Texts in Computer Science

Editors
David Gries
Fred B. Schneider

For other titles published in this series, go to
http://www.springer.com/3191

Texts in Computer Science

Editors
David Gries
Fred B. Schneider

Luiz Velho · Alejandro C. Frery ·
Jonas Gomes

Image Processing
for Computer Graphics
and Vision

Second Edition

 Springer

Luiz Velho, BE, MS, PhD
IMPA - Instituto de Matematica Pura e
    Aplicada,
Rio de Janeiro, Brazil

Alejandro Frery, BSc, MSc, PhD
Universidade Federal de Alagoas,
Maceió, Brazil

Jonas Gomes, PhD
IMPA - Instituto de Matematica Pura e
    Aplicada,
Rio de Janeiro, Brazil

Series Editors

David Gries
Department of Computer Science
415 Boyd Graduate Studies
    Research Center
The University of Georgia
Athens, GA 30602-7404, USA

Fred B. Schneider
Department of Computer Science
Upson Hall
Cornell University
Ithaca, NY 14853-7501, USA

Translated by Silvio Levy

British Library Cataloguing in Publication Data
A catalogue record for this book is available from the British Library

ISBN 978-1-4471-6015-1 2nd edition
978-0-387-94854-6 1st edition
DOI 10.1007/978-1-84800-193-0

ISBN 978-1-84800-193-0 (eBook)

© Springer-Verlag London Limited 2009
Softcover re-print of the Hardcover 2nd edition 2009
First published 1997
Second edition 2009

Apart from any fair dealing for the purposes of research or private study, or criticism or review, as permitted under the Copyright, Designs and Patents Act 1988, this publication may only be reproduced, stored or transmitted, in any form or by any means, with the prior permission in writing of the publishers, or in the case of reprographic reproduction in accordance with the terms of licences issued by the Copyright Licensing Agency. Enquiries concerning reproduction outside those terms should be sent to the publishers.

The use of registered names, trademarks, etc., in this publication does not imply, even in the absence of a specific statement, that such names are exempt from the relevant laws and regulations and therefore free for general use.

The publisher makes no representation, express or implied, with regard to the accuracy of the information contained in this book and cannot accept any legal responsibility or liability for any errors or omissions that may be made.

Printed on acid-free paper

9 8 7 6 5 4 3 2 1

Springer Science+Business Media
springer.com

To Solange and Daniel

To Noni and Alice

To Enilson

Preface

A escrita é a forma mais duradoura de conservar nossos pensamentos. Através dela, nos é permitido transmitir, de geração em geração, a essência de nossas reflexões sobre a vida, a humanidade e, sobretudo, o amor.

Solange Visgueiro

This book originated when we noticed, several years ago, that the importance of image processing in the area of visualization and computer graphics was not reflected in either the existing curricula or the current textbooks.

On the one hand, traditional image processing books do not cover important topics for computer graphics such as warping, morphing, digital compositing, color quantization, and dithering. Often even basic facts about signals are not adequately discussed in the context of graphics applications. This kind of knowledge is now more important than ever for computer graphics students, given the interactions between audio, images, and models in most applications.

Computer graphics books, on the other hand, emphasize primarily modeling, rendering, and animation, and usually do not contain a proper exposition of signal processing techniques.

We have adopted a conceptual approach, with emphasis on the mathematical concepts and their applications. We introduce an abstraction paradigm that relates mathematical models with image processing techniques and implementation methods. This paradigm is used throughout the book, and helps the reader understand the mathematical theory and its practical use. At the same time, we keep the presentation as elementary as possible by sacrificing mathematical rigor, when necessary, for an intuitive description.

This book is intended to be useful either as a textbook or as a reference book. In draft form and after publication, the Portuguese edition has been used since 1992 at a course taught at Instituto de Matemática Pura e Aplicada (IMPA) in Rio de Janeiro, attended by undergraduate and master's students in mathematics and computer science. Chapters 1 through 7 correspond to the course's contents; the remaining chapters have been used as topics for discussion and seminars with the students. The English version has been in use outside Brazil since the fall of 1996. The initial edition of the book had a strong emphasis on deterministic image models. The current extended edition of the book includes also stochastic image models, as well as applications in Computer Vision. The book in its present form can thus be used more flexibly for teaching or research.

Acknowledgments

We wish to thank our friend Silvio Levy, who accepted the invitation to translate the book and did an excellent job. Our interaction with him during the translation was very rewarding, and his comments greatly influenced several changes we have made for this English edition.

We are grateful to numerous colleagues who contributed their comments and criticism to this work: André Antunes, Bruno Costa, Lucia Darsa, Marcelo Dreux, Luiz Henrique de Figueiredo, and Valéria Iório, among others. We also thank Siome Goldenstein and Paulo Roma, who helped us revise the translation. We acknowledge the editorial work of Martin Gilchrist and Wayne Wheeler of Springer-Verlag, who respectively agreed to publish the first and second editions of this book.

Rio de Janeiro, March 2008

Luiz Velho
Alejandro Frery
Jonas Gomes

Contents

1

Introduction

Images are the final product of most processes in computer graphics. This book is devoted to the study of images and of image manipulation systems by computer. It also covers important mathematical concepts in the analysis and understanding of images.

1.1 Computer Graphics

The International Standards Organization (ISO) defines computer graphics as the sum total of "methods and techniques for converting data for a graphics device by computer." This definition would probably not help a reader totally unfamiliar with the field to understand what it's all about. In fact, the best way to understand a field is to grasp what its main problems are. From this point of view, the ISO definition can be said, with goodwill, to define the main problem of computer graphics: *converting data into images.*

The process of converting data into images is known as *visualization.* It is schematically illustrated in Figure 1.1.

In order to understand computer graphics, then, we must study the methods for creating and structuring data in the computer as well as methods for turning these data into images. These two steps correspond to the two main areas of research in computer graphics: *modeling* and *visualization.*

In this book we will not study modeling or data visualization. Instead, we will focus on a more fundamental and very important problem: understanding

Fig. 1.1. Computer graphics: converting data into images.

L. Velho et al., *Image Processing for Computer Graphics and Vision,*
Texts in Computer Science, DOI 10.1007/978-1-84800-193-0_1,
© Springer-Verlag London Limited 2009

the notion of an image and also the techniques of image manipulation by computer—in other words, *image processing*. At the same time, this is not a typical image processing book, because it covers primarily the aspects of image processing used most often in computer graphics.

Since its inception, computer graphics has sought methods to allow the visualization of information stored in computer memory. Since there are practically no limitations on the nature or origins of such data, researchers and professionals use computer graphics today in the most diverse fields of human knowledge. Its use is important whenever it is necessary to have a visual representation of *objects, actions, relations,* or *concepts*. The importance of this visual representation is reflected in the huge number of computer graphics applications, ranging from scientific visualization to special effects in the entertainment industry.

Partly because computer graphics has so many applications, there are no sharp boundaries between it and related fields. However, we can take as a working criterion in differentiating among these fields the nature of the input and output of the process in question, as shown in Figure 1.2.

In *data processing,* the system takes in data and, after processing, returns data of more or less the same nature. For example, a bank account management system processes input transactions and yields output data such as a daily balance, interest earned, and so on.

In *computer graphics*, the input data are (typically) nonvisual, and the output is an *image* that can be seen through some graphics output device. For instance, the account management system of the preceding paragraph might plot a graph of the daily balance over a period.

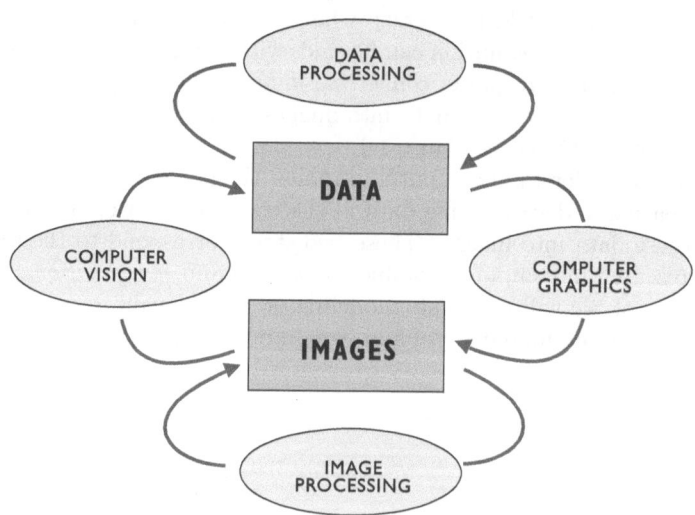

Fig. 1.2. Computer graphics and kindred disciplines.

In *digital image processing*, the input is already an image, which gets processed to yield another image, the output. The latter can again be seen through a graphics output device. An example would be the processing of data sent by an orbiting satellite, with the purpose of coloring or enhancing the image.

The goal of *computer vision* is to, given an input image, obtain information about the physical, geometric, or topological properties of the objects that appear in it. This is used, for example, in robotics, to endow machines with a sense of sight.

In most applications, two or more of these areas act in concert. Thus, the data output by an image processing or computer vision system might be further subjected to computer graphics techniques, to give the user better qualitative information. For instance, height information can be extracted from satellite images and processed to yield a relief map, which can further be combined with enhanced and colorized image data to yield a realistic three-dimensional model of the area. It is exactly the joint use of techniques in these various areas that holds the greatest potential for applications.

Sometimes these cross-disciplinary links are so vigorous that they can spawn new disciplines. The joint use of geometric modeling and computer vision is the basis of the discipline known as *visual modeling*, which allows the creation of models starting from the images of a scene. Medical imaging, likewise, uses (almost transparently) techniques from image processing, computer graphics, and computer vision.

Another way to summarize these distinctions is by saying that computer vision is interested in *analyzing* images, while computer graphics is interested in *synthesizing* them. These two areas, plus image processing, are the triple foundation of all computational processes involving images. In computer graphics, the image is the end product; image processing plays a role in the early phase of image generation and in a later phase called postprocessing. In computer vision, the image is the input data, and in general the early, preprocessing phase involves image processing. Figure 1.3 illustrates this idea.

In all these areas, of course, the study of images as abstractions is of paramount importance. We therefore turn to the general ideas that underlie this study.

1.2 Abstraction Paradigms

In any area of applied mathematics, one needs to model the objects under study mathematically. To set things on the right conceptual footing, one must create a hierarchy of abstractions, and at each level of abstraction one must apply the most appropriate models.

In applied areas that involve computational methods, and in particular in computer graphics, one abstraction paradigm that is generally applicable consists in establishing four universes (sets), shown schematically in Figure 1.4: the physical universe P, the mathematical universe M, the representation

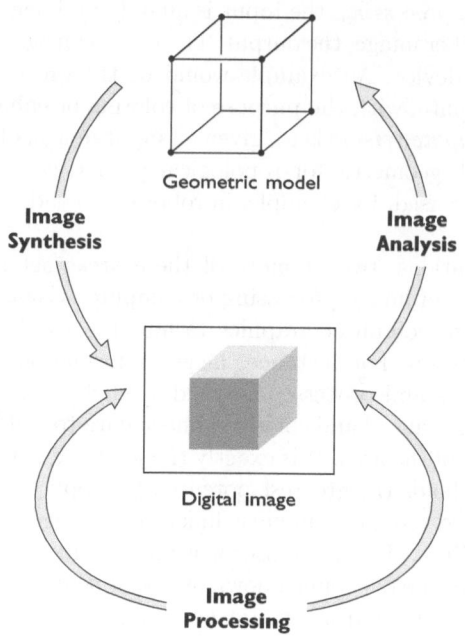

Fig. 1.3. Image synthesis, processing and analysis.

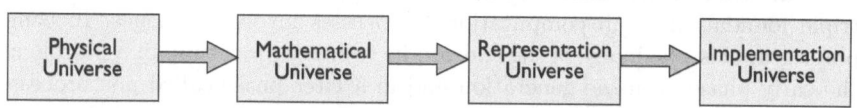

Fig. 1.4. Conceptual levels of abstraction.

universe R, and the implementation universe I. The *physical universe* contains the real-world objects we intend to study; the *mathematical universe* contains an abstract description of the objects from the physical world; the *representation universe* is made up of the discrete representations associated with the objects from the mathematical universe; and, finally, in the *implementation universe*, we map the entities from the representation universe to concrete data structures, in order to actually represent the objects on the computer. The implementation universe is designed to separate the discretization step (representation) from the particularities of the programming language used in the implementation.

We call this conceptual layering the *four-universe paradigm*. In short, in order to study a given phenomenon or object on the computer, we must associate to it a mathematical model and then find a discrete representation of the model that can be implemented in the computer.

Based on this paradigm, here are the main types of problems that we can expect to encounter in an area of study:

- defining the elements of the universe M;
- relating the universes P, M, R, and I with one another;
- defining representation relations from M to R;
- studying the properties of the possible representations from M to R;
- converting among the different representations;
- devising good data structures for implementation.

Clearly, once the elements of M are defined, other specific problems can be posed; for instance, one may create additional abstraction sublevels, in a process similar to top-down structured programming. Creating abstraction levels allows one better to encapsulate, pose, and solve the problems at each level, much as object-oriented programming does. We will use the paradigm above many times in this book; in each case we will discuss how it applies to a particular area and how the problems itemized above translate to that area.

1.3 About This Book

We now turn to the contents of each chapter of this book. Our goal throughout has been to present the subject from a theoretically consistent point of view, and we have adopted innovative formulations whenever we felt that this was necessary for clarity of exposition. We have likewise included a great many examples and illustrations, as an aid to understanding.

To the extent possible, we have avoided the discussion of implementation details. There is no "pseudocode" in this book. Throughout the work, the emphasis is on describing and analyzing underlying concepts rather than on presenting algorithms or discussing optimization questions. A good conceptual understanding is exceedingly important as a basis for further studies, and it is also a prerequisite for understanding existing algorithms and creating new ones.

The chapters of the book are as follows.

Chapter 2: Signal Theory

Discusses signal theory and so introduces the reader to the various ideas and results of digital signal processing. It also prepares the reader for the study of color and images, two particular cases of signals used in computer graphics, and which are studied in detail in subsequent chapters.

Chapter 3: Random Processes

Introduces the basic notions of Probability and Random Processes. These concepts will be used to formulate stochastic image models and design their applications.

Chapter 4: Fundamentals of Color

Introduces color theory from the point of view of signal processing. Discusses the various mathematical models of color and develops the notion of color space representation, or discretization of the visible spectrum, as a means of allowing the representation of color on the computer.

Chapter 5: Color Systems

Discusses the different color systems used for the specification and computation of color, emphasizing the RGB and XYZ standards from the International Commission for Illumination (CIE). Also covers video component systems, which have acquired great importance in video and computer graphics applications.

Chapter 6: Digital Images

Introduces the main object of study of this book: the mathematical model for an image, and its digital representation. The problem of image discretization is extensively discussed.

Chapter 7: Operations on Images

Covers the important topic of operations on an image's domain and color space. Emphasizes image filtering, a topic of great importance in computer graphics. Discusses a variety of linear filters, both in the spatial and the frequency domain, and gives applications to reconstruction. There is some overlap between this chapter and Chapter 2, for the benefit of readers who have already studied signal processing and who may want to read this chapter independently.

Chapter 8: Sampling and Reconstruction

This chapter uses the theory and concepts developed in the previous chapters to discuss in more detail the important problem of image sampling and reconstruction.

Chapter 9: Multiscale Analysis and Wavelets

Discusses the concept of multiscale representation of images and computation with wavelets. Presents the Multiresolution framework to build wavelets and describes the implementation of the Fast Wavelet Transform.

Chapter 10: Probabilistic Image Models

Introduces stochastic models for images. Relates these models with the process of image acquisition from sensors. Analyzes the role of noise in images. Probabilistic methods for classification and inference are also considered.

Chapter 11: Color Quantization

This chapter is devoted to the discretization of image attributes, with emphasis on color quantization. Uniform and adaptive quantization algorithms are presented.

Chapter 12: Digital Halftoning

Dithering is often treated together with quantization, in a very superficial way. We treat it separately, to stress the fact that dithering is a type of nonlinear filtering. The decision to devote a whole chapter to the subject was based on the importance of dithering in electronic publishing. The chapter contains a detailed study of dithering algorithms, including some stochastic screening techniques.

Chapter 13: Image Compression

This chapter gives an overview of image compression techniques. It includes methods based on image transforms and wavelets.

Chapter 14: Combining Images

Studies several operations that allow one to combine into one image elements from several. This is a very important topic in applications of image processing to computer graphics, especially in the creation of special effects for video and movies.

Chapter 15: Warping and Morphing

Discusses in detail, from a conceptual point of view, image warping and morphing. These techniques are used to obtain smooth transitions between two images and involve a correspondence of geometric elements simultaneously with color interpolation.

Chapter 16: Imaging Systems

Discusses a number of problems related to image analysis and processing systems. As an example, it treats in some detail the case of an electronic publishing system with the ability to produce color separations for offset printing.

Appendix: Radiometry and Photometry

Covers in detail necessary background material on these classical areas of physics, which, while useful for understanding parts of Chapter 3, would break the conceptual flow of ideas if included in the main text.

Naturally, the study of images takes up most of the book. However, the contents are not identical with those of a traditional course in image processing: as already remarked, our interest is to exploit the aspects of image processing that have importance in computer graphics. For this reason we stress more color quantization than, say, the different methods of image encoding and compression. This also has led us to devote whole chapters to operations on images, dithering, warping and morphing, and image composition, topics of great importance in computer graphics applications, which are not covered in most books devoted to image processing.

1.4 Comments and References

Until recently, many computer graphics books included a historical synopsis of the field's evolution. This started in early days, when the field was new and relatively little-known, as a way to acquaint the public with the potential of computer graphics; later the tradition was maintained, in large measure because the explosive growth in the body of knowledge and applications demanded constant updating of the literature.

Today, although still a young discipline in comparison with other areas of science, computer graphics has developed to the point where the historical dimension plays a different role. A history of computer graphics must cover not only applications but also the evolution of mathematical and physical models, the algorithms, and even the hardware. Such a history, or even a bare chronology, would be far too long to be adequately dealt with in a single chapter. What is needed is a book entirely devoted to the history of computer graphics.

Nonetheless, here are some highlights of the literature, from a historical point of view.

The seminal work of Ivan Sutherland, in his Ph.D. thesis (Sutherland 1963), marked a watershed between early, rudimentary uses of the computer for graphics and the modern notion of interactive computer graphics. Sutherland's "Sketchpad" allowed the user to interactively manipulate plane geometric figures.

The first texts that refer to computer graphics as such were connected with computer-aided design (CAD) in high-technology industries, especially the automobile, aircraft and shipbuilding industries. See, for example, (Parslow 1969) and (Prince 1971).

Many introductory articles about computer graphics have appeared in popular scientific magazines. Among the oldest, and yet most interesting for its historical perspective, is the article (Sutherland 1970) in *Scientific American*. We mention also (Crow 1978), (Whitted 1982), and (van Dam 1984).

An excellent way to get an overall view of the evolution of the state of the art in computer graphics is to watch the videos put out every year since 1978 by SIGGRAPH, the Special Interest Group in Computer Graphics of the ACM (Association for Computing Machinery). These videos contain the animations selected for display at the annual SIGGRAPH meeting in the United States, itself a showcase of the most important current work in the field.

Among the many introductory computer graphics books available, we mention two classics: (Newman and Sproull 1979) and (Foley et al. 1990). Both stress the interactive aspects of the field and have appeared in a second, revised, edition. Other general texts are (Giloi 1978), (Magnenat-Thalmann and Thalmann 1987), and (Watt 1990).

Another general textbook, with good coverage of implementation aspects of algorithms, is (Rogers 1985). Along the same lines, (Harrington 1983, pp. 345–352) is geared toward the implementation of a graphics system in the spirit of the CORE system proposed by the ACM (Michener and Van Dam 1979) as a possible ISO standard.

There is a series of "Graphics Gems" books devoted to the implementation of graphics algorithms, covering many problems in image synthesis, processing and analysis: (Glassner 1990), (Arvo 1991), (Kirk 1992), (Heckbert 1994), and (Paeth 1995). They can be used (selectively) to complement the present book, which does not stress the implementation side.

Although it can be appropriately applied to all areas of applied mathematics that involve computational methods, the four-universe paradigm first appeared explicitly in the literature in (Requicha 1980), in the context of geometric modeling. More details on the use of this paradigm in various areas of computer graphics can be found in (Gomes and Velho 1995).

As the subject of computer graphics matured, textbooks started to appear on specific subfields, such as ray tracing (Glassner 1989) and lighting (Hall 1989). The book (Fiume 1989), devoted to raster graphics, is an effort to lay a solid conceptual foundation for this subject.

Scientific visualization, and its importance to scientific and technological development, are well documented in (McCormick 1987).

A historical, if anecdotal, perspective on computer graphics can be gleaned from (Machover 1978) and (Rivlin 1986).

References

[Arvo 1991]Arvo, J. (1991). *Graphics Gems II*. Academic Press, New York.
[Crow 1978]Crow, F. C. (1978). Shaded computer graphics in the entertainment industry. *Computer*, 11(3):11–22.

[Fiume 1989]Fiume, E. L. (1989). *The Mathematical Structure of Raster Graphics*. Academic Press, New York.

[Foley et al. 1990]Foley, J. D., van Dam, A., Feiner, S. K., and Hughes, J. F. (1990). *Fundamentals of Interactive Computer Graphics*, second ed. Addison-Wesley, Reading, MA.

[Giloi 1978]Giloi, W. K. (1978). *Interactive Computer Graphics. Data Structures, Algorithms, Languages*. Prentice-Hall, Englewood Cliffs, NJ.

[Glassner 1990]Glassner, A. (1990). *Graphics Gems I*. Academic Press, New York.

[Glassner 1989]Glassner, A. (editor) (1989). *An Introduction to Ray Tracing*. Academic Press, New York.

[Gomes and Velho 1995]Gomes, J. and Velho, L. (1995). Abstract paradigms for computer graphics. *The Visual Computer*, 11:227–239.

[Hall 1989]Hall, R. A. (1989). *Illumination and Color in Computer Generated Imagery*. Springer-Verlag, New York.

[Harrington 1989]Harrington, S. (1983). *Computer Graphics: a Programming Approach*. McGraw-Hill, New York.

[Heckbert 1994]Heckbert, P. (1994). *Graphics Gems IV*. Academic Press, New York.

[Kirk 1992]Kirk, D. (1992). *Graphics Gems III*. Academic Press, New York.

[Machover 1978]Machover, C. (1978). A brief, personal history of computer graphics. *Computer*, 11:38–45.

[Magnenat-Thalmann and Thalmann 1987]Magnenat-Thalmann, N. and Thalmann, D. (1987). *Image Synthesis*. Springer-Verlag, New York.

[McCormick 1987]McCormick, B. (1987). Visualization in scientific computing. *Computer Graphics*, 21(6).

[Michener and Van Dam 1979]Michener, J. C. and Van Dam, A. (1978). A functional overview of the Core System with glossary. *ACM Computing Surveys*, 10:381–387.

[Newman and Sproull 1979]Newman, W. M. and Sproull, R. F. (1979). *Principles of Interactive Computer Graphics*. McGraw-Hill, New York.

[Paeth 1995]Paeth, Alan W. (1995). *Graphics Gems V*. Academic Press, New York.

[Parslow 1969]Parslow, R. D., Prowse, R. W., and Elliot Green, R. (editors), (1969). *Computer Graphics, Techniques and Applications*. Plenum Press, New York.

[Prince 1971]Prince, D. (1971). *Interactive Graphics for Computer Aided Design*. Addison Wesley, New York.

[Requicha 1980]Requicha, A. A. G. (1980). Representations for rigid solids: Theory methods, and systems. *ACM Computing Surveys*, 12:437–464.

[Rivlin 1986]Rivlin, R. (1986). *The Algorithmic Image*. Microsoft Press, Redmond, WA.

[Rogers 1985]Rogers, D. F. (1985). *Procedural Elements for Computer Graphics*. McGraw-Hill, New York.

[Sutherland 1963]Sutherland, I. (1963). *A man-machine graphical communication system*. Ph.D. thesis, MIT, Dept. of Electrical Engeneering.

[Sutherland 1970]Sutherland, I. (1970). Computer displays. *Scientific American*, June 1970.

[van Dam 1984]van Dam, A. (1984). Computer software for graphics. *Scientific American*, September 1984, 146–161.

[Watt 1990]Watt, A. (1990). *Fundamentals of Three-Dimensional Computer Graphics*, second ed. Addison-Wesley, Wokingham, England.

[Whitted 1982]Whitted, T. (1982). Some recent advances in computer graphics. *Science*, 215:764–767.

2

Signal Theory

Our everyday interaction with the environment takes place by means of signals of many types. The sense of sight is based on light signals; the sense of hearing and the ability to speak are based on sound signals; and electromagnetic signals open the doors to the fantastic world of telecommunications.

With the advent of computers, and their ever-increasing role in managing the many activities of everyday life, the representation and processing of signals in digital form have acquired great importance. The representation of signals plays a central role in computer graphics, especially in image synthesis, processing, and analysis. The aim of this chapter is to give the reader a conceptual understanding of the problems arising from signal representation and processing.

The study of signal theory can involve sophisticated mathematics. Here we hope to give the reader a firm grasp of the mathematical foundations of the theory. In order to make the chapter accessible to as wide a readership as possible, we have not adhered to a rigorous mathematical formulation. A mathematically sophisticated reader may consider it a good exercise to spot the places where we left aside mathematical rigor in favor of simplicity of exposition.

We hope this chapter will be useful to readers who are somewhat acquainted with digital signal processing, as well as to those who are approaching the subject for the first time.

2.1 Abstraction Paradigms

An application of the four-universe paradigm of Chapter 1 to the study of the multitude of signals that surround us leads to the following breakdown of our task: we must search for mathematical descriptions of these signals, find effective means to construct discrete signal representations, and seek algorithms to implement signal synthesis, analysis, and processing operations in the computer. See Figure 2.1.

L. Velho et al., *Image Processing for Computer Graphics and Vision*,
Texts in Computer Science, DOI 10.1007/978-1-84800-193-0_2,
© Springer-Verlag London Limited 2009

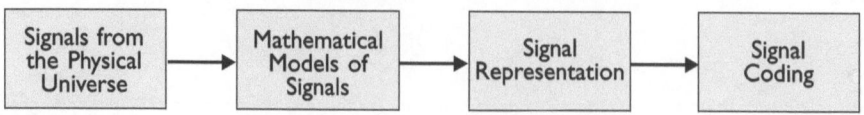

Fig. 2.1. Applying the four-universe paradigm to the study of signals.

2.1.1 Levels of Abstraction

Based on the paradigm just described, we think in terms of three abstraction levels in the study of signals: *continuous signals*, *discrete signals*, and *encoded signals*. Each level corresponds to one form of signal description, appropriate for the formulation and solution of a particular set of problems. The transition between the three abstraction levels is carried out by operations of four types: *discretization*, *encoding*, *decoding*, and *reconstruction*. See Figure 2.2.

Discretization and Reconstruction

Discretization is the task of converting a continuous signal into a discrete representation. The opposite task, converting a discrete representation into a continuous signal, is called *reconstruction*.

Ideally, the operation of reconstruction should be inverse to that of discretization; that is, given a continuous signal s, with discrete representation s_d, the process of reconstruction should recover s from s_d:

$$s \rightarrow \text{discretization} \rightarrow s_d \rightarrow \text{reconstruction} \rightarrow s.$$

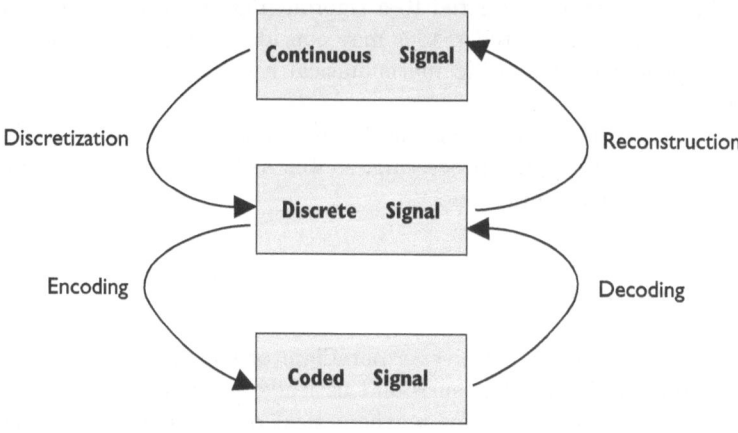

Fig. 2.2. Levels of abstraction in signal representation.

In general, however, the result of reconstruction is not exactly the original signal s, but some other signal s_r. One of the central problems in signal processing is finding discretization and reconstruction methods such that the reconstructed signal s_r is a good approximation of the original signal s. The meaning of "good approximation" depends, of course, on the application.

Encoding and Decoding

Encoding is the step that goes from a discrete representation of the signal to a finite representation, one that can be described by a finite set of symbols organized according to some data structure. *Decoding* allows one to go from the encoded data back to a discrete representation. Given a discrete representation s_d of a signal s, we have

$$s_d \to \text{encoding} \to s_c \to \text{decoding} \to \tilde{s}_d.$$

If the decoded signal $\tilde{s}_d$ equals the discrete signal s_d, we have *lossless* encoding; otherwise the process is *lossy*. In addition to lossiness, there are many other questions to be considered in the creation or choice of structures for a finite signal representation, the most important of which are the space occupied by the code and the speed of encoding and decoding.

2.2 Mathematical Models for Signals

According to our plan for following the abstraction paradigm, we now turn to the mathematical models that can be used in the study of the signals of our physical universe.

A signal arises when some physical magnitude varies in time or space. Thus a sound signal corresponds to the variation of air density (or pressure) with time. A static visual image corresponds to variations of color in space, say among the points of a photograph. A video signal conveys color variation in time as well as in space.

It follows that a signal should be represented by a mathematical object that records the variation of the magnitude in question. If the variation is deterministic, we can use a function to describe the signal. If it is nondeterministic, we can use a *stochastic process*. In the first case we have a *functional model* for the signal, and in the second a *stochastic model*.

Functional models are sufficient for the purposes of computer graphics and therefore are the ones we adopt here. In such a model, a signal is represented by a function $f : U \subset \mathbb{R}^m \to V$, where V is an arbitrary vector space; that is, the physical magnitude in question is represented by a vector, varying in a space with m degrees of freedom. A *signal space* is a vector subspace of the space of functions $\{f : U \subset \mathbb{R}^m \to V\}$, where U, m, and V are fixed. In other words, a signal space is a space of functions with a natural vector space structure given by the usual operations of addition of functions and multiplication of a function by a real number:

$$(f + g)(t) = f(t) + g(t),$$
$$(\lambda f)(t) = \lambda \cdot f(t) \quad \text{for } \lambda \in \mathbb{R}.$$

It is also important to consider signals whose values are complex numbers. The definition and the theory are essentially the same, and it will be clear from the context whether we are referring to real- or complex-valued signals.

Two final remarks about our mathematical model for signals should be made.

We defined a signal as a function assuming values in an arbitrary vector space. We might have simplified our mathematical model for signals by allowing a signal to assume values only in a finite-dimensional space, that is, $f : U \subset \mathbb{R}^m \to \mathbb{R}^n$. In fact, most signals are covered by this model; but the image signal, a very important one for us, assumes values in an infinite-dimensional vector space, as we will see later.

The second remark concerns how general our mathematical model for signals is. Is it possible to represent any signal from the physical universe by a function? The answer, unfortunately, is no, as exemplified by an impulse signal.

An *impulse signal* is characterized by having an *instantaneous variation in magnitude, a very large intensity, and finite energy*. One can attempt to represent such a signal by a "function" f with the following properties:

- $f(t_0) \neq 0$, and $f(t) = 0$ if $t \neq t_0$ (instantaneous variation);
- $f(t_0) = +\infty$ (very large intensity at t_0);
- $\int_{-\infty}^{+\infty} f(t) \, dt < \infty$ (finite energy).

Obviously, the second property means that f is not, in fact, a function.

One could use an appropriate extension of the notion of a function in order to cover such signals: for example, distributions would be adequate. However, in this book we have opted for a more elementary approach, sacrificing mathematical rigor when necessary. Later we will show how we can use the functional model to give an approximate description of an impulse signal.

2.2.1 Approximation of Signals

In many problems involving signals it is very important to have a metric (notion of distance) to measure how close two signals are to one another. Several metrics can be chosen, depending on the application. Our perception of signals takes place, directly or indirectly, through our senses: the sense of sight perceives electromagnetic signals within a certain range of frequencies, the sense of hearing perceives sound signals, and so on. A metric d on a space of signals is called a *perceptual metric* if two signals f and g satisfy $d(f, g) = 0$ if and only if they are perceptually indistinguishable. Strictly speaking, if

$d(f,g) = 0$ for two distinct functions $f \neq g$, we should call d a *pseudometric* rather than a metric, but we won't make that distinction, since all perceptual metrics are of this type, as well as many other nonperceptual metrics that are important for reasons of computational efficiency.

We now consider two commonly used metrics.

The Uniform Metric

In the *uniform metric* the distance $d(f,g)$ between two signals f and g is

$$d(f,g) = \sup\{|f(u) - g(u)| : u \in U\},$$

where sup indicates the supremum (least upper bound) of a set of real numbers. We must, of course, assume that the signals are bounded in U. Figure 2.3 shows a neighborhood of radius $\varepsilon > 0$ of a signal f in the uniform metric, and a signal g in this neighborhood.

The L^p Metric

When the signals take values in $\mathbb{R}$ or $\mathbb{C}$, we can introduce the L^p metric, in which the distance between two signals is

$$d(f,g) = \left(\int_{-\infty}^{+\infty} |f(u) - g(u)|^p \, du \right)^{1/p},$$

assuming the integral exists. A particular case of great importance in signal theory is $p = 2$. In this case the metric arises from the L^2 inner product in signal space given by

$$\langle f, g \rangle = \int_{-\infty}^{+\infty} f(u)\overline{g(u)} du,$$

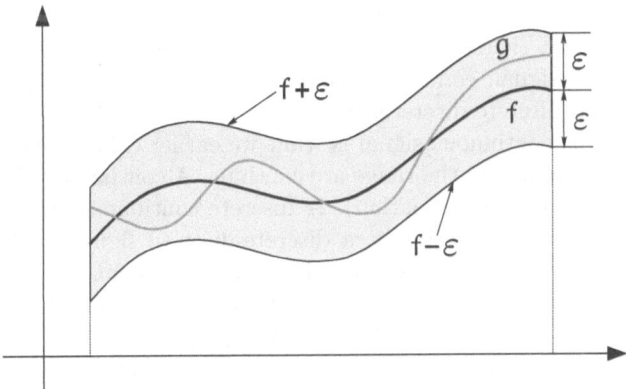

Fig. 2.3. The ε-neighborhood of a signal f in the uniform metric.

where the bar indicates complex conjugation. Clearly, the L^2 metric is defined only if

$$\int_{-\infty}^{+\infty} |f|^2 < \infty$$

for all f in the space of signals. Physically, this condition means that all signals have finite energy. The assumption of finite energy is physically meaningful and mathematically very convenient. In this book we will assume that all signals have finite energy, unless we say otherwise.

While the uniform metric measures a point-by-point difference between the two signals, the L^p metric gives an average difference between them. We leave it to the reader to provide an intuitive discussion of the perceptual characteristics of these two metrics.

2.2.2 Functional Models and Abstraction Levels

A function $f : U \subset \mathbb{R}^m \to \mathbb{R}^n$ on a space of signals is called a *continuous signal*. "Continuous" here contrasts with "discrete" and means simply that the domain and range of f are "continua" (spaces parametrized by real numbers, rather than discrete spaces). We are *not* assuming that f is necessarily a continuous function in the sense of topology or analysis. Continuous signals are also known as *analog signals* in engineering.

On a computer, the set of real numbers is replaced by a finite set of numbers, namely those that can be represented by computer words in some floating-point arithmetic scheme. Thus, in practice, we consider a continuous signal to be one that is defined using floating-point arithmetic.

We now return to the three abstraction levels for signal representation (Section 2.1) in the context of a functional model. Given a signal defined by a function f, the representation process consists of discretizing the domain of f, while encoding requires the discretization of both the domain and the range of f. Discretization of the domain is traditionally known as *sampling*, and discretization of the range as *quantization*.

More precisely, there are four conceptual variants for the functional representation of a signal: continuous-continuous, continuous-discrete, discrete-continuous, and discrete-discrete.

A continuous-continuous signal is what we earlier called a continuous signal: both the domain and the range are continua. A continuous-discrete signal has a discrete range (quantization). A discrete-continuous signal has a discrete domain (sampling). Finally, a discrete-discrete signal is both sampled and quantized. Such a signal is also called a *digital signal*.

Intuitively, sampling a signal f amounts to computing and preserving the values taken by f at a finite number of points $p_1, p_2, \ldots, p_k$ of the original domain U (such values are known as *samples*) and ignoring all other information about f.

Reconstruction of the signal amounts to using an interpolation process to obtain the original signal f, or an approximation f_r to f, starting from the

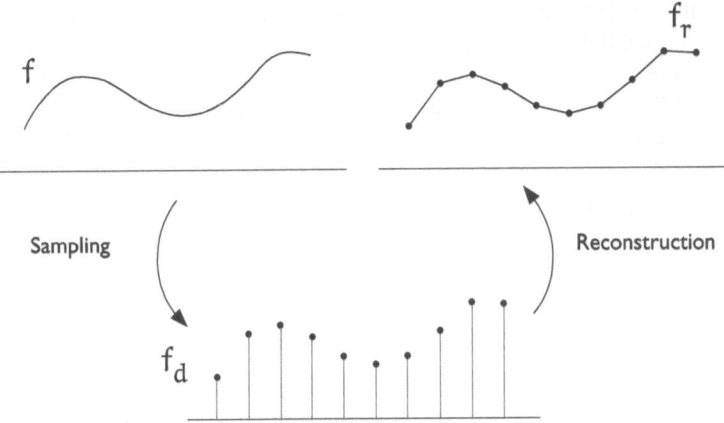

Fig. 2.4. Sampling and reconstruction of a signal.

samples $f(p_1), f(p_2), \ldots, f(p_k)$ of f. Figure 2.4 illustrates this idea, using a particular reconstruction method (linear interpolation).

If the interpolation method recovers the original signal, that is, if $f_r = f$, we say the reconstruction is *ideal* or *exact*. In this book we will extend and elaborate on the notions of sampling and reconstruction for images; we will also briefly review the problem of image encoding and decoding.

Different functional models arise from different interpretations for the domain and range variables of the function representing the signal. The physical interpretation of these variables, although mathematically irrelevant, is of course what gives a representation its meaning, and it is very important in applications where different functional models can be used. We will examine here two functional models: *spatial* and *spectral*.

2.2.3 The Spatial Model

In the *spatial model*, the domain U of the function $f : U \subset \mathbb{R}^m \to V$ representing the signal stands for a region in physical space or an interval of time (in which case, $m = 1$). For this reason U is called the *space domain* or *time domain*; the two expressions are often used interchangeably when $m = 1$. The function expresses a physical magnitude, a vector in V, varying according to position or time. The dimension of the domain is called the dimension of the signal: for $m = 1$ we have a *one-dimensional* signal, and so on.

Returning to our earlier examples: an audio signal (such as recorded sound) is one-dimensional, varying in time. A photographic image is a two-dimensional signal; U is a subset of the Euclidean plane $\mathbb{R}^2$, and the function f associates to each point $p \in U$ a vector $f(p) \in \mathbb{R}^n$ representing color information at p, where $\mathbb{R}^n$ is the color space. For a video signal we also have variation in time; therefore we have a three-dimensional signal $f : U \times \mathbb{R} \subset \mathbb{R}^3 \to \mathbb{R}^n$.

The theory developed in this chapter is valid for m-dimensional signals, with $m \geq 1$ arbitrary. To simplify the exposition and the notation, we will introduce our definitions and most examples in the context of one-dimensional signals. While reading the chapter, you should mentally translate the ideas also to the m-dimensional case—especially for $m = 2$, the case of images.

Pulse and Impulse Signals

Let $a > 0$. The one-dimensional *pulse signal* $p_a(t)$ is represented by

$$p_a(t) = \begin{cases} 1 & \text{if } |t| \leq a, \\ 0 & \text{if } |t| > a. \end{cases}$$

Clearly, this signal has constant intensity and finite duration. Its graph is given in Figure 2.5(a).

It can easily be generalized to m dimensions: the m-dimensional pulse signal $p_a : \mathbb{R}^n \to \mathbb{R}$ is defined as

$$p_a(x_1, \ldots, x_n) = p_a(x_1)p_a(x_2)\ldots p_a(x_n)$$

(that is, it is the tensor product of m one-dimensional signals). The graph of p_a for $m = 2$ appears in Figure 2.5(b).

Sometimes it is convenient to consider a normalized pulse signal,

$$p(t) = \frac{p_a(t)}{2a},$$

where the area under the graph equals 1.

As we mentioned in Section 2.2, the impulse signal cannot be represented by a function. Now we will show how it is possible to use the pulse signal to obtain an approximate mathematical description of an impulse in the functional model.

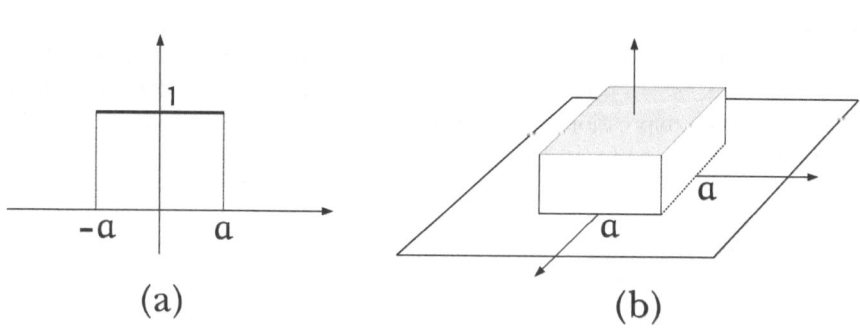

(a)   (b)

Fig. 2.5. (a) One-dimensional pulse signal of duration $2a$. (b) The two-dimensional analogue.

The *impulse "function"* $\delta(t)$, also called a *Dirac delta function*, can be represented in the functional model by the limit

$$\delta(t) = \lim_{n \to +\infty} \tfrac{1}{2}n \cdot p_{1/n}(t), \tag{2.1}$$

where $p_{1/n}$ is the pulse function defined above. Note that, as $n \to +\infty$, the support of the signals $\frac{1}{2}np_{1/n}$ converges to 0, and the value at $t = 0$ converges to $+\infty$, as shown in Figure 2.6. (Recall that the *support* of a function is the smallest closed set outside of which the function is zero.)

In the limit, we have a signal with instantaneous variation at the origin $(f(0) \neq 0$, and $f(t) = 0$ for $t \neq 0)$ and also with a very large intensity at the origin $(f(0) = \infty)$. The energy of this signal is finite and equal to 1. This can be checked as follows:

$$\int_{-\infty}^{+\infty} \delta(t)\,dt = \int_{-\infty}^{+\infty} \lim_{n \to +\infty} \tfrac{1}{2}np_{1/n}(t)\,dt = \lim_{n \to +\infty} \int_{-\infty}^{+\infty} \tfrac{1}{2}np_{1/n}(t)\,dt = 1.$$

The above limit-of-pulses signal represents an impulse signal δ at the origin. An impulse at an arbitrary point t_0 would be represented by $\delta(t - t_0)$. However, we should point out that this representation of the impulse signal is not mathematically correct, because the limit in (2.1) has no meaning for $t = 0$.

The above "approximation" can be used to justify most of the properties of the impulse signal, but be aware that manipulations involving this approximation are not mathematically correct. However, they have heuristic value and can be rigorously formalized in the context of a distribution model for signals.

The Dirac δ plays a central role in signal theory, both in theory and in applications. One of its most important properties is given by the equation

$$f(x) = \int_{-\infty}^{+\infty} f(t)\,\delta(x - t)\,dt, \tag{2.2}$$

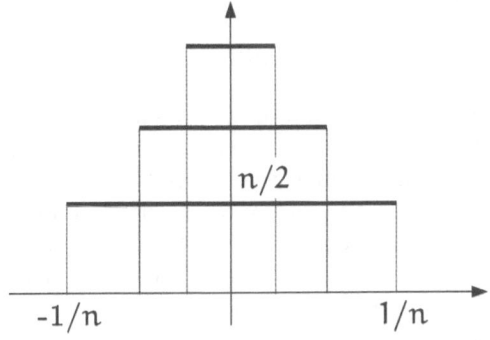

Fig. 2.6. A sequence of pulses converging toward an impulse at the origin.

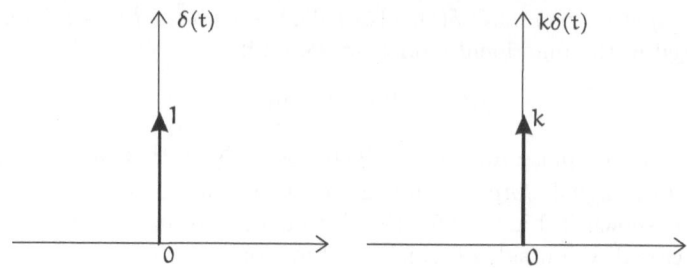

Fig. 2.7. Graphical representation of the impulse function.

which shows how every signal can be "reconstructed" as an infinite sum of impulses, translated and scaled according to the value of the signal. We leave it to the reader to show the plausibility of (2.2), using the definition of δ as a limit of pulse signals (2.1).

Figure 2.7, left, shows the graphical representation of the impulse δ. The "graph" of a scaled version $k\delta$, for $k \in \mathbb{R}$, is shown on the right. In these figures, the finite height of the arrow indicates finite energy, and not finite value.

2.2.4 The Frequency Model

We recall that, in the spatial model, a signal is determined by a function $f : U \subset \mathbb{R}^m \to V$ that defines the variation of the signal in the space or time domain U. This is the model that relates directly to the signal magnitudes of the physical world. The frequency model, to be introduced in this section, is closely related with our perception of the signal. But instead of motivating it with a perceptual example, we prefer to start with a very simple and well-known mathematical example.

Consider the signal defined in time by a sine curve

$$f(t) = a \sin(2\pi\omega_0 t + \Phi).$$

Such a signal is completely characterized by its amplitude a, its frequency ω_0, and its phase angle Φ. The frequency measures how fast the signal changes: namely, ω_0 cycles per unit of time. See Figure 2.8.

Based on this example, we can try to characterize an arbitrary signal in terms of its frequency components. That is, we can consider a functional model that associates to each frequency in a signal the corresponding amplitude and phase. Consider, for example, the periodic signal $f(t) = a\cos(2\pi\omega_0 t)$, whose graph is shown in Figure 2.9(a). This signal has a single frequency component ω_0, with amplitude a and phase 0. Thus its functional representation in terms of frequencies can be considered to be

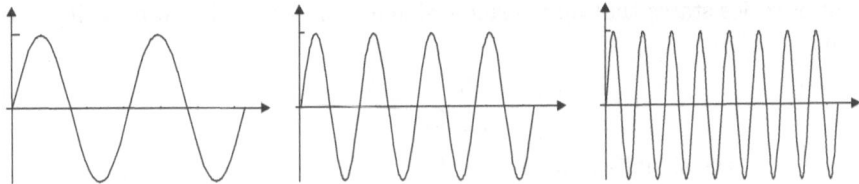

Fig. 2.8. Sinusoidal signals with frequencies 2, 4, and 8.

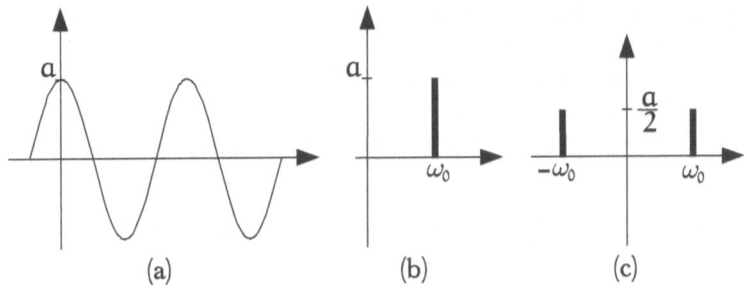

Fig. 2.9. A periodic signal and two alternative frequency models for it.

$$\mathrm{Freq}(f)(s) = \begin{cases} a & \text{if } s = \omega_0, \\ 0 & \text{if } s \neq \omega_0. \end{cases}$$

The graphical representation of $\mathrm{Freq}(f)$ is shown in Figure 2.9(b).

We now make an important observation with respect to the definition of $\mathrm{Freq}(f)(s)$. Using complex notation, we can write

$$\cos t = \tfrac{1}{2}(e^{it} + e^{-it});$$

therefore,

$$f(t) = a\cos(2\pi\omega_0 t) = \tfrac{1}{2}a(e^{2\pi i\omega_0 t} + e^{2\pi i(-\omega_0)t}).$$

We see that in this formulation we have two symmetric frequency components, with frequencies ω_0 and $-\omega_0$, both having amplitude $\tfrac{1}{2}a$, so that we can instead write

$$\mathrm{Freq}(f)(s) = \begin{cases} \tfrac{1}{2}a & \text{if } s = \omega_0 \text{ or } s = -\omega_0, \\ 0 & \text{otherwise.} \end{cases}$$

The graph of this $\mathrm{Freq}(f)$ is shown in Figure 2.9(c). As we will see, allowing components with negative frequencies is a natural way to take into account the phase information of a periodic signal.

The analysis just made can be generalized, yielding a functional model in the frequency variable for an arbitrary periodic signal. Indeed, the theory of

Fourier series states that any periodic signal f of period T_0 can be written as a sum

$$f(t) = \sum_{k=-\infty}^{+\infty} c_k e^{i 2\pi k \omega_0 t}, \tag{2.3}$$

where ω_0 is the signal's *fundamental frequency*, and the c_k are complex numbers.

This Fourier series development shows that a periodic signal has (potentially) all the frequencies that are multiples of the fundamental frequency ω_0, and no others. Thus a frequency-domain functional model for a periodic real- or complex-valued signal is a function that associates to each multiple $k\omega_0$, for $k \in \mathbb{Z}$, a complex amplitude c_k.

Usually f is real-valued. Then the coefficients c_k occur in complex conjugate pairs. That is, $c_{-k} = \bar{c}_k$ for each k (where the bar stands for complex conjugation), and Equation (2.3) can be written in the form

$$f(t) = c_0 + 2\sum_{k=1}^{+\infty} \left(\operatorname{Re} c_k \, \cos(2\pi k \omega_0 t) - \operatorname{Im} c_k \, \sin(2\pi k \omega_0 t) \right). \tag{2.4}$$

Thus, for real-valued f, we can think in terms of positive frequencies only, if we so desire. The component with frequency $k\omega_0$ is the linear combination of $\sin(2\pi k\omega_0 t)$ and $\cos(2\pi k\omega_0 t)$ that appears in (2.4). We see that the modulus of c_k is the amplitude of this component, and the argument of c_k is its phase. In the complex exponential formulation (2.3), this combined component of frequency $k\omega_0$ splits into two, with frequencies $\pm k\omega_0$.

Example 2.1 (Sawtooth signal). The periodic signal $f(t)$ whose graph is shown in Figure 2.10(a) is called a *sawtooth signal*. With $\omega_0 = 1/T_0$, the Fourier series representation of f is

$$f(t) = \frac{1}{T_0} + \sum_{\substack{k>0 \\ k \text{ odd}}} \frac{4}{\pi^2 k^2 T_0} \left(e^{i \cdot 2\pi k \omega_0 t} + e^{-i \cdot 2\pi k \omega_0 t} \right)$$

$$= \frac{1}{T_0} + \frac{8}{\pi^2 T_0} \left(\cos(2\pi\omega_0 t) + \frac{1}{3^2} \cos(6\pi\omega_0 t) + \frac{1}{5^2} \cos(10\pi\omega_0 t) + \cdots \right).$$

The frequency representation of f is shown in Figure 2.10(b).

Fourier series allowed us to define a frequency functional model for periodic signals. We now want to extend this model to nonperiodic signals. We need a tool to measure the contribution of a given frequency to an arbitrary signal. We cannot use Fourier series, since a nonperiodic signal can contain arbitrary frequencies, not just multiples of a fixed fundamental frequency—in other words, we have a continuum of frequencies.

The classical technique to measure this contribution is the *Fourier transform*, which we now explain.

The Fourier Transform

Given a signal $f : \mathbb{R} \to \mathbb{R}$, the *Fourier transform* $F(f)$ of f is defined by

$$F(f)(s) = \hat{f}(s) = \int_{-\infty}^{+\infty} f(t)e^{-2\pi its}\, dt. \qquad (2.5)$$

The reader should be aware that sometimes the transformed signal $\hat{f}(s)$ is denoted by $F(s)$. We will avoid this notation because it confuses the operator F with the transformed function.

One can intuitively see that the Fourier transform detects frequencies in the signal f. Indeed, the kernel $e^{-2\pi its}$ is a periodic signal with frequency s. Thus the modulation $f(t)e^{-2\pi its}$ detects the frequency values s of the signal f that are in resonance with the frequencies of the kernel $e^{-2\pi its}$. The integral in (2.5) measures the "density" of the frequency s in the signal f throughout its domain. Thus $\hat{f}(s)$ shows with what intensity the frequency s occurs in the signal f, but it does not say where it occurs in the domain of the signal.

One immediately checks that the Fourier transform defines a linear operator $F : \mathcal{S} \to \mathcal{S}'$, with $F(f) = \hat{f}$, between two signal spaces. The Fourier transform operator F is invertible, and its inverse F^{-1} is given by

$$f(t) = F^{-1}(\hat{f}(s)) = \int_{-\infty}^{+\infty} \hat{f}(s)e^{2\pi ist}\, ds. \qquad (2.6)$$

Intuitively, (2.6) says that the signal f can be "reconstructed" as an infinite sum of signals with frequency s (for $s \in \mathbb{R}$) and amplitude $\hat{f}(s)$.

Thus a signal can be characterized either by its spatial model f or by its frequency model $\hat{f}$. The Fourier transform operator and its inverse allow us to go back and forth between the two models. Thus the spatial model gives us information on the variation of the signal in the space domain, whereas the frequency model records the variation of the signal in the frequency domain. Generally speaking, the spatial model is used for signal synthesis, while the frequency model is more useful for signal analysis.

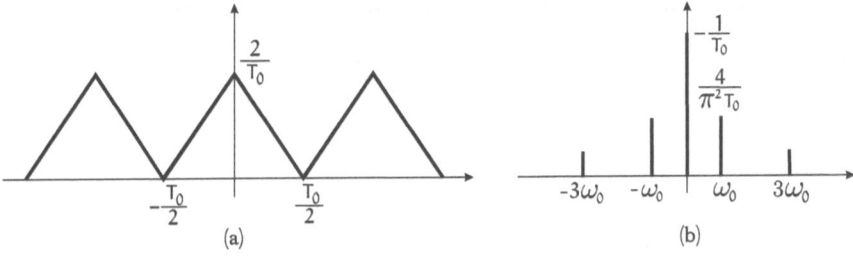

Fig. 2.10. A sawtooth signal in a spatial model (a) and frequency model (b) representation.

We remark that there are three notation conventions in the literature for writing the Fourier transform and its inverse. These conventions differ from each other in the placement of the factor 2π. We opted for placing the factor 2π on the exponent, as shown in Equations (2.5) and (2.6), and nowhere else. In this convention, if t represents time or space, the variable s of the transformed signal indeed represents the frequency with no scale factor.

All these considerations generalize to m-dimensional signals. The Fourier transform of a signal $f : \mathbb{R}^m \to \mathbb{R}$ is defined by

$$\hat{f}(U) = \int_{\mathbb{R}^m} f(X)e^{-2\pi i \langle X, U \rangle} \, dX, \tag{2.7}$$

where $U = (u_1, u_2, \ldots, u_m)$, $X = (x_1, x_2, \ldots, x_m)$, $\langle X, U \rangle = \sum_{i=1}^m x_i u_i$, and $dX = dx_1 \, dx_2 \ldots dx_m$.

Other Frequency Models

There are other frequency-variable functional models for signals in addition to the one based on the Fourier transform. Each such model has a corresponding transform operator, which fills the role played by the Fourier transform in the frequency model described in this section. We mention the *cosine transform* and the *window Fourier transform*. Also, the *wavelet transform* is associated with a scale-space model of a signal. The existence of different functional models for signals, and of transform operators that allow one to convert between them, is very important for image synthesis and analysis. For each application we must devise the model most appropriate to the problem.

Fourier Transform of the Pulse

Consider the pulse function

$$p_a(t) = \begin{cases} 1 & \text{if } |t| \leq a, \\ 0 & \text{if } |t| > a \end{cases}$$

defined in Section 2.2.3. A short calculation shows that its Fourier transform is

$$\hat{p}_a(s) = 2a \, \frac{\sin(2\pi a s)}{2\pi a s}.$$

Using the classical notation

$$\text{sinc}(t) = \begin{cases} \dfrac{\sin t}{t} & \text{if } t \neq 0, \\ 1 & \text{if } t = 0, \end{cases}$$

we can write

$$\hat{p}_a(s) = 2a \, \text{sinc}(2\pi a s).$$

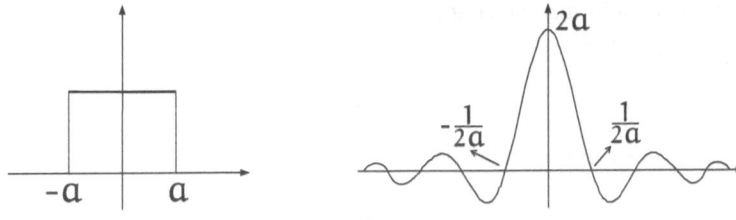

Fig. 2.11. The pulse signal and its Fourier transform.

Figure 2.11 shows the graphs of the pulse function and of its Fourier transform sinc. As we will see, this function plays an important role in the study of signals.

Looking at the Fourier transform of the pulse signal, we see that, as the width $2a$ of the support of the pulse function approaches 0, the presence of high frequencies in the signal increases: $1/(2a) \to \infty$. Conversely, if $a \to \infty$, the presence of high frequencies in the signal decreases. This is a particular case of a general result that relates the standard deviation of the signal in the space domain with that in the frequency domain. The smaller the standard deviation is in one of the domains, the greater it is in the other. This observation is the basis of the *uncertainty principle*, which is very important when relating signals in the spatial and frequency models.

Fourier Transform of the Impulse

The Fourier transform of an impulse signal can be obtained by approximating the impulse by pulse functions, as explained in Section 2.2.3. The calculations we make here are not mathematically rigorous, but they have heuristic value; a rigorous calculation using distributions would lead to the same result.

$$\hat{\delta}(s) = F(\delta) = F\left(\lim_{n \to \infty} \tfrac{1}{2} n p_{1/n}(t)\right)$$

$$= \lim_{n \to \infty} F\left(\tfrac{1}{2} n p_{1/n}(t)\right) = \lim_{n \to \infty} \frac{\sin(2\pi s/n)}{2\pi s/n} = 1.$$

Thus the Fourier transform of a unit impulse signal, or Dirac delta, at the origin is the constant function $\hat{\delta}(s) = 1$.

2.3 Linear Representation of Signals

Now that we have defined our signal models in the mathematical universe, we tackle the second level in our abstraction hierarchy: the discretization problem. As already said, discretization of a signal is classically known as sampling. We introduce here the concept of *representation*, which extends, in a sense to be made precise, the notion of sampling.

We start by defining the space ℓ^2 of *square-summable sequences*. This consists of all sequences

$$(\ldots, c_{-2}, c_{-1}, c_0, c_1, c_2, \ldots)$$

of real or complex numbers, such that

$$\sum_{i=-\infty}^{+\infty} |c_i|^2 < \infty. \tag{2.8}$$

The convergence condition in (2.8) is made so we can define an inner product in ℓ^2. The inner product of two sequences $\boldsymbol{b} = (b_i)$ and $\boldsymbol{c} = (c_i)$ is

$$\langle \boldsymbol{c}, \boldsymbol{b} \rangle = \sum_{i=-\infty}^{+\infty} c_i b_i.$$

Note that Euclidean spaces $\mathbb{R}^n$ have a natural isometric embedding in ℓ^2 (i.e., they extend each n-tuple to an infinite sequence by adding zeros).

A *linear representation* of a signal space $\mathcal{S} = \{f : U \subset \mathbb{R}^m \to V\}$ is a continuous linear operator $R : \mathcal{S} \to \ell^2$. The image $V = R(\mathcal{S})$ of the space $\mathcal{S}$ under R is a subspace of ℓ^2, called the *representation space* of $\mathcal{S}$. If V is finite-dimensional, we have a *finite representation*. When the signal f can be recovered from its representation $R(f)$, we say that f is *represented exactly*. If every $f \in \mathcal{S}$ is represented exactly, we have an *exact representation*. If the operator R is invertible, the representation is obviously exact. But it is possible to obtain exact reconstruction even for noninvertible representation operators.

The purpose of the representation step is to obtain a discrete signal from its continuous model.

Example 2.2 (Finite point sampling). Given a signal space $\mathcal{S} = \{f : U \subset \mathbb{R}^m \to V\}$, fix k points $u_1, u_2, \ldots, u_k$ in the domain U. One immediately verifies that the map $R : \mathcal{S} \to \mathbb{R}^k \subset \ell^2$ defined by

$$R(f) = \big(f(u_1), f(u_2), \ldots, f(u_k)\big)$$

is a linear representation of $\mathcal{S}$, which is also finite. The elements $f(u_i)$ of the representation vector are called *samples* of the signal f. This representation is traditionally known as *finite point sampling*. It is an exact representation if we can devise an interpolation method that reconstructs f from its samples.

The purpose of a linear representation is to replace the initial signal space, which is generally infinite-dimensional, by a subspace of ℓ^2. In general, we try to work with finite-dimensional subspaces of ℓ^2, so we can later obtain an encoding of the represented signal. A linear representation induces a natural relationship between signal space and representation space. Indeed, given a representation $R : \mathcal{S} \to \ell^2$, we define an equivalence relation in the signal space $\mathcal{S}$ by

$$f \simeq g \quad \text{if and only if} \quad R(f) = R(g).$$

The quotient map $\bar{R} : \mathcal{S}/{\simeq} \to V$, making the diagram

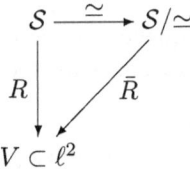

commute, is a linear isomorphism. We can then use the representation space V instead of the signal space. However, unless the representation is exact— that is, unless the equivalence relation $\simeq$ is the trivial one—there is a loss of information in the representation process. This is illustrated in Figure 2.12, for the case of finite point sampling (Example 2.2): two distinct signals have the same representation if their values happen to coincide at all sampling points (here $f(u_i) = g(u_i)$ for $i = 1, \ldots, 4$).

In principle, the isomorphism $\bar{R}$ in the diagram above establishes only an algebraic equivalence between the quotient space and the representation space. For each representation one must make a detailed analysis of the loss of information incurred in passing to the quotient.

2.3.1 Existence of Exact Representations

Consider an enumerable set of signals $\{g_k : k \in \mathbb{Z}\}$ from a signal space $\mathcal{S}$ with an inner product $\langle \, , \, \rangle$. We define the *representation operator* R by

$$R(f) = (c_k)_{k \in \mathbb{Z}} \in \ell^2, \quad \text{where} \quad c_k = \langle f, g_k \rangle.$$

From the properties of the inner product, it follows that R is indeed a linear representation of the signal space $\mathcal{S}$. The elements g_k are called *atoms* of the representation. Here is a well-known example of this representation.

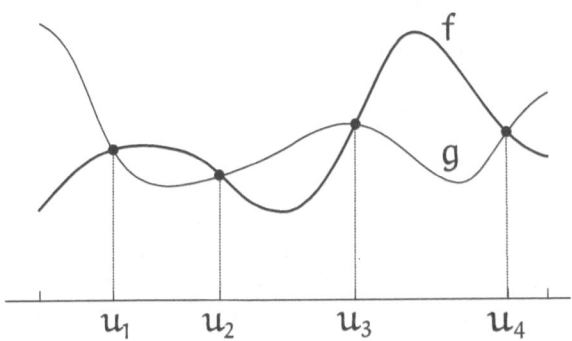

Fig. 2.12. Distinct signals can have the same point sampling representation.

Example 2.3 (Fourier sampling). The Fourier series of a periodic signal f, given by (2.3), defines in fact a linear representation

$$f \mapsto (\ldots, c_{-2}, c_{-1}, c_0, c_1, c_2, \ldots)$$

of the signal f, where $c_k = \langle f, e^{i2\pi k\omega_0 t} \rangle$. This representation gives a discrete model of the signal in the frequency domain. It is called *Fourier sampling*. The exactness of this representation is related to the convergence properties of the Fourier series.

In Example 2.2 we introduced the finite point representation of a signal. Clearly, every signal has representations of this type, so a given signal space always has a linear representation. However, as already seen, this representation is not necessarily exact.

Using the representation by atoms introduced in the beginning of this section, one can look for conditions on the set $\{g_k\}_{k \in \mathbb{Z}}$ that guarantee the exactness of the representation operation R. This would give us a method to obtain exact representations. The following theorem is a result in this direction.

Theorem 2.4 (Representation theorem). *Every space $\mathcal{S}$ of signals of finite energy admits an exact representation.*

Proof (Outline of proof). Consider $\mathcal{S}$ as a (not necessarily closed) subspace of L^2. It can be proved that there exists a complete orthonormal set $\{\ldots, e_{-2}, e_{-1}, e_0, e_1, e_2, \ldots\}$, such that

$$f = \sum_{k=-\infty}^{\infty} c_k e_k$$

for every signal $f \in \mathcal{S}$, where $c_k = \langle f, e_k \rangle$. Moreover,

$$\sum_{k=-\infty}^{\infty} |c_k|^2 = \|f\|^2 < \infty,$$

so $(c_k) \in \ell^2$. Now define a representation by setting

$$f \mapsto L(f) = (\ldots, c_{-2}, c_{-1}, c_0, c_1, c_2, \ldots).$$

It is clear that L is linear.

In practice, Theorem 2.4 requires that we devise methods to obtain complete orthonormal sets of signal spaces. This is not an easy task in general, especially when the orthonormal set must satisfy additional properties.

Complete orthonormal sets constitute the proper generalization of a basis in a signal space with inner product. Theorem 2.4 says that whenever we have a basis we obtain a method for exact representation, at least theoretically.

We should point out, however, that even when the family $\{g_k; k \in \mathbb{Z}\}$ of atoms is not a basis, the representation operator can be of great value in applications. Indeed, if the set generates the space, the only property that we miss is the uniqueness of the representation, but this is not important for some applications.

Here is another example of a representation based on an orthonormal family of signals.

Example 2.5 (Area sampling). Consider a signal space

$$\mathcal{S} = \{f : U \subset \mathbb{R}^m \to \mathbb{R}\},$$

and take a countable partition of the domain U, say $\mathcal{P} = \bigcup_i U_i$ for $i \in \mathbb{Z}$, such that the area of the set U_i satisfies $\text{Area}(U_i) > \varepsilon > 0$ (see Figure 2.13). Define a family of functions $\chi_i : U \to \mathbb{R}$ on U by setting

$$\chi_i(p) = \begin{cases} \dfrac{1}{\text{Area}(U_i)} & \text{if } p \in U_i, \\ 0 & \text{if } p \notin U_i. \end{cases}$$

The family $\{\chi_i\}_{i \in \mathbb{Z}}$ forms an orthogonal set in the space of signals, with respect to the L^2 inner product. Given a signal f, we have

$$f = \sum_i \text{Area}(U_i) \, c_i \chi_i, \qquad (2.9)$$

where

$$c_i = \langle f, \chi_i \rangle = \frac{1}{\text{Area } U_i} \int_{U_i} f(u) \, du. \qquad (2.10)$$

That is, the value of the coefficient c_i in this representation is exactly the mean value of the signal over the set U_i. This representation is known in the literature as *area sampling*. We remark that, since the orthonormal set of functions $\{\chi_i\}_{i \in \mathbb{Z}}$ is not complete in $L^2(U)$, the representation arising from (2.9) is not exact: distinct signals can have the same mean values over each U_i.

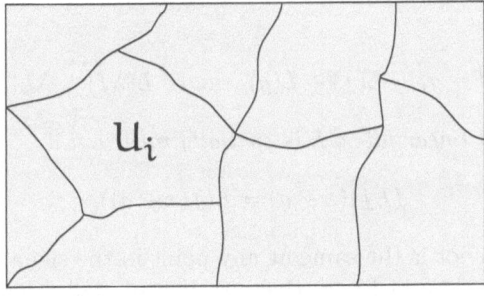

Fig. 2.13. A partition of U.

2.4 Operations on Signals

An *operation* on a signal space $\mathcal{S}$ is a map

$$T : \mathbb{R}^m \times \mathcal{S}^n \to \mathcal{S}'$$

into a (possibly different) signal space $\mathcal{S}'$.

As already remarked, since a signal space is a vector space, there are two natural operations: *addition* of signals, an operation $\mathcal{S} \times \mathcal{S} \to \mathcal{S}$ defined by

$$(f + g)(t) = f(t) + g(t),$$

and *multiplication* of a signal by a scalar, an operation $\mathbb{R} \times \mathcal{S} \to \mathcal{S}$ defined by

$$(\lambda f)(t) = \lambda \cdot f(t),$$

where $\lambda \in \mathbb{R}$. Combining these operations, we can, for example, define the linear interpolation between two signals f and g, obtaining a one-parameter family h_u of signals

$$h_u(t) = (1 - u)f(t) + ug(t), \quad \text{for } u \in [0, 1].$$

This family gives a continuous transition between the signals f and g, as the parameter u varies from 0 to 1.

For real-valued signals $f : U \subset \mathbb{R}^m \to \mathbb{R}$ (or for signals taking values in another space having a product structure, such as the complex numbers $\mathbb{C}$), we can define the *multiplication* of two signals, an operation $\mathcal{S} \times \mathcal{S} \to \mathcal{S}$, by setting

$$(fg)(x) = f(x)g(x).$$

2.4.1 Filters

A unary operation $L : \mathcal{S} \to \mathcal{S}$ on a signal space is called a *filter*. If L is a linear map, so that

$$L(f + g) = L(f) + L(g) \quad \text{and} \quad L(\lambda f) = \lambda L(f),$$

we say that L is a *linear filter*. L is *spatially invariant* if

$$(Lf)(x - a) = L(f(x - a)),$$

that is, if its behavior is the same at any point in the signal's space domain.

An important class of filters that are not spatially invariant consists of *adaptive filters*, where the value of the filter at a point depends on the behavior of the signal in a neighborhood of the point.

Convolution and Impulse Response

The *impulse response* of a filter L is the image $L(\delta)$ of the impulse signal δ under the filter. From (2.2) we see that a signal f can be written as an infinite sum of impulse signals, appropriately translated and modulated. Thus, if we know the impulse response $h(t) = L(\delta)$ of a linear and spatially invariant filter L, we can say that the response $L(f)$ of the filter to an arbitrary signal f is an infinite sum of copies of h, appropriately translated and modulated. More precisely:

$$Lf(x) = \int_{-\infty}^{+\infty} f(t)L\big(\delta(x-t)\big)\,dt = \int_{-\infty}^{+\infty} f(t)h(x-t)\,dt. \tag{2.11}$$

The second integral in (2.11) defines an operation $\mathcal{S} \times \mathcal{S} \to \mathcal{S}$ called the *convolution product*, or simply *convolution*. In general, the convolution of two signals f and g is denoted $f * g$ and is defined by

$$f * g(x) = \int_{-\infty}^{+\infty} f(t)g(x-t)\,dt.$$

We therefore have the following result.

Theorem 2.6. *A spatially invariant linear filter L is entirely determined by its impulse response function h. More precisely, for every signal f we have $L(f) = f * h$.*

The impulse response function is also called the *kernel* of the filter. When using linear and spatially invariant filters, we will often identify the operation of filtering with the operation of convolution with the appropriate kernel.

2.4.2 Transforms

A linear transformation $T : \mathcal{S} \to \mathcal{S}'$ between two functional models of a signal space is called a *transform operation*, or just a *transform*. In general, we will require our transform operations to be invertible, so that they provide a faithful conversion between the two functional signal models. An important example is the *Fourier transform*, introduced in Section 2.2.4. It relates the spatial model to the frequency model, providing information about the frequencies present in each signal.

The use of transforms allows us to switch among various signal models, so we can choose the one most appropriate to the solution of a given problem. It is therefore important to know the correspondence between operations for different signal spaces under a given transform. As a generic example, consider a binary operation $\oplus : \mathcal{S} \times \mathcal{S} \to \mathcal{S}$ and a transform $T : \mathcal{S} \to \mathcal{S}'$ into another space. It is useful to know what binary operation on $\mathcal{S}'$ corresponds to $\oplus$ on $\mathcal{S}$ under the transform T. More precisely, we want to know what operation $\ominus$ on $\mathcal{S}'$ makes the diagram

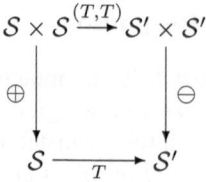

$$\mathcal{S} \times \mathcal{S} \xrightarrow{(T,T)} \mathcal{S}' \times \mathcal{S}'$$

commute (this means $T(f) \ominus T(g) = T(f \oplus g)$ for all $f, g \in \mathcal{S}$).

A particular case of great importance consists in finding the operation on the frequency domain that corresponds to a certain filter on the space domain. If the filter is linear and spatially invariant, the answer is simple and comes from a classical connection between the Fourier transform and the convolution product. Namely, given two real-valued (or complex-valued) signals f and g in the space domain, we have

$$F(f * g) = F(f)F(g), \tag{2.12}$$

where on the right we have the product of signals. Thus, filtering with a kernel h in the space domain corresponds, in the frequency model, to multiplying by the Fourier transform $F(h) = \hat{h}$ of the kernel. We call $\hat{h}$ the *transfer function* of the filter.

This result is fundamental, in theory as well as in applications. The transfer function allows us, for example, to analyze the action of a filter on the various frequencies present in a signal. We will discuss this in greater detail in the next section.

2.4.3 Filtering and Frequencies

It is usual to divide the frequency domain of a signal into two regions: a *low-frequency region*, near the origin, and a *high-frequency region*, away from the origin. Obviously, what counts as "low" or "high" depends on the application. For some applications it is useful to subdivide the frequency domain further, partitioning it into several disjoint regions, so as to obtain a decomposition of the signal into frequency *bands*.

We say that a signal f is *bandlimited* if the support $\text{supp} \hat{f}$ of its Fourier transform is a bounded set. In the one-dimensional case this means that $\text{supp} f \subset [-\Omega, \Omega]$, where Ω is finite. When the kernel of a filter has bounded support, we say the filter has *finite impulse response*, or that it is an FIR filter. Otherwise we say the filter has *infinite impulse response*, or that it is an IIR filter.

Let L be a filter and f a signal. We say that L is a *highpass filter* if the action of L removes or attenuates the low frequencies in the signal, leaving the high frequencies mostly unchanged. We say it is a *lowpass* or *smoothing* filter in the opposite situation, when high frequencies are removed. A *bandstop* filter is one that selectively eliminates the frequencies in a certain band of the spectrum, while allowing to pass frequencies above and below this band.

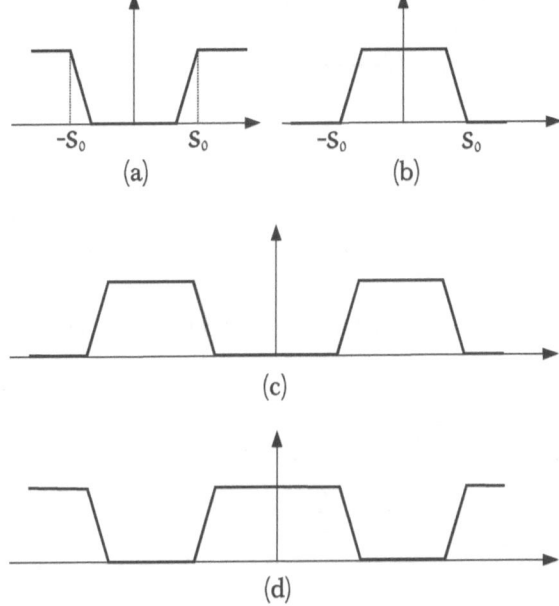

Fig. 2.14. Transfer function for filters: (a) highpass; (b) lowpass; (c) bandpass; (d) bandstop.

Finally, a *bandpass* filter eliminates frequencies outside a given band. The transfer functions of these filters are shown schematically in Figure 2.14.

The frequency value that separates (approximately) the low- and high- frequency regions in lowpass and highpass filters is called the *cutoff frequency*. Bandpass and bandstop filters can be considered to have two cutoff frequencies, one at each end of the band in question.

An ideal lowpass filter allows frequencies under the cutoff frequency to pass and completely rejects other frequencies. The transfer function for such a filter in dimension 1 is the pulse function, shown in Figure 2.15.

In higher dimensions the transfer function of an ideal filter can assume different shapes. Figure 2.16 shows a cylinder-shaped and a square-shaped ideal reconstruction filters.

2.5 Sampling Theory

In this section we return to sampling. We start by defining point sampling, a notion we discussed briefly in Example 2.2.

Let $f : U \subset \mathbb{R}^m \to \mathbb{R}^n$ be a signal, and let $\bar{U} = \{\ldots, u_{-2}, u_{-1}, u_0, u_1, u_2, \ldots\}$ be an enumerable subset of U. *Point sampling* f at $\bar{U}$ consists in taking the sequence

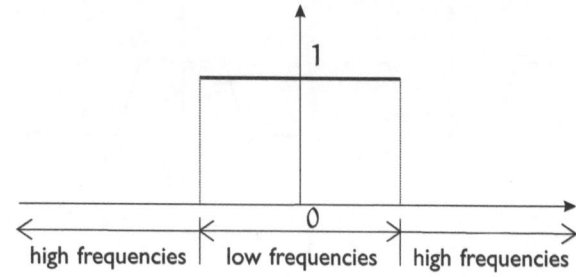

Fig. 2.15. One-dimensional ideal lowpass filter.

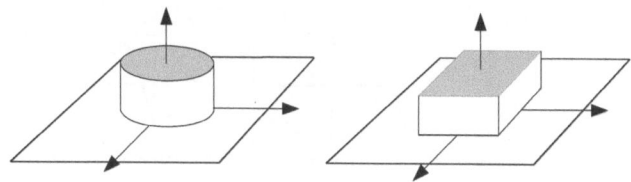

Fig. 2.16. Transfer function of ideal two-dimensional lowpass filters.

$$(\dots, f(u_{-2}), f(u_{-1}), f(u_0), f(u_1), f(u_2), \dots). \qquad (2.13)$$

Each value $f(u_i)$ is a *sample* of the signal, and the sequence (2.13) is the *sample sequence*. See Figure 2.17.

2.5.1 Uniform Point Sampling

Uniform point sampling is very common and important in signal theory. It is amenable to a relatively simple mathematical treatment and is used in many types of digital hardware for signal manipulation.

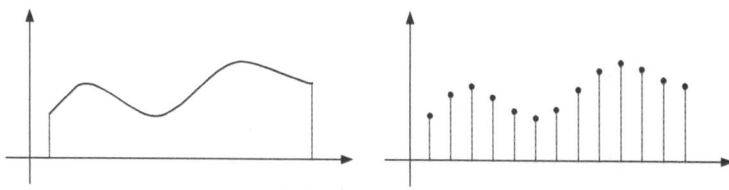

Fig. 2.17. Point sampling of a signal.

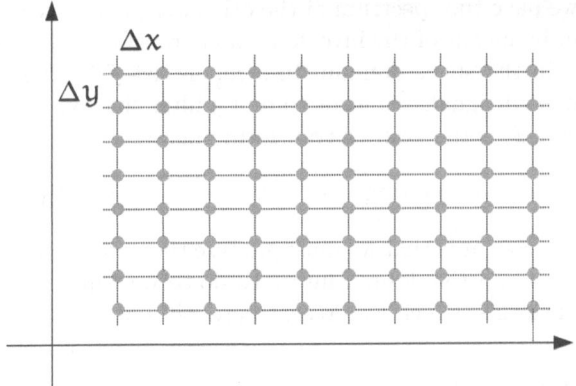

Fig. 2.18. A plane lattice (lines are not part of the lattice).

We start by defining lattices. For each axis x_i of $\mathbb{R}^n$, take a nonzero length Δx_i. The set of points of $\mathbb{R}^n$ given by

$$L_\Delta = L_{\Delta x_1, \ldots, \Delta x_n} = \{(m_1 \, \Delta x_1, \, m_2 \, \Delta x_2, \, \ldots, \, m_n \, \Delta x_n) : m_i \in \mathbb{Z}\}$$

is called a *lattice* in $\mathbb{R}^n$. Figure 2.18 shows a lattice on the plane $\mathbb{R}^2$.

Each point of L_Δ is a *vertex* of the lattice. We will often refer to the lattice by its defining vector $\Delta = (\Delta x_1, \ldots, \Delta x_n)$.

A point sampling is *uniform* when the set $\bar{U}$ of points at which samples are taken in (2.13) forms a lattice. The number of samples per unit of length in each direction of space is called the *sampling rate* or *sampling frequency* in that direction.

As we saw in Section 2.3 (Figure 2.12), there is usually loss of information under point sampling. Here we are interested in two problems:

- Under what conditions can we avoid loss of information when performing point sampling? That is, when is it possible to reconstruct the signal f from its sample sequence (2.13)?
- Assuming that the sampling was lossless, how can the original signal be reconstructed from the samples? What interpolation method should be used?

These two questions can be subsumed under one: *under what conditions does point sampling give rise to an exact linear representation?* We will study this question for uniform sampling.

Intuitively, point sampling replaces the signal with a sum of "finite impulses" (Figure 2.17). This process introduces high frequencies in the sampled signal. These high frequencies, combined with the frequencies present in the original signal, form the spectral model of the sampled signal. Thus exact reconstruction is possible whenever we can find an appropriate filter to extract the frequencies of the original signal from the frequencies of the sampled signal. Once we have the spectrum of the original signal, we can return to the spatial domain by means of the inverse Fourier transform, thus recovering the original signal in the space domain. In symbols, let f be the original signal and f_d the sampled signal. We want to find a filter that will transform $F(f_d)$ into $F(f)$, where F is the Fourier transform operator:

$$F(f_d) \;\rightarrow\; \text{filtering} \;\rightarrow\; F(f) \;\rightarrow\; f = F^{-1}(F(f)).$$

From this we conclude that we must analyze the frequencies of the sampled signal in order to find the desired filter. To do so we will need several results on Fourier transforms, which we present here without proof.

2.5.2 Point Sampling and the Fourier Transform

A very interesting way of analyzing point sampling is by using an *impulse train* associated to the sampling lattice. In the one-dimensional case the impulse train is given by

$$\delta_{\Delta t} = \sum_{n=-\infty}^{\infty} \delta(t - n\,\Delta t),$$

as illustrated in Figure 2.19.

The lattice interval length Δt is called the *sampling period*, and $\omega = 2\pi/\Delta t$ is the *sampling frequency*. The discretization of the signal is a two-step process. First we obtain the signal

$$f_\Delta(t) = f \cdot \delta_{\Delta t} = \sum_{n=-\infty}^{\infty} f(n\,\Delta t)\delta(t - n\,\Delta t) \qquad (2.14)$$

by multiplying the signal and the impulse train. This is illustrated in Figure 2.20.

From the "train signal" $f_\Delta = f \cdot \delta_\Delta$, the discretized signal is obtained by replacing $\delta(t)$ by the *unit discrete impulse*

$$\delta[n] = \begin{cases} 1 & \text{if } n = 0, \\ 0 & \text{if } n \neq 0, \end{cases}$$

and changing the variable to n. This is illustrated in Figure 2.21.

The unit discrete impulse plays the role of a Dirac delta impulse when we work with signals in the discrete domain.

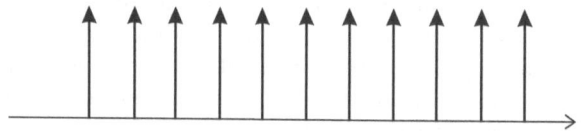

Fig. 2.19. One-dimensional impulse train.

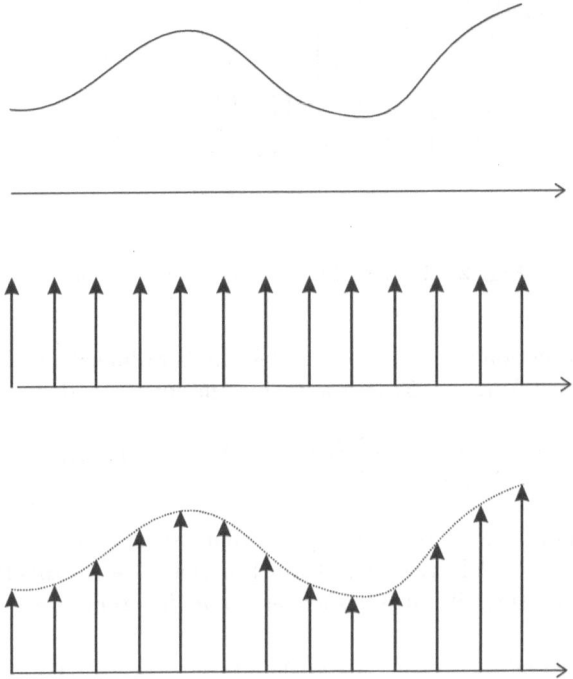

Fig. 2.20. Product of the signal by an impulse train.

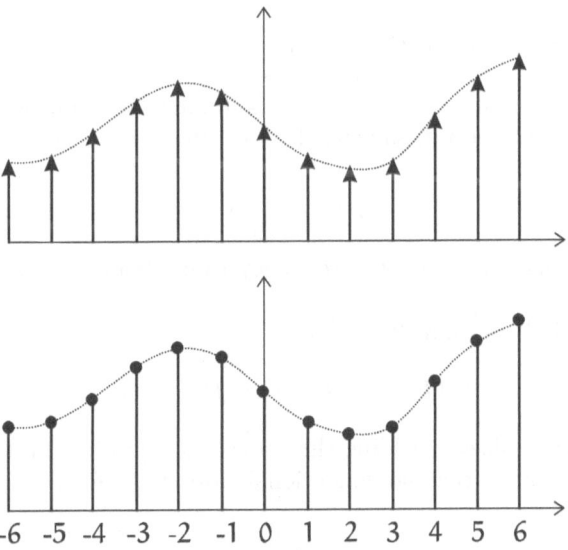

Fig. 2.21. From an impulse train to a discrete signal.

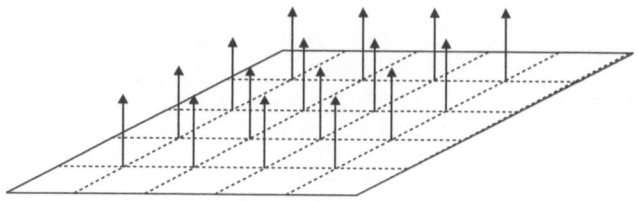

Fig. 2.22. Train function in two dimensions.

The train function can be generalized for higher dimensions. Given a lattice L_Δ, where $\Delta = (\Delta x_1, \dots, \Delta x_n)$, the train function δ_Δ is defined by

$$\delta\Delta = \sum_{i_1 \in \mathbb{Z}} \cdots \sum_{i_n \in \mathbb{Z}} \delta(x - (i_1\,\Delta x_1, \dots, i_n\,\Delta x_n))$$

for $x \in \mathbb{R}^n$. Figure 2.22 shows a graphical representation of the train function for a two-dimensional lattice. The train function receives different names in the signal processing literature, such as *comb function, spike train,* or *sha function.*

As we discussed for the one-dimensional case, the importance of the comb function is that *performing uniform point sampling with respect to a lattice L_Δ is equivalent to multiplying the signal by the comb function $comb_\Delta$,* in the sense that all the information in the product is retained at the lattice vertices and abandoned in between.

More Properties of the Comb Function

Another important step in our analysis is finding the frequency model of the comb function. The result, somewhat surprisingly, is another comb function! More precisely,

$$F(\delta_{\Delta t}(t)) = \frac{1}{\Delta t}\,\delta_{1/\Delta t}(s).$$

Notice the change in the lattice frequency from Δt in the spatial domain to $1/\Delta t$ in the frequency domain.

Finally, it follows from (2.2) that

$$(f * \delta(t - t_0))(t) = f(t - t_0). \tag{2.15}$$

This can be generalized, yielding the convolution product of a signal f with $\delta_{\Delta t}(t)$. The result, in the one-dimensional case, is the following:

$$(f * \delta_{\Delta t}) = \sum_{k=-\infty}^{+\infty} f(t - k\,\Delta t). \tag{2.16}$$

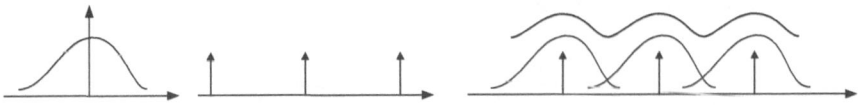

Fig. 2.23. Convolution of a signal with a comb function.

In other words, to obtain the convolution $f * \delta_{\Delta t}$, the function f is translated to the vertices of the lattice, and the result is obtained by adding all these translates. (In fact, this is true for higher-dimensional signals as well.) See Figure 2.23.

2.5.3 The Sampling Theorem

Using the results above, we can now obtain information on the frequency model of a uniformly sampled signal. We do this in Figure 2.24: in the space domain, sampling is obtained by multiplying the signal by the comb function; in the frequency domain, this corresponds to convolving (taking the convolution product) with the transform of the comb function, which is given by (2.16). In Figure 2.24(c), notice the sampled signal and its Fourier transform.

We conclude that the spectral model of a signal sampled at a lattice $L_{\Delta t}$ (in one dimension) is obtained by adding translates of the frequency model of the original signal, the translation distances being all multiples of $1/\Delta t$. Figure 2.25 is an enlarged version of the frequency model of the sampled signal of Figure 2.24. One can see that the translates of the original spectrum overlap. In particular, high frequencies present in the original signal can give

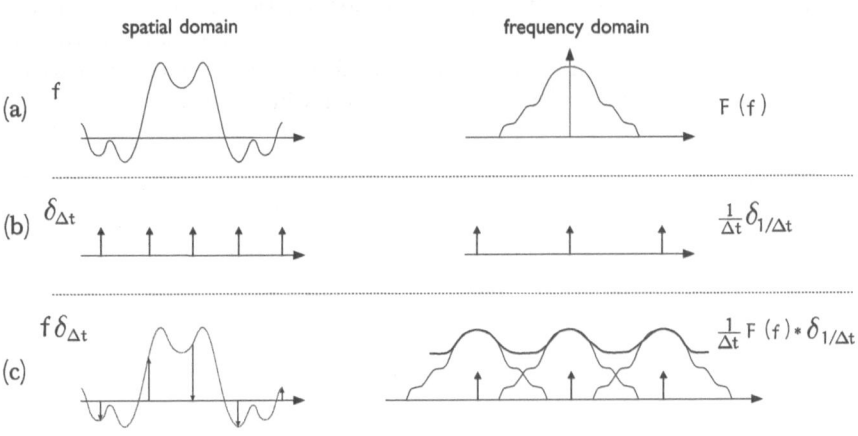

Fig. 2.24. Effect of sampling in the space domain and in the frequency domain.

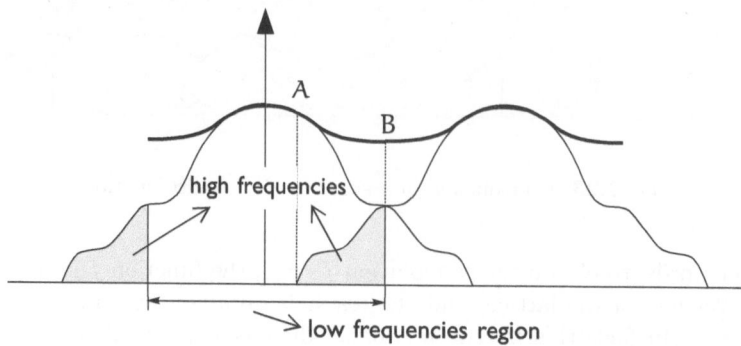

Fig. 2.25. Sampling causes translates of the spectrum of a signal to be superimposed.

rise to low frequencies in the spectrum of the sampled signal. In the final spectrum, within the segment AB in Figure 2.25, the high frequencies get scrambled with the low frequencies, which makes it impossible to recover the correct frequencies of the original signal. In fact, in the reconstruction process these high frequencies are reconstructed as low frequencies.

Now consider a bandlimited signal f, with frequencies contained in the interval $[-\Omega, \Omega]$. As we can see in Figure 2.26, if

$$\frac{1}{\Delta t} > 2\Omega, \quad \text{or, equivalently,} \quad \Delta t < \frac{1}{2\Omega},$$

there will be no overlap of translates in the frequency model of the sampled signal. Then the frequency model of the original signal can be obtained from that of the sampled filter by applying a lowpass filter with cutoff frequency Ω and unit gain, that is, a filter whose transfer function is the pulse function $p_\Omega(s)$ (this is also shown in Figure 2.26). The original signal can then be recovered by taking the inverse Fourier transform of the frequency model of the signal. This argument demonstrates the following result.

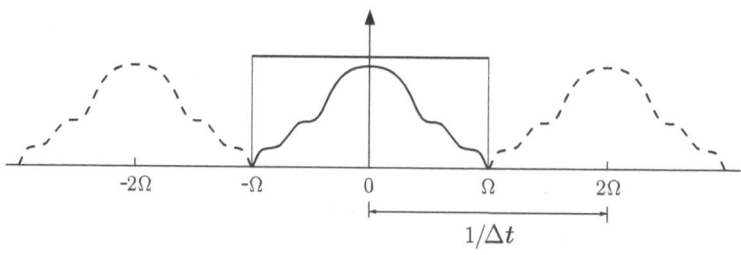

Fig. 2.26. Frequency model of a sampled bandlimited signal.

Theorem 2.7 (The Shannon–Whittaker sampling theorem). *Let f be a bandlimited signal and Ω the smallest frequency such that* $\operatorname{supp} \hat{f} \subset [-\Omega, \Omega]$. *Then f can be exactly recovered from the uniform sample sequence $\{f(m\,\Delta t) : m \in \mathbb{Z}\}$ if $\Delta t < 1/(2\Omega)$.*

In other words, if the signal is bandlimited to a frequency band going from 0 to Ω cycles per second, it is completely determined by samples taken at uniform intervals at most $1/(2\Omega)$ seconds apart. Thus we must sample the signal at least two times every full cycle.

The sampling rate limit $1/(2\Omega)$ is known as the *Nyquist limit*, in honor of H. Nyquist, who pointed out in the 1920s the importance of this limit in telegraphy.

The Shannon–Whittaker theorem relates the high frequencies in the signal with the sampling rate. Intuitively speaking, the higher the frequencies present in the signal, the higher the sampling rate must be if we want to have a faithful reconstruction. Later we will mention extensions of the theorem to m-dimensional signals.

Ideal Reconstruction

The Shannon–Whittaker theorem does not explain what interpolation method should be used to reconstruct the original signal from the signal sampled according to the Nyquist limit. However, we have already seen in the previous section how we can do this (see especially Figure 2.26). We now go through the calculations that led to the proof of the Shannon–Whittaker theorem more carefully, proving that this method leads to the desired ideal reconstruction.

Given a signal f with $\operatorname{supp} \hat{f} \subset [-\Omega, \Omega]$, the signal sampled at the lattice $L_{\Delta t}$ has the expression

$$f(t)\delta_{\Delta t}(t).$$

Using the Fourier transform, we obtain the spectral model of the sampled signal:

$$F\big(f(t)\delta_{\Delta t}(t)\big) = \hat{f}(s) * F\big(\delta_{\Delta t}(t)\big) = \frac{1}{\Delta t}\hat{f}(s) * \delta_{1/\Delta t}(t).$$

The original frequencies of the original signal can be obtained using a lowpass filter whose transfer function is the pulse $p_\Omega(s)$, modulated by the constant Δt to compensate for the amplitude distortion of the transformed comb signal. Then

$$\hat{f}(s) = \Big(\frac{1}{\Delta t}\hat{f}(s) * \delta_{1/\Delta t}(t)\Big)p_\Omega(s)\,\Delta t = \hat{f}(s) * \delta_{1/\Delta t}(t) \cdot p_\Omega(s).$$

To get the spatial model of the signal, we must apply the inverse Fourier transform to the above equation. We get

$$f(t) = F^{-1}(\hat{f}(s)) = F^{-1}\Big(\hat{f}(s) * \delta_{1/\Delta t}(t)\Big) * F^{-1}\Big(p_\Omega(s)\Big).$$

Using (2.12) and the fact that

$$F^{-1}\Big(p_\Omega(s)\Big) = 2\Omega \, \mathrm{sinc}(2\pi\Omega t),$$

we get

$$f(t) = 2\Omega\Delta t f(t)\delta_{\Delta t}(t) * \mathrm{sinc}(2\pi\Omega t).$$

From Equation (2.14) we obtain

$$f(t) = \sum_{k=-\infty}^{+\infty} 2\Omega\Delta t \, f(k\Delta t)\delta(t - k\Delta t) * \mathrm{sinc}\,(2\pi\Omega t).$$

From (2.15) we obtain

$$f(t) = \sum_{k=-\infty}^{+\infty} 2\Omega\Delta t f(k\Delta t) \, \mathrm{sinc}\big(2\pi\Omega(t - k\Delta t)\big). \tag{2.17}$$

Equation (2.17) is precisely the expression of the exact reconstruction of the signal f from its samples $f(k\Delta t)$, for $k \in \mathbb{Z}$. One can show that the series in (2.17) converges absolutely, and the convergence is uniform on compact parts of the domain.

In fact, one can show that, by normalizing the elements of the set

$$\{\mathrm{sinc}\big(2\pi\Omega(t - k\Delta t)\big)\}, \quad \text{for } k \in \mathbb{Z},$$

one obtains a complete orthonormal basis in an appropriate signal space. This is called the *Shannon basis*. In the notation of the preceding proof, the Shannon–Whittaker sampling theorem can be restated as follows:

Theorem 2.8 (Ideal reconstruction theorem). *Let f be a bandlimited signal. If f is point sampled within the Nyquist limit, the resulting sample sequence is an exact representation of f, and f can be reconstructed from this representation using the Shannon basis.*

2.5.4 Extensions of the Sampling Theorem

The Shannon–Whittaker sampling theorem was derived under several assumptions, three of which are quite restrictive:

- the signal is one-dimensional;
- sampling is uniform; and
- the signal is bandlimited.

Extensions of the theorem to nonbandlimited signals and to nonuniform sampling exist, but they will not be discussed here. The reader can consult the references mentioned in Section 2.12.

The extension to n-dimensional signals is of paramount importance, especially in view of the fact that image signals, those that most concern us in this book, are two-dimensional. An extension to higher dimensions can also be applied to the problem of volume reconstruction, which is important in scientific visualization.

There are several possibilities of extension to n-dimensional signals. One of the difficulties is the geometry and topology of the signal's domain. One very natural extension, sufficient for our purposes, will now be given.

Suppose the signal is defined in a rectangular domain $U \in \mathbb{R}^n$, given by

$$U = \prod_{i=1}^{n} [a_i, b_i].$$

Samples are taken at the vertices of a uniform n-dimensional lattice $\Delta = (\Delta x_1, \ldots, \Delta x_n)$.

We saw in (2.7) that the Fourier transform extends naturally. The notion of a bandlimited signal, therefore, still applies. The Nyquist limit must be satisfied for each coordinate separately. That is, we consider the vector $\Omega = (\Omega_1, \ldots, \Omega_n)$ of upper bounds Ω_i for the frequencies of the signal in each coordinate. Then we must have

$$\Delta x_1 < \frac{1}{2\Omega_1}, \ldots, \Delta x_n < \frac{1}{2\Omega_n}. \tag{2.18}$$

The sinc function can be extended to n dimensions by "separability":

$$\mathrm{sinc}(x_1, \ldots, x_n) = \mathrm{sinc}(x_1) \ldots \mathrm{sinc}(x_n).$$

Figure 2.27 shows the graph of sinc in the two-dimensional case, on the square $[-12, 12] \times [-12, 12]$. It also shows a cross section along the x-axis.

Under the conditions just stated, the theorem's extension is immediate: *Suppose a signal $f : U \subset \mathbb{R}^n \to \mathbb{R}$ is bandlimited and is sampled on a uniform lattice above the Nyquist rate—that is, according to (2.18). Then f is completely determined by its samples.* More precisely,

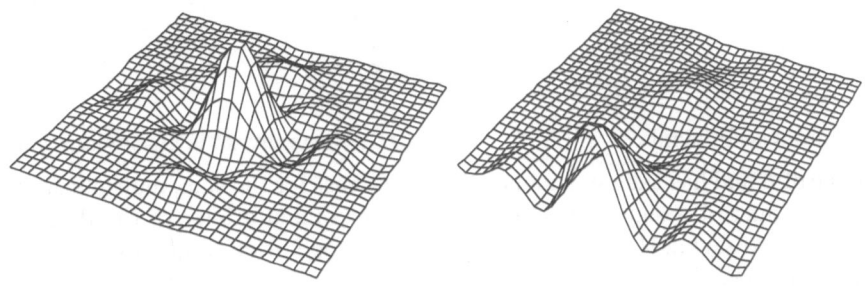

Fig. 2.27. Graph of the sinc function in two dimensions.

$$f(X) = \sum_{k_1=-\infty}^{+\infty} \cdots \sum_{k_n=-\infty}^{+\infty} 2\Omega \Delta X f(K\Delta X) \operatorname{sinc}\big(2\pi\Omega(X - K\,\Delta X)\big), \quad (2.19)$$

where $X = (x_1, \ldots, x_n)$, $K\,\Delta X = (k_1\,\Delta x_1, \ldots, k_n\,\Delta x_n)$,

$$\operatorname{sinc}[2\pi\Omega(X - K\,\Delta X)] = \operatorname{sinc}[2\pi\Omega_1(x_1 - k_1\,\Delta x_1), \ldots, 2\pi\Omega_n(x_n - k_n\,\Delta x_n)],$$

and

$$2\Omega\,\Delta X = 2\Omega_1 \ldots \Omega_n\,\Delta x_1\,\Delta x_2 \ldots \Delta x_n.$$

The proof of this n-dimensional version of the Shannon–Whittaker sampling theorem is analogous to that of the one-dimensional version. The comments we made earlier about sampling, reconstruction, and aliasing are also valid here. When we study image signals in Chapter 6, we will return to these problems and give several examples.

2.6 Operations in the Discrete Domain

In this section we extend the operations on signals studied in Section 2.4 to the discrete universe. Here a signal f is given by a representation sequence

$$(\ldots, f_{-2}, f_{-1}, f_0, f_1, f_2, \ldots)$$

in ℓ^2. Sometimes we will use a functional notation $f(i)$ instead of the positional one f_i. The usual operations of addition and multiplication of signals, and multiplication of a signal by a scalar, extend easily to discrete signals. In general, given an operation $L : S \to S'$ in the continuous domain, and a representation $R : S \to S_d$, we must define an operation $L' : S_d \to S_d$ in the discrete domain, such that the diagram

$$
\begin{array}{ccc}
S & \xrightarrow{\;L\;} & S \\
\downarrow{\scriptstyle R} & & \downarrow{\scriptstyle R} \\
S_d & \xrightarrow[\;L'\;]{} & S_d
\end{array}
$$

commutes. In symbols, $L'(R(f)) = R(L(f))$. Normally L' is an obvious analogue of L and has the same name. This is the case for the sum, multiplication, and product of a signal by some scalar. The commutativity of the diagram amounts to saying that the order in which the operation in question and the discretization operation are applied to the signal is immaterial. An important particular case of this problem occurs when L is a linear and spatially invariant filter and R is the Fourier transform.

2.6.1 Discrete Convolution

As we saw in Section 2.4, the application of a spatially invariant linear filter reduces to taking a convolution product of the signal with the filter kernel. Given two signals f and g, and their discretization sequences (f_m), $m \in \mathbb{Z}$, and (g_n), $n \in \mathbb{Z}$, the *convolution product* of (f_m) and (g_n) is the discrete signal (h_q), $q \in \mathbb{Z}$, defined by

$$h_k = ((f_m) * (g_n))_k = \sum_{j=-\infty}^{+\infty} f_j g_{k-j}. \tag{2.20}$$

This assumes the series converges, which is always the case if the two original sequences are in ℓ^2.

If the signals (f_m) and (g_n) are defined only for finitely many values of m and n, say $m = 0, 1, \ldots, M-1$ and $n = 0, \ldots, N-1$, and we want to compute h_k for the same range of values of k, the signal (g_n) must be extended beyond its original domain, since we need to have g_{k-j} for $k = 0, \ldots, N-1$. There are several (inequivalent) ways to extend a signal. We will discuss this problem in more detail in Chapter 6, in the context of image operations.

An important case of convolution is when we have a discrete representation $(f(t_k))$ (where $k \in \mathbb{Z}$) of a signal f and a continuous signal h. The convolution is given by

$$g(t) = (f(t_k)) * h) = \sum_{j=-\infty}^{+\infty} f(t_j) h(t - t_j). \tag{2.21}$$

That is, the convolution works as an interpolation technique using the translates $h(t - t_j)$ as an "interpolation basis." This is illustrated in Figure 2.28.

2.6.2 The Discrete Fourier Transform

Consider the expression that defines the Fourier transform,

$$\hat{f}(s) = \int_{-\infty}^{+\infty} f(t) e^{-2\pi i s t} \, dt. \tag{2.22}$$

Our purpose is to obtain a method to discretize this transform. We will consider the discrete transform for a finite, uniform point sampling of the signal.

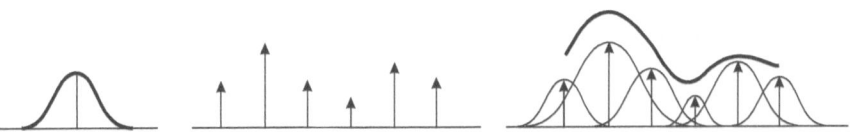

Fig. 2.28. Convolution of a continuous signal with a discrete one.

Therefore, we suppose that the continuous signal is defined on a bounded, closed interval of length A and that the signal domain in centered at the origin, that is, the signal is defined on the interval $[-A/2, A/2]$, where $A > 0$.

We define a grid on this interval with $N + 1$ points, for N even, using subintervals of length Δx. Therefore, $N\,\Delta x = A$, and the grid vertices are defined by $x_k = k\,\Delta x$, for $k = -N/2, \ldots, N/2$. The samples of the signal f on the grid are computed by $f_k = f(k\,\Delta x)$, for $k = -N/2, \ldots, N/2$.

A discrete version of the Fourier transform could be obtained by using some numerical technique associated to the above partition to solve the integral in (2.22). The discrete Fourier transform, DFT, is obtained when we use the trapezoid rule to compute the numerical approximation to the integral. This is illustrated in Figure 2.29. The integrand is approximated by a line segment inside each subinterval of the grid (linear interpolation), and the integral is approximated by the sum of the areas of the resulting trapezoid.

To obtain the expression for the DFT we suppose that $f(-A/2) = f(A/2)$. Using the trapezoid rule to approximate the integral, we get

$$\hat{f}(s) = \int_{-\frac{A}{2}}^{\frac{A}{2}} f(t)e^{-2\pi i s t}dt = \frac{A}{N} \sum_{k=-\frac{N}{2}+1}^{\frac{N}{2}} f_k e^{-2\pi i s x_k}. \qquad (2.23)$$

The above sum is a discretized version of the Fourier transform that allows us to compute $\hat{f}(s)$ for any frequency value s of the spectrum. We need to obtain a discrete sequence $(\hat{f}_n)$ that is the image of the discrete signal (f_n) by the discrete transform. We will suppose that $\hat{f}(s)$ is defined on a domain $[-\Omega/2, \Omega/2]$ and we need to sample $\hat{f}$ on a uniform grid $k\,\Delta s$, with N intervals. Before proceeding, we must obtain the relationship between Δx, A, Ω, and Δs. The equations relating these parameters are called *reciprocity relations*. We can obtain them using a very simple and intuitive argument.

The longest period existing in the signal equals at most the length A of the time domain interval $[-A/2, A/2]$. The corresponding frequency is $1/A$;

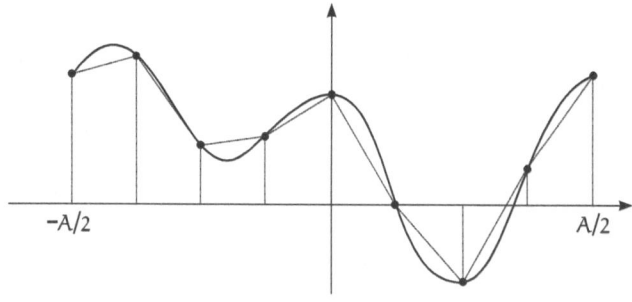

Fig. 2.29. Trapezoidal approximation of the integral.

therefore, in order for it to be detected in the frequency domain sampling, we should use a grid of length

$$\Delta s = \frac{1}{A}.$$

Since we have N subintervals on $[\Omega/2, \Omega/2]$, we get

$$\Omega = N \, \Delta s = \frac{N}{A}, \quad \text{or} \quad A\Omega = N.$$

This is the *first reciprocity relation*.

Now, it is easy to verify that

$$\Delta x \, \Delta s = \frac{A}{N} \cdot \frac{1}{A} = \frac{1}{N},$$

the *second reciprocity relation*.

From the second reciprocity relation we obtain the following correspondence between the vertices $x_k = k \, \Delta x$ in the space domain and the vertices $s_j = j \, \Delta s$ in the frequency domain:

$$s_k x_j = (k \, \Delta s)(j \, \Delta x) = \frac{kj}{N}. \tag{2.24}$$

Evaluating Equation (2.23) on the grid vertex $s_k = k \, \Delta s$, we obtain

$$\hat{f}(s_k) \approx \frac{A}{N} \sum_{j=-\frac{N}{2}+1}^{\frac{N}{2}} f_j e^{-2\pi i s_k x_j}.$$

Substituting (2.24) into this equation, we get

$$\hat{f}(s_k) = A \cdot \frac{1}{N} \sum_{j=-\frac{N}{2}+1}^{\frac{N}{2}} f_j e^{-2\pi i k j/N}.$$

From this we extract the expression that defines the *discrete Fourier transform* of the finite sequence (f_n),

$$(\mathrm{DFT}(f_n))_k = F_k = \hat{f}_k = \frac{1}{N} \sum_{j=-\frac{N}{2}+1}^{\frac{N}{2}} f_j e^{-2\pi i k j/N}. \tag{2.25}$$

Notice the relation

$$\hat{f}_k = \hat{f}(s_k) \approx A \cdot (\mathrm{DFT}(f_n))_k = F_k$$

between the discrete Fourier transform and the discretized version of it.

2.7 The Inverse Discrete Transform

Consider an even positive integer number N, and let F_k be a sequence of N complex numbers, for $k = -(N/2) + 1, \ldots, N/2$. The *inverse discrete Fourier transform*, IDFT, of the sequence (F_n) is another sequence (f_n), defined by

$$(\text{IDFT}(F_n))_k = f_k = \sum_{j=-\frac{N}{2}+1}^{\frac{N}{2}} F_k e^{2\pi i k j/N},$$

for $k = -(N/2) + 1, \ldots, N/2$.

It is possible to show that the inverse discrete Fourier transform and the discrete Fourier transform are inverse to each other, that is,

$$\text{IDFT}(\text{DFT}(f_k))_n = f_n \quad \text{and} \quad \text{DFT}(\text{IDFT}(F_n))_k = F_k.$$

The proof of this fact is a long algebraic computation involving the definition of each transform.

It is also possible to show that if the sequence (F_n) represents a sampling of the Fourier transform $\hat{f} = F(f)$ of some signal f, then the $\text{IDFT}(F_n)$ are samples from an approximation to the function f.

2.7.1 Properties of the DFT

In this section we will briefly describe several properties of the discrete Fourier transform. We will not demonstrate these properties, but we do point out that since we are dealing with the discrete transform, the proofs reduce to algebraic computations.

Periodicity

Consider the sequences (F_n) and (f_n) defined by the direct and inverse discrete Fourier transforms. These sequences are N-periodic, that is,

$$f_{n+N} = f_n \quad \text{and} \quad F_{k+N} = F_k$$

for all integers n and k.

Linearity

The discretization of the Fourier transform preserves the linearity of the continuous transform. More precisely, we have

$$\text{DFT}[\alpha(f_n) + (h_n)] = \alpha\text{DFT}(f_n) + \text{DFT}(h_n).$$

Obviously, the inverse discrete transform is also linear.

Shift

The shift property is very useful when we need to compute the discrete Fourier transform of a translated (shifted) sequence. It can be stated as

$$(\text{DFT}(f_{n-j}))_k = F_k e^{-2\pi ijk/N}.$$

Geometrically, this means that each term F_k of the transformed sequence $\text{DFT}(f_n)$ is rotated by an angle $e^{-2\pi ijk/N}$.

DFT and Discrete Convolution

The relation between the Fourier transform and the operation of convolution holds in the discrete domain. More precisely,

$$(\text{DFT}[(f_n) * (g_n)])_k = NF_k \cdot G_k,$$
$$(\text{DFT}[(f_n) \cdot (g_n)])_k = F_k * G_k.$$

2.8 The Discrete Transform on the Interval $[0, A]$

We defined the DFT for a symmetric domain $[-A/2, A/2]$, centered at the origin. What happens if the signal does not have this domain symmetry? The periodicity property of the DFT allows us to extend the signal periodically and compute the DFT of this extended signal. This is illustrated in Figure 2.30.

The reader should remember when making this extension that we must enforce the condition $f(-A/2) = f(A/2)$ when making the signal extension. Also, at jump discontinuities we should take the average of the limits from each side at the discontinuity point.

There are different formulations of the DFT on the literature. A very common definition, used in several software packages, is

$$F_k = \frac{1}{N} \sum_{j=0}^{N} f_j - 2\pi ikj/N.$$

This formulation is particularly suitable if the signal is defined over the interval $[0, A]$.

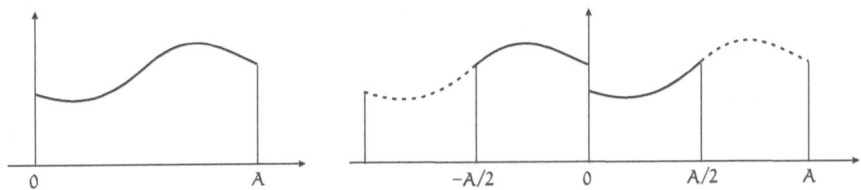

Fig. 2.30. Periodic extension of the signal.

2.9 Matrix Representation of the DFT

We have seen that the discrete Fourier transform is linear. If we represent the sequence (f_n) by $\mathbf{f} = (f_0, f_1, \ldots, f_{N-1})$ and its discrete Fourier transform by $\mathbf{F} = (F_0, F_1, \ldots, F_{N-1})$, the DFT can be written in matrix form:

$$\mathbf{F} = \mathbf{W}\mathbf{f}.$$

In fact, using the notation $\omega_N = e^{-2\pi i/N}$, the matrix $\mathbf{W}$ is easily computed from Equation (2.25), which defines the DFT:

$$\mathbf{W} = \frac{1}{N}\begin{pmatrix} \omega_N{}^0 & \omega_N{}^0 & \omega_N{}^0 & \cdots & \omega_N{}^0 \\ \omega_N{}^0 & \omega_N{}^1 & \omega_N{}^2 & \cdots & \omega_N{}^{(N-1)} \\ \omega_N{}^0 & \omega_N{}^2 & \omega_N{}^4 & \cdots & \omega_N{}^{2(N-1)} \\ \vdots & \vdots & \vdots & \ddots & \vdots \\ \omega_N{}^0 & \omega_N{}^{(N-1)} & \omega_N{}^{2(N-1)} & \cdots & \omega_N{}^{(N-1)(N-1)} \end{pmatrix}. \tag{2.26}$$

It is easy to see that the matrix $\mathbf{W}$ in invertible, and its inverse is the matrix of the inverse discrete Fourier transform. The inverse matrix is easily computed by

$$\mathbf{W}^{-1} = N\mathbf{W}^*,$$

where $\mathbf{W}^*$ is the complex conjugate matrix.

2.10 The Fast Fourier Transform

The computation of the discrete Fourier transform of a sequence with N points has complexity N^2. This follows easily from the matrix representation of the DFT.

It is possible to factor the matrix in (2.26) in order to obtain faster computations of the discrete Fourier transform. There are several algorithms in the literature that achieve this optimization, reducing the computational complexity of the DFT from N^2 to $N \log N$. These algorithms are generically known by the name of *fast Fourier transform*, or simply FFT.

The fast Fourier transform and the relation between the product operation, on the frequency domain, the convolution operation, on the spatial domain, and the discrete Fourier transform can save substantial computational time when performing filtering operations. In fact, filtering a signal using a linear spatial invariant filter reduces to a convolution operation of complexity $O(N^2)$, where N is the number of samples. If we transform the signal and the filter to the frequency domain, convolution reduces to a point-by-point product. Therefore, the whole operation of filtering has complexity $2N \log N + N$. This complexity corresponds to the use of the FFT twice (direct and inverse transform) and to the point-by-point product, which has complexity N.

2.11 Finite Transform

All calculations so far presupposed that the support of the signal f is a compact (bounded) interval. If this is not the case, we must clip f, replacing it by the signal

$$h(x) = \begin{cases} f(x) & \text{if } x \in [a,b], \\ 0 & \text{if } x \notin [a,b] \end{cases}$$

(for a and b appropriately chosen). For this reason some texts refer to the Fourier transform introduced above as the *finite Fourier transform*. It is clear that, unless the signal has compact support to begin with, the Fourier transform computed above is not the true Fourier transform of the signal.

Clipping is equivalent to multiplying the spatial domain model of the signal by a pulse function. It introduces high frequencies in the spectral model of the signal. Perceptually, these high frequencies give rise to a ringing effect on the reconstructed image. This is the spatial domain version of the well-known Gibbs phenomenon that takes place when we truncate a Fourier series (clipping in the frequency domain).

The operation of clipping a signal to a finite domain is called *windowing*. There is a vast literature on how to choose the correct windows to mitigate the above-mentioned problems (see references below).

2.12 Comments and References

The present chapter is included in the book because it is hard to find in the literature a concise, conceptual exposition of the various aspects of signal theory as they apply to computer graphics. Also, the chapter creates a common basis of knowledge about signal processing that suffices for the understanding of the whole book. Later on, we will return to study the problems of signal filtering, aliasing, and reconstruction, in the context of digital images.

Of course, it is impossible to cover the whole theory of signals in one chapter. Several topics were omitted. A very important one is multiresolution signal theory, which is of first importance in signal synthesis, analysis, transformation, and encoding. We refer the reader to the introductory text (Chui 1992). An elementary and quite complete reference for signal processing is (Oppenheim et al. 1983).

An exposition of sampling theory from the viewpoint of measure theory can be found in (Fiume 1989). That book includes a proof of the fact that point supersampling with uniform distribution converges to area sampling as the number of samples increases.

An elementary but quite complete exposition of the Fourier transform and its applications can be found in the classical (Bracewell 1986). For an introductory yet rigorous introduction to the subject using distribution theory, see (Weaver 1989).

The representation theorem (Section 2.3.1) is based on the fact that every finite-energy signal can be decomposed as a (possibly infinite) sum of "building blocks" (atoms) in the space in question. This result can be extended to other signal spaces. There are also many signal representation methods where the building blocks don't form a basis of the signal space. See (Zayed 1993).

Brief discussions of the fast Fourier transform can be found in several books. We recommend in particular (Bracewell 1986) and (Weaver 1989). An implementation in C is listed in the appendix of (Wolberg 1990).

Two recent books about the Fourier transform are likely to become classics. One is (Briggs and Henson 1995), which brings a wide coverage of the theory, computation, and applications of the discrete Fourier transform. Our exposition of the DFT was greatly inspired by it. The other book is (Van Loan 1986), which discusses computational frameworks for the fast Fourier transform from the point of view of matrix factorization.

Many works deal with extensions of the Shannon–Whittaker theorem to dimensions greater than one. A very good reference for dimension 2 is (Lim 1990). Another fairly complete source is (Zayed 1993), which includes extensions to nonbandlimited signals and nonuniform sampling. The same reference has a synopsis of the theorem's history.

A concise, but complete, coverage of the history of point sampling theory and the extensions of the Shannon theorem to non-bandlimited signals can be found in (Butzer and Stens 1992).

The Shannon basis can be used as an efficient discretization method for the solution of many numerical problems. The reader interested in this topic can consult (Lund and Bowers 1992), which contains a proof that the Shannon basis is a complete orthonormal set in an appropriately chosen signal space.

References

[Bracewell 1986]Bracewell, R. (1986). *The Fourier Transform and its Applications*, second ed. McGraw-Hill, New York.

[Briggs and Henson 1995]Briggs, W. L. and Henson, V. E. (1995). *The DFT: An Owner's Manual for the Discrete Fourier Transform*. SIAM, Philadelphia.

[Butzer and Stens 1992]Butzer, P. L. and Stens, R. L. (1992). *Sampling theory for not necessarily band-limited functions: a historical overview*. SIAM Review, 34(1):40–53.

[Chui 1992]Chui, C. (1992). *An Introduction to Wavelets*. Academic Press, Boston.

[Fiume 1989]Fiume, E. L. (1989). *The Mathematical Structure of Raster Graphics*. Academic Press, Boston.

[Hamming 1983]Hamming, R. W. (1983). *Digital Filters*. Second Edition. Prentice-Hall, Englewood Cliffs, NJ.

[Lim 1990]Lim, J. S. (1990). *Two-Dimensional Signal and Image Processing.*
Prentice-Hall, Englewood Cliffs, NJ.

[Lund and Bowers 1992]Lund, J. and Bowers, K. (1992). *Sinc Methods for
Quadrature and Differential Equations.* SIAM Books, Philadelphia.

[Oppenheim et al. 1983]Oppenheim, A. V., Willsky, A. S., and Young, I. T.
(1983). *Signals and Systems.* Prentice-Hall, Englewood Cliffs, NJ.

[Van Loan 1986]Van Loan, C. (1986). *Computational Frameworks for the Fast
Fourier Transform.* SIAM, Philadelphia.

[Weaver 1989]Weaver, J. (1989). *Theory of Discrete and Continuous Fourier
Analysis.* John Wiley & Sons, New York.

[Wolberg 1990]Wolberg, G. (1990). *Digital Image Warping.* IEEE Computer
Society Press, Los Alamitos, CA.

[Zayed 1993]Zayed, A. (1993). *Advances in Shannon's Sampling Theory.* CRC
Press, Boca Raton, FL.

3

Random Processes

This chapter presents some basic definitions on probability and stochastic processes that will be used later, mainly in Chapter 10. Among the numerous excellent textbooks on this field, the reader is referred to the works by (Dekking et al 2005), by (Grinstead and Snell 1997) and by (Resnick 1999).

3.1 Random Variables

The main idea of this Chapter is that of a *stochastic model*. By this, we understand a mathematical means of describing a phenomenon with the following properties:

1. It is **observable**, so it can be measured somehow.
2. It can be **repeated** as many times as desired under the same conditions.
3. The result of any trial is **unknown** beforehand.
4. The set of all possible outcomes is **well defined**; it is called *sample space*, and it will be denoted Ω.

The subsets of Ω are usually called *events*.

A *probability* is a measure, denoted here 'Pr', defined on subsets of Ω with the following properties:

A1) $\Pr(A) \geq 0$ for every $A \subset \Omega$
A2) $\Pr(\Omega) = 1$
A3) For every sequence of disjoint sets $A_1, A_2, \ldots \subset \Omega$ holds that $\Pr(\cup_i A) = \sum_i \Pr(A_i)$

These, in fact, are the *probability axioms*.

This measure aims at providing a nice mathematical description of the plausibility of the set $A \subset \Omega$. Other measures, with different properties but with similar scope, can be defined leading to, for instance, fuzzy sets and related fields (Banon 1981).

L. Velho et al., *Image Processing for Computer Graphics and Vision*,
Texts in Computer Science, DOI 10.1007/978-1-84800-193-0_3,
© Springer-Verlag London Limited 2009

Knowing the distribution of a random experiment is the capacity of being able to express the probability of all interesting events. The notion of "interesting event" will be made clear after the definition of random variables.

Observing and/or recording random events may be expensive, cumbersome, dangerous or even impossible. For this reason, and for the convenience of working with real numbers, random variables are defined.

A real random variable is a transformation between Ω and the real line $\mathbb{R}$. The term *variable* should not lead to confusion, albeit random variables are functions of the form $X : \Omega \to \mathbb{R}$. More general, i.e., not necessarily real random variables will be used in Chapter 10.

Knowing the distribution of a random variable X amounts to being able to compute the probability of all events of the form of enumerable unions and intersections of real intervals. The events in Ω that are mapped onto such sets by X are the ones we referred to as "interesting".

Every well-defined random variable belongs to a *probability space* given by three elements: the sample space Ω, the set of interesting events $\mathcal{A}$ and a probability Pr for every element of $\mathcal{A}$.

Example 3.1 (Tossing a coin). Consider the experiment of tossing a coin. The possible outcomes are either heads or tails, so $\Omega = \{\text{heads}, \text{tails}\}$. Assume that $\Pr(\text{heads}) = 2/3$ and that $\Pr(\text{tails}) = 1/3$, so the coin is biased to heads; we have a probability defined on Ω. The mapping $X : \{\text{heads}, \text{tails}\} \to \{-1, +1\}$ is a real random variable, and the information provided about the probabilities on Ω is enough to compute the distribution of X.

Example 3.2 (Measuring a continuous function). Consider the situation of making a continuous measurement of the sea level at a certain site. Each record is the height during 24 hours. In this case, Ω is the set of all continuous functions defined on the $[0, 24)$ interval, i.e., $\Omega = \{H : [0, 24) \to \mathbb{R}$ such that H is continuous$\}$ and there is no obvious way of attaching probabilities to them since they form an uncountable space. Also, making a complete record of each outcome $h = H(\omega)$ would demand an infinite amount of storage. A convenient random variable could be defined as $X : \Omega \to \mathbb{R}$ in the following manner: $X(h) = \int_{[0,24)} h(t) \, dt$ for every $h \in \Omega$, i.e., a measure of the mean height.

Example 3.3 (Still measuring a continuous function). In the example above, consider being interested in detecting flood situations. A more convenient random variable would be, for that case, $X(h) = \max\{h(t) : t \in [0, 24)\}$.

The sample space of the random variable X is the set of all possible outcomes. If a distinction between it and the original sample space Ω is needed, which is seldom the case, it can be denoted Ω_X.

The distribution of a random variable can be specified in several manners. One of the most convenient ways to do it is through the *cumulative distribution function*. This function is simply defined as $F(t) = \Pr(X \leq t)$, and it has the following properties:

1. F is non decreasing, i.e., if $t_1 < t_2$ then $F(t_1) \leq F(t_2)$
2. F is right-continuous, i.e., if $t_n \uparrow t$ when $n \to \infty$ then $F(t_n) \to F(t)$
3. $F(t_n) \to 0$ if $t_n \to -\infty$ and $F(t_n) \to 1$ if $t_n \to \infty$.

Any function satisfying properties 1, 2 and 3 above is a cumulative distribution function and, therefore, it characterizes the distribution of a random variable.

There are three basic types of random variables, namely discrete, continuous and singular; we will only see the two first.

Definition 3.4 (Discrete random variable). *A discrete random variable has a non-empty finite or countable sample space. It this case, attaching a probability p_i to every possible outcome ω_i completely specifies the distribution.*

This specification is called *probability function*.

Definition 3.5 (Continuous random variable). *If there exists a function f such that $f(t) = dF(t)/dt$, with F the cumulative distribution function of the random variable X, then we say that X is a continuous random variable and that f is its density.*

The density of a continuous random variable characterizes its distribution. It exhibits two properties: it is non-negative $f \geq 0$, and it integrates to one $\int f = 1$.

The set $A \subset \mathbb{R}$ such that $f(t) > 0$ for every $t \in A$ is known as the *support* of the random variable or of the distribution characterized by f.

The distribution of a set of random variables defined on the same probability space requires the specification of the probability of all possible events, i.e., if $X = (X_1, \ldots, X_n)$ are random variables defined on $(\Omega, \mathcal{A}, \mathrm{Pr})$, the distribution of X is known if one is able to compute $\mathrm{Pr}(X_1 \in A_1, \ldots, X_n \in A_n)$ for any $A_1, \ldots, A_n \in \mathcal{A}$. Sets of random variables are often referred to as "multivariate random variables", as seen in Example 3.7.

The random variables $X_1, \ldots, X_n$, defined on the same probability space $(\Omega, \mathcal{A}, \mathrm{Pr})$, are jointly independent if for every $A_1, \ldots, A_n \in \mathcal{A}$ holds that $\mathrm{Pr}(X_1 \in A_1, \ldots, X_n \in A_n) = \prod_{i=1}^{n} \mathrm{Pr}(X_i \in A_i)$. Pairwise independence does not grant joint independence.

The simplest discrete distribution is the Bernoulli. The random variable X is said to follow a Bernoulli law if $\Omega_X = \{0, 1\}$ and $p_1 = p = 1 - p_0$, with $p \in (0, 1)$. Usually, ω_1 is called "success". A sum of n independent identically distributed random variables following Bernoulli laws with probability p of success follows a Binomial distribution with parameters p and n.

Examples of discrete distributions with countable support are the Geometric, for which $\Omega = \{1, 2, \ldots\}$ and $p_i = p(1-p)^{i-1}$ with $p \in (0, 1)$, and the Poisson, for which $\Omega = \{0, 1, 2 \ldots\}$ and $p_i = e^{-\lambda}\lambda^i/i!$ with $\lambda > 0$.

Two important continuous random variables are the standard versions of the Uniform and the Gaussian (or Normal) laws. They are characterized, respectively, by the densities

$$f(t) = \mathbb{I}_{(0,1)}(t) \text{ and } f(t) = \frac{1}{\sqrt{2\pi}} \exp\{-t^2/2\},$$

where $\mathbb{I}_A$ denotes the indicator function of the set A, i.e.,

$$\mathbb{I}_A(t) = \begin{cases} 1 \text{ if } t \in A, \\ 0 \text{ otherwise.} \end{cases}$$

It is often important to compute certain functionals on random variables. In the following we will define the expected value of a transformation of the continuous real random variable $X : \Omega \to \mathbb{R}$. The discrete case follows by replacing integrals by summations.

Definition 3.6 (Expected value). *Consider the real random variable X and the function $\Psi : \mathbb{R} \to \mathbb{R}$. If the integral exists, the expected value of $\Psi(X)$ is given by*

$$E(\Psi(X)) = \int_{\mathbb{R}} \Psi(t) f(t) \, dt,$$

where f is the density that characterizes the distribution of X.

The following particular cases are of interest:

- When $\Psi(t) = t$ we have the mean of X, denoted by $E(X)$.
- When $\Psi(t) = t^p$, with $p \neq 0$, we have the moment of order p of X.
- When $\Psi(t) = |t - E(X)|^p$, with $p \neq 0$, we have the central moment of order p of X.

Clearly, if well defined, the first central moment of X is zero; the second central moment is known as "variance":

$$\text{Var}(X) = E(X - E(X))^2 = E(X)^2 - (E(X))^2.$$

The mean value is related to the long term tendency of averages of samples, while the square root of the variance, called "standard deviation", describes the width of typical fluctuations around the mean.

The transformation of a random variable used in Definition 3.6 may derive from more than a single random variable, provided the elements are defined on the same probability space. This leads us to the need of defining multivariate random variables.

It is easy to build a multivariate real random variable starting from Example 3.2, as will be seen in the following.

Example 3.7 (A multivariate random variable). Consider the situation of improving the measures of the continuous process presented in Examples 3.2 and 3.3. Instead of recording a single value, it would be much informative to register three: the minimum, the mean and the maximum heights. With this, $X : \Omega \to \mathbb{R}^3$ is given by

$$X(h) = \left(\min\{h(t) : t \in [0, 24]\}, \int_{[0,24)} h(t) \, dt, \max\{h(t) : t \in [0, 24]\} \right).$$

Multivariate random variables play a central role in image modelling.

One can also compute expected values of functions of several random variables. Consider, for instance, the vector of continuous random variables $X_1, \ldots, X_n$ defined on $(\Omega, \mathcal{A}, \mathrm{Pr})$, whose distribution is characterized by the density $f : \mathbb{R}^n \to \mathbb{R}_+$, and the real-valued function $\Psi : \mathbb{R}^n \to \mathbb{R}$.

Definition 3.8 (Expected value of real functions of random variables). *The expected value of $\Psi(X_1, \ldots, X_n)$ is*

$$\mathrm{E}(\Psi(X_1, \ldots, X_n)) = \int_{\mathbb{R}} \Psi(x_1, \ldots, x_n) f(x_1, \ldots, x_n) d(x_1, \ldots, x_n),$$

if the integral is well defined.

This definition will be particularly useful for computing the covariance and correlation matrices of random vectors.

The relationships between the elements of a multivariate random variable can be quantified by means of Covariance and Correlation matrices, among other measures of association.

Definition 3.9 (Covariance). *The covariance between X and Y, real random variables defined on the same probability space, is*

$$\mathrm{Cov}(X, Y) = \mathrm{E}(XY) - \mathrm{E}(X)\,\mathrm{E}(Y),$$

if the integrals are well defined.

Definition 3.10 (Correlation). *The correlation between X and Y, real random variables defined on the same probability space, is*

$$\rho(X, Y) = \frac{\mathrm{Cov}(X, Y)}{\sqrt{\mathrm{Var}(X)\,\mathrm{Var}(Y)}},$$

if the integrals and the ratio are well defined.

Definition 3.11 (Covariance matrix). *The covariance matrix of the random vector $X = (X_1, \ldots, X_n) : \Omega \to \mathbb{R}^n$ is the matrix $\Sigma = (\mathrm{Cov}(X_i, X_j))$, $1 \leq i, j \leq n$, provided every element is well defined.*

Definition 3.12 (Correlation matrix). *The correlation matrix of the random vector $X = (X_1, \ldots, X_n) : \Omega \to \mathbb{R}^n$ is the matrix $\varrho = (\rho(X_i, X_j))$, $1 \leq i, j \leq n$, provided every element is well defined.*

As we will see in Chapter 10, these matrices play a central role in the analysis and simulation of image data.

This brief presentation would not be complete without the definition of conditional probabilities and distributions.

Definition 3.13 (Conditional probability of A given B). *Consider the probability space $(\Omega, \mathcal{A}, \Pr)$ and two events $A, B \subset \Omega$. The following quantity*

$$\Pr(A \mid B) = \begin{cases} \dfrac{\Pr(A \cap B)}{\Pr(B)} & \text{if } \Pr(B) > 0, \\ \Pr(A) & \text{otherwise} \end{cases}$$

is a probability, i.e., it satisfies the probability axioms for every A, provided B fixed.

The following theorems are useful probability tools.

Theorem 3.14 (Product of probabilities). *The probability of a compound event can be computed as the product of conditional probabilities. For arbitrary events $A_1, \ldots, A_n$ on the same probability space holds that*

$$\Pr(A_1 \cap \cdots \cap A_n) =$$
$$\Pr(A_1) \Pr(A_2 \mid A_1) \Pr(A_3 \mid A_1 \cap A_2) \cdots \Pr(A_n \mid A_1 \cap A_2 \cap \cdots \cap A_{n-1}).$$

Definition 3.15 (Partition of the sample space). *The events $B_1, \ldots, B_k$ form a partition of the sample space Ω if they exhibit the following properties:*

1. *They are pairwise disjoint, i.e., $B_j \cap B_k = \varnothing$ whenever $j \neq k$.*
2. *Their union is the sample space, i.e., $\cup_{i=1}^{k} B_i = \Omega$.*
3. *They have positive probability, i.e., $\Pr(B_i) > 0$ for every $1 \leq i \leq k$.*

Theorem 3.16 (Total probability). *Consider the event $A \in \mathcal{A}$ and the partition $B_1, \ldots, B_k$ both defined on the same probability space $(\Omega, \mathcal{A}, \Pr)$, then*

$$\Pr(A) = \sum_{1 \leq i \leq k} \Pr(A \mid B_i) \Pr(B_i).$$

This theorem is useful for obtaining the probability of events (A) whose direct computation is cumbersome, or impossible, but for which it is possible to collect conditional information with respect to a partition of the sample space.

The following is one of the most famous theorems in probability theory (Bayes 1763).

Theorem 3.17 (Bayes' theorem). *Consider again the event $A \in \mathcal{A}$ and the partition $B_1, \ldots, B_k$ both defined on the same probability space $(\Omega, \mathcal{A}, \Pr)$, then*

$$\Pr(B_i \mid A) = \frac{\Pr(A \mid B_i) \Pr(B_i)}{\sum_{1 \leq i \leq k} \Pr(A \mid B_i) \Pr(B_i)}.$$

This theorem, that gave birth to a whole branch in statistics, namely Bayesian inference, allows using prior knowledge as an ingredient in the process of drawing conclusions from data. This is further commented in Section 10.6, when dealing with image classification.

3.2 Stochastic Processes

We will now turn our attention sets of random variables with a relatively complex structure: the so-called *stochastic processes*.

One of the most important stochastic processes involves both discrete and continuous random variables: the Poisson process. It serves as a basis for other relevant stochastic processes, and it is tractable.

The Poisson process can be used to describe a huge variety of phenomena, among them: the arrival times of customers in a bank, the instants at which a radioactive substance emits particles, the failure of electronic devices and so forth. Loosely speaking, the Poisson process is a good model for those situations in which the probability of a single event is very small when compared to the number of trials. This stochastic process is also referred to as the one with complete randomness. In the following we will present it for the one-dimensional case, and later it will be used to build two-dimensional point processes.

Consider we are observing the arrival of customers in a bank, starting at $t = 0$. A Poisson process can be built *ab initio* from the following hypothesis:

P1) Stationary increments: the number of events in interval $(t, t+\Delta t)$ depends only on Δt.

P2) Independent increments: the number of events on disjoint intervals are independent.

P3) Finiteness: the probability of more than one event take place on a small interval is negligible when compared to the probability of observing just one event on the same interval.

In technical terms, if $E_{t,\Delta t}^k$ denotes "k events took place in interval $(t, t+\Delta t]$", these hypotheses can be written as

H1) $\Pr(E_{t,t+\Delta t}^k) = \Pr(E_{0,\Delta t}^k)$ for every k, t and Δt.

H2) $\Pr(E_{t_1,\Delta t_1}^{k_1} \cap E_{t_2,\Delta t_2}^{k_2}) = \Pr(E_{t_1,\Delta t_1}^{k_1}) \Pr(E_{t_2,\Delta t_2}^{k_2})$ if $(t_1, \Delta t_1] \cap (t_2, \Delta t_2] = \varnothing$.

H3) $\lim_{t\to 0}(\Pr(E_{0,t}^{2\ \text{or more}}) / \Pr(E_{0,t}^{\text{at least }1})) = 0$.

With these hypotheses, and denoting $\lambda = -\log \Pr(E_{0,1}^0)$ holds that, for every $t > 0$, $\Pr(E_{0,t}^0) = \exp\{-\lambda t\}$ and, more generally, that

$$\Pr(E_{0,t}^k) = \frac{(\lambda t)^k}{k!} \exp\{\lambda t\}. \tag{3.1}$$

Consider the time of the first event or arrival T_1; it is possible to verify that $\Pr(T_1 \le t) = 1 - \exp\{-\lambda t\}$. Denoting now $T_2, T_3, \ldots$ the time of the second arrival after the first arrival, the time of the third arrival after the second arrival and so on, also holds that

$$\Pr(T_n \le t) = 1 - \exp\{-\lambda t\} \tag{3.2}$$

for every $n \geq 2$ and $t > 0$. In this manner, the Poisson process can be defined either in terms of the number of events or in terms of the times between them. Equation (3.2) implies that the density of the inter arrival times is

$$f_{T_n}(t) = \lambda \exp\{-\lambda t\} \mathbb{I}_{\mathbb{R}_+}(t).$$

In other words, the times between arrivals in a Poisson process obeys an exponential law.

A very convenient property that stems from the definition of Poisson processes concerns the location of the events, given the knowledge of their number. As can be seen in (Dekking et al 2005), given that there are n points in the interval $[a, b]$, the locations of these points are independently distributed, each with a uniform distribution on $[a, b]$. This property will be useful for building Poisson point processes on $\mathbb{R}^2$.

In the following we will recall the basic terminology and definitions for the characterization of stochastic processes.

The main stochastic processes categories are related to

Space state: integer valued (discrete space state) or continuous. Section 10.4 discusses a class from the latter, while Section 10.6 comments an important model from the former.

Index parameter: When describing random variables indexed on subsets of $\mathbb{Z}^p$, integer p, we call them a discrete-time stochastic process. Otherwise, if the index belongs to $\mathbb{R}^p$ we are dealing with a continuous time process.

Consider now the random variables $(X_t)_{t \geq 0}$ (unless otherwise stated, this is a continuous time stochastic process). Some of the classical types of stochastic processes are defined in the following (Karlin and Taylor 1975)

Definition 3.18 (Stationary independent increments). *If the difference random variables $D_1, D_2, \ldots$ defined as $X_{t_2} - X_{t_1}, X_{t_3} - X_{t_2}, \ldots$ are independent regardless of the times $t_1 < t_2 < t_3 < \cdots$, then $(X_t)_{t \geq 0}$ is a process with stationary increments. If the time index is discrete, the differences D_i, $i \geq 1$, are independent random variables. If the distribution of the differences $X_{t_i + h} - X_{t_i}$ depends solely on h, then the process is said to have stationary increments.*

Definition 3.19 (Markov processes). *If*

$$\Pr(X_{t_{n+1}} \in (a, b] \mid X_{t_1} = x_1, X_{t_2} = x_2, \ldots, X_{t_n} = x_n) =$$
$$\Pr(X_{t_{n+1}} \in (a, b] \mid X_{t_n} = x_n),$$

we have a Markov process. In words, Markov processes have limited memory; when conditioned on the whole past, they only depend on the last event.

Technically speaking, Definition 3.19 describes a *first order* Markov process. If the dependence extends back to ℓ terms, the process is known as Markov of order ℓ.

Definition 3.20 (Stationary processes). *If for every $h > 0$ and every n the random variables $(X_{t_1}, X_{t_2}, \ldots, X_{t_n})$ and $(X_{t_1+h}, X_{t_2+h}, \ldots, X_{t_n+h})$ have the same joint distribution , then it is a stationary stochastic process.*

3.3 Point Processes

Though point processes are just a kind of stochastic processes, it is our understanding that they deserve a separate treatment. They can be used for simulating spatial patterns and, therefore, they should be one of the working tools in image synthesis.

One of the best references for point processes, regarding completeness and readability, is the work by (Baddeley 2006). It also provides a comprehensive review of more specialized literature, and it is the reference we will follow in the remaining of this chapter.

This section is divided into two parts: the first deals with the basic definitions and examples, and presents homogeneous process governed by independent laws; the second shows situations for wich either homogeneity or independence or both are dropped. All the examples were produced by functions available in R's spatstat library, available from http://www.spatstat.org (Baddley and Turner 2005).

3.3.1 Homogeneous Processes with Independence

The most important point processes has already been defined: it stems from the Poisson process. Without loss of generality, consider $\lambda = 1$ in the times at which events take place in a Poisson process (equation (3.2)). A sample of size ten from such random variable is $T_1 = 0.041$, $T_2 = 0.287$, $T_3 = 0.773$, $T_4 = 0.786$, $T_5 = 3.892$, $T_6 = 0.443$, $T_7 = 0.619$, $T_8 = 0.853$, $T_9 = 0.904$ and $T_{10} = 2.979$. Figure 3.1 depicts the position of these arrival times.

Such process is regarded as the one with complete randomness, and can be used to form other types of point processes.

A formal definition of point processes is well beyond the scope of this text. Instead of that, we will provide a few constructive examples in two dimensions which give an idea of their potential in image simulation and analysis.

In the following, consider the window $W = [0, 1] \times [0, 1]$, bearing in mind that the point processes we will present can be defined on bounded regions of $\mathbb{R}^2$.

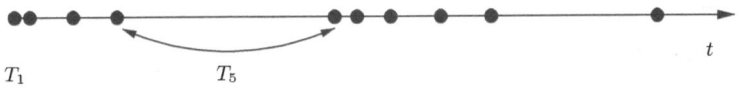

Fig. 3.1. Exponential times from a Poisson process.

Definition 3.21 (Binomial point process). *A fixed number of points $n \geq 1$ is said to obey a Binomial point process if each of their coordinates in W are outcomes of independent identically distributed random variables following uniform laws on $[0, 1]$.*

Figure 3.2 shows three Binomial point processes, with $n = 10$, $n = 50$ and $n = 100$. They were all obtained by taking $2n$ independent samples from the Uniform distribution on $[0, 1]$, and placing a mark at each pair. Note that any perceived pattern in Figures 3.2(a), 3.2(b) and 3.2(c) is mere chance.

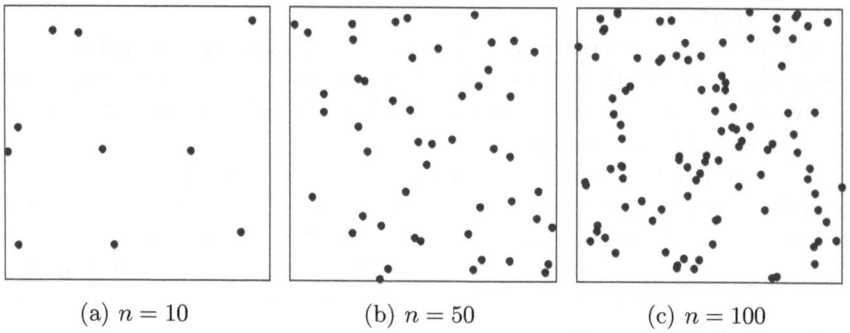

(a) $n = 10$ (b) $n = 50$ (c) $n = 100$

Fig. 3.2. Three Binomial point processes.

Introducing randomness on the number of points, i.e., considering that n is an outcome of the discrete random variable N, we can define the Poisson point process in a constructive manner.

Definition 3.22 (Poisson point process). *Firstly, observe n, the outcome of N, discrete random variable following the Poisson distribution presented in equation (3.1) with intensity $\eta = \lambda t$ (assume $t = 1$ for simplicity). Secondly, place these n points on W following a Binomial point process. The resulting is an outcome of the homogeneous Poisson point process with intensity η.*

Figure 3.3 presents three outcomes of a Poisson process with $\eta = 50$. In order to build these events, three outcomes of $N : \Omega \to \mathbb{N}$ following a Poisson law with mean 50 were observed; these outcomes were $n_1 = 37$, $n_2 = 66$ and $n_3 = 47$. Three Binomial processes were then obtained, with n_1 (Figure 3.3(a)), n_2 (Figure 3.3(b)) and n_3 points (Figure 3.3(c))

A more formal definition of the Poisson point process than the one given in Definition 3.22 can be stated as

Definition 3.23 (Poisson point process). *A Poisson point process with intensity $\eta > 0$ is a collection $P_1, P_2, \ldots$ of points located in a compact set $W \subset \mathbb{R}^p$, $p \geq 1$, such that the number of points in any set $A \subset W$, denoted $N(A)$, has the following properties:*

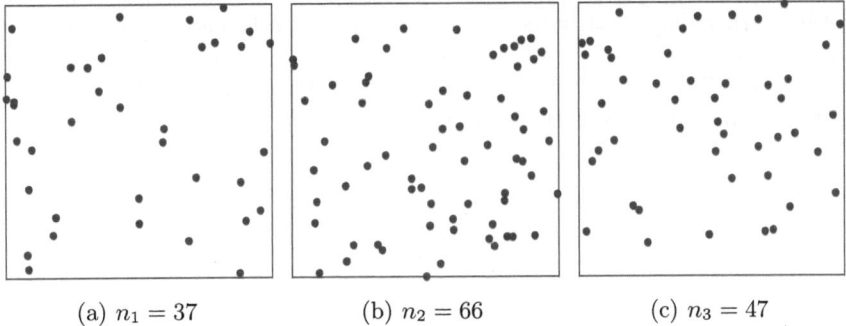

(a) $n_1 = 37$ (b) $n_2 = 66$ (c) $n_3 = 47$

Fig. 3.3. Three Poisson point processes with $\eta = 50$.

1. $N(A)$ *follows a Poisson distribution with mean* $\eta\mu(A)$, *where* $\mu(A)$ *is the volume of the set* A, *and*
2. *the random variables* $N(A_1), N(A_2), \ldots$ *are collectively independent if the sets* $A_1, A_2, \ldots$ *are disjoint.*

A Binomial point process is a Poisson point process, conditioned on the number of points n.

The Poisson point process is used as a reference for complete randomness, i.e., for contrasting a process with unknown properties against one where there is no interaction between disjoint sets, and for building other types of processes. See, for instance, Figure 3.4. Three Binomial processes with $n = 100$ where simulated and used as the centers (not shown in the figure) of circumferences. Each circumference, independently of its position, has a radius of size R, a random variable uniformly distributed on $(0, 1/20)$. A different color is used for each process.

Poisson processes with varying intensities and different (deterministic or stochastic) rules for the marks are possible.

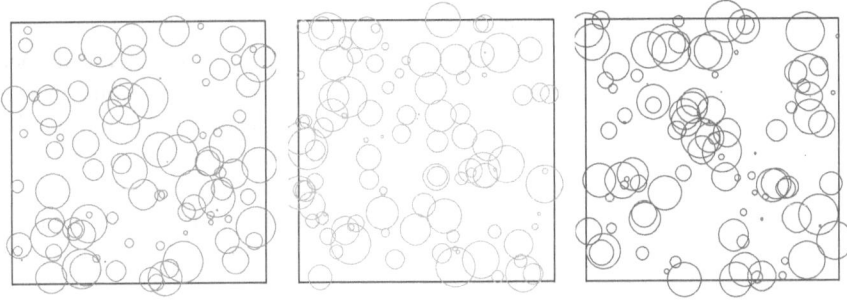

Fig. 3.4. Three marked Poisson point processes.

An interesting spatial process is the one formed by segments. Consider, for instance, $2n$ points distributed on W according to a Binomial process. Draw the segment from i to $i+1$, with $i = 1, 3, \ldots, 2n-1$. This is a Binomial n-lines process, and Figure 3.5 shows outcomes for $n = 10$, 50 and 100. Such process, again, can be enhanced with other point and marks distributions.

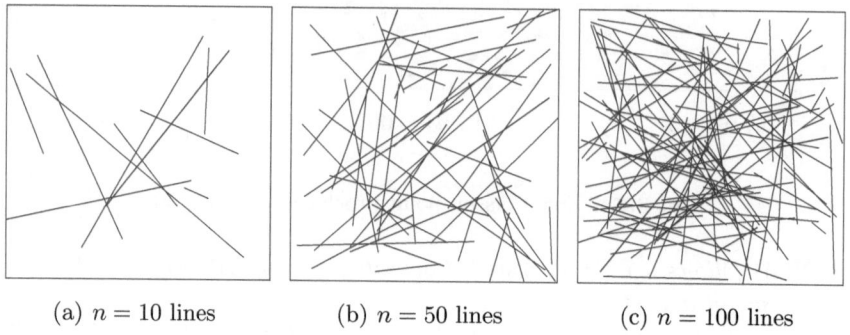

(a) $n = 10$ lines   (b) $n = 50$ lines   (c) $n = 100$ lines

Fig. 3.5. Three line processes on the unit square.

An outcome of the uniform Poisson line process is formed by lines crossing a convex region of the plane W. Such process is governed by a positive paramenter, λ: the expected number of lines intersecting W is equal to lambda times the perimeter length of W. The expected total length of the lines crossing W is equal to $\pi\lambda$ times the area of W.

In order to simulate outcomes form the uniform Poisson line process, start observing n, sample from the random variable N that follows a Poisson law with parameter $\lambda\varrho(W)$, where $\varrho(W)$ is the perimeter of the convex set W. Simulate now a Binomial point process on W with $2n$ points, and join each adjacent pair of points by a straight line. Figure 3.6 presents three outcomes of such process.

3.3.2 Inhomogeneity and/or Dependence

One of the basic properties of the Poisson point process, as presented in Definitions 3.22 and 3.23 is that the probability of observing n points in sets $A_1, A_2 \subset W$ is the same if their areas are equal, regardless their relative location in W. This may be a limitation when, for instance, it is desirable to observe more densely packed points in selected areas of the window.

A simple way of defining such process is using an additional function $\eta : W \to \mathbb{R}_+$, such that $\int_A \eta < \infty$ in any $A \subset W$, the intensity of the process. With this, we can formulate the following.

Definition 3.24 (Inhomogeneous Poisson point process). *Such process with intensity function η is a collection $P_1, P_2, \ldots$ of points located in*

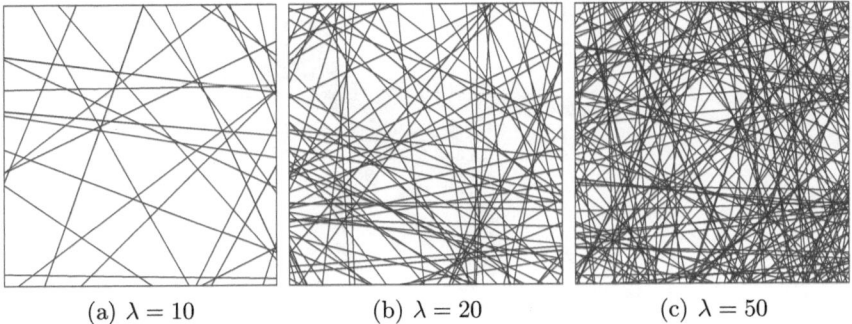

(a) $\lambda = 10$ (b) $\lambda = 20$ (c) $\lambda = 50$

Fig. 3.6. Uniform Poisson line processes.

a compact set $W \subset \mathbb{R}^p$, $p \geq 1$, such that the number of points in any set $A \subset W$, denoted $N(A)$, has the following properties:

1. *$N(A)$ follows a Poisson distribution with mean $\int_A \eta$, and*
2. *the random variables $N(A_1), N(A_2), \ldots$ are collectively independent if the sets $A_1, A_2, \ldots$ are disjoint.*

Note that the basic Poisson point process is obtained making η constant over W.

Figure 3.7 presents an application of such model in nonphotorealistic rendering. Figure 3.7(a) shows an image in shades of gray. As will be seen in Chapter 10, such image can be described as a function $f : S \to \mathbb{R}$; in our case $f : \{0, \ldots, 259\} \times \{0, \ldots, 169\} \to \{0, \ldots, 255\}$. The values $\{0, \ldots, 255\}$ are presented associating 0 to black, 255 to white and gray levels in between. Figures 3.7(b) and 3.7(c) show outcomes of inhomogeneous Poisson point processes with intensity $\eta = (256 - f)/s$, where $s > 0$ is a scale. Notice that darker regions are associated by this transformation to areas where points are more likely, and that the smallest probability of observing a point is strictly positive in any area.

Such inhomogeneous processes capture the main features of the original image and, at the same time, introduce a stochastic component in its rendering. Provided η, the intensity, the points are independent. In the sequel, point processes with dependence structures will be presented.

Definition 3.25 (Simple Sequential Inhibition (SSI)). *Starting with an empty window W, an outcome from the SSI process is built adding points one-by-one. While (i) less than n points are drawn, or (ii) less than T trials have been made without changing the previous configuration, generate uniformly a new point independently of preceding points; if the new point lies closer than r units from an existing point, then it is rejected and another random point is generated.*

(a) Original image

(b) Inhomogeneous Poisson point
process, scale = 50

(c) Inhomogeneous Poisson point
process, scale = 100

Fig. 3.7. Original image and stochastic inhomogeneous derived Poisson point processes.

As defined, provided the window W, a SSI process is characterized by n, r and T. If $r = 0$, the SSI becomes a Binomial point process, while the bigger r, the more repulsive the process.

Figure 3.8 presents three SSI processes with $n = 50$ and increasing values of the exclusion parameter r.

While the point process presented in Definition 3.25 is intuitive and relatively easy to build, it describes a strict exclusion rule; for this reason, it is also known as *hardcore process*. A generalization of such process is the Strauss point process, which ranges from the Poisson point process to the Simple Sequential Inhibition process, i.e., from independence among points to exclusion. This process is defined in terms of an interaction radius r and two parameters: a normalizing constant β and the dependence parameter $0 \leq \gamma \leq 1$.

Definition 3.26 (Strauss point process). *The points $\boldsymbol{x} = x_1, \ldots, x_n \subset W$ are outcomes of the Strauss process with parameters r, β and γ if they obey a law characterized by the density*

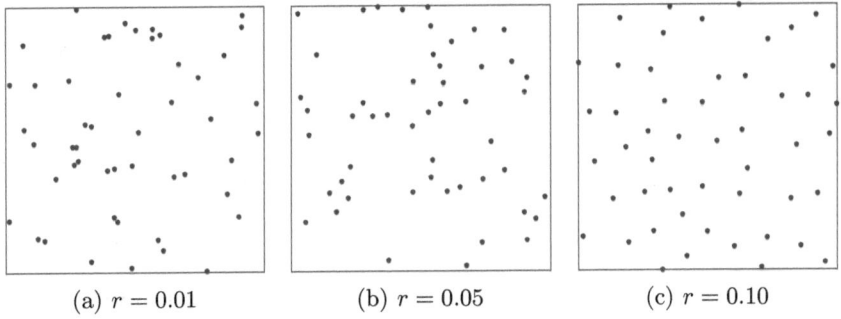

(a) $r = 0.01$ (b) $r = 0.05$ (c) $r = 0.10$

Fig. 3.8. Simple Sequential Inhibition processes with $n = 50$.

$$f(x_1, \ldots, x_n) = \alpha \beta^n \gamma^{s(\boldsymbol{x})},$$

where $s(\boldsymbol{x})$ is the number of (distinct unordered) pairs of points that are closer than r units apart in W.

If $\gamma = 1$ it reduces to a Poisson process with intensity β, whereas if $\gamma = 0$ it is a "hard core process" with exclusion radius $r/2$, since no pair of points is permitted to lie closer than r units apart. (Kelly and Ripley 1975) prove that, as presented in Definition 3.26, the process is well defined. Its exact simulation is discussed by (Berthelsen and Møller 2002).

Figure 3.9 presents three outcomes of the Strauss process for the same window W, same intensity $\beta = 100$, same interaction radius $r = 0.1$ and varying dependence paramenter γ.

Another point process that is built upon the notion of stochastic dependence is the Matérn Cluster process, which is characterized on a window W by the parent intensity κ, the disc r and the cluster intensity μ as follows.

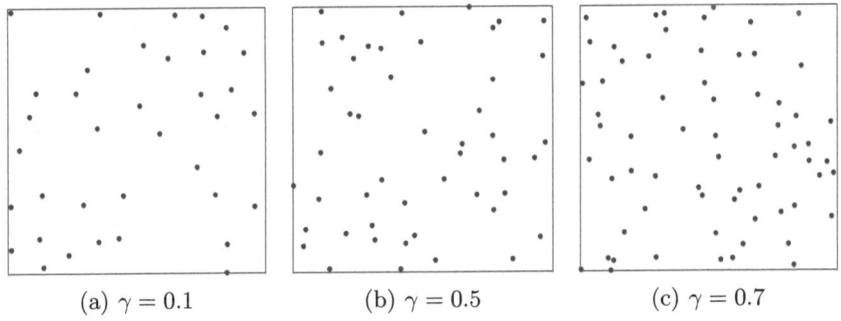

(a) $\gamma = 0.1$ (b) $\gamma = 0.5$ (c) $\gamma = 0.7$

Fig. 3.9. Three outcomes of the Strauss point process with $\beta = 100$, $r = 0.1$ and varying γ.

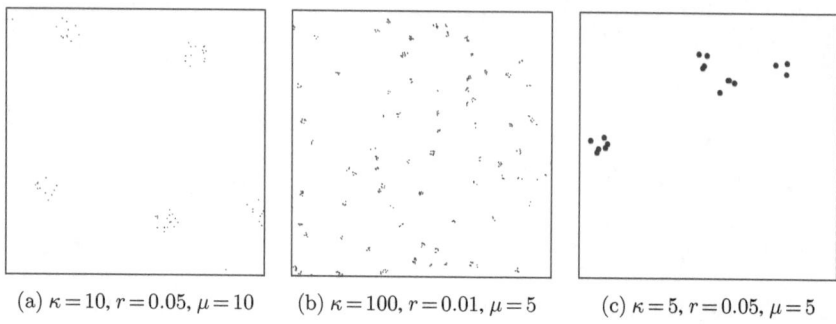

(a) $\kappa = 10, r = 0.05, \mu = 10$ (b) $\kappa = 100, r = 0.01, \mu = 5$ (c) $\kappa = 5, r = 0.05, \mu = 5$

Fig. 3.10. Matérn Cluster processes.

Definition 3.27 (Matérn Cluster process). *First generate a Poisson point process on W with intensty κ, the parent process. Then replace each point in the parent process by an outcome of a Poisson point process on a disc of radius r centered on the parent point, with intensity μ, the cluster process.*

Points from the cluster process outside W can be either discarded, or the Matérn Cluster process can be defined on W', the dilation of W by a disc of radius r. This process can be extended for other types of supports for the cluster process.

Sparse parent processes and dense cluster processes are significantly different from a simple Poisson point process, as presented in Figure 3.10. Figures 3.10(a) and 3.10(b) show the points with a small symbol, while in Figure 3.10(c) they are depicted with solid ones.

It is noteworthy that every point process can be turned into a marked point process, by defining the set of properties of each point.

3.4 Comments and References

The book by (Grinstead and Snell 1997) is highly recommended because it attains a hard-to-find equilibrium between mathematical rigor and intuitiveness. The main ideas are illustrated by means of simulation exercises. (Dekking et al 2005) provide a nice balance between probability and statistics, where the former is used to explain application of the latter; the books ranges from basic probability theory to testing statistical hypothesis. The book by (Resnick 1999) is rigorous, it uses Measure Theory, and is an excellent introduction to the mathematical details of probability theory.

Discovering the richness of point processes through `spatstat` (Baddley and Turner 2005) is one of the best ways to be introduced to these powerful stochastic models.

The work by (Johnson et al 1993; Johnson et al 1994) is one of the most complete references for distributions of, respectively, discrete and continuous random variables. The textbook by (Johnson and Kotz 1972) is a compendium on continuous multivariate distributions. More references on multivariate random variables are discussed in Chapter 10.

References

[Baddeley 2006]A. Baddeley. Spatial point processes and their application. In W. Weil, editor, *Stochastic Geometry*, volume 1892 of *Lecture Notes in Mathematics*, pages 1–75. Springer, Belin, 2006.

[Baddley and Turner 2005]A. Baddeley and R. Turner. spatstat: An R package for analyzing spatial point patterns. *Journal of Statistical Software*, 12(6):1–42, 2005.

[Banon 1981]G. J. F. Banon. Distinction between several subsets of fuzzy measures. *Fuzzy Sets and Systems*, 5(3):291–305, May 1981.

[Bayes 1763]T. R. Bayes. An essay towards solving a problem in the doctrine of chances. *Philosophical Transactions of the Royal Society*, 53:370–418, 1763.

[Berthelsen and Møller 2002]K. K. Berthelsen and J. Møller. A primer on perfect simulation for spatial point processes. *Bulletin of the Brazilian Mathematical Society*, 33(3):351–367, 2002.

[Dekking et al 2005]F. M. Dekking, C. Kraaikamp, H. P. Lopuhaä, and L. E. Meester. *A Modern Introduction to Probability and Statistics: Understanding Why and How*. Springer, 2005.

[Grinstead and Snell 1997]C. M. Grinstead and J. L. Snell. *Introduction to Probability*. American Mathematical Society, 2 edition, 1997.

[Karlin and Taylor 1975]S. Karlin and H. M. Taylor. *A First Course in Stochastic Processes*. Academic, New York, 1975.

[Johnson and Kotz 1972]N. L. Johnson and S. Kotz. *Distributions in Statistics: Continuous Multivariate Distributions*. Wiley, New York, 1972.

[Johnson et al 1994]N. L. Johnson, S. Kotz, and N. Balakrishnan. *Continuous Univariate Distributions*, volume 1 of *Wiley Series in Probability and Mathematical Statistics*. John Wiley & Sons, New York, 2 edition, 1994.

[Johnson et al 1993]N. L. Johnson, S. Kotz, and A. W. Kemp. *Univariate Discrete Distributions*. Wiley Series in Probability and Mathematical Statistics. John Wiley & Sons, New York, 2 edition, 1993.

[Kelly and Ripley 1975]F. P. Kelly and B. D. Ripley. On Strauss's model for clustering. *Biometrika*, 63:357–360, 1975.

[Resnick 1999]S. Resnick. *A Probability Path*. Birkhäuser, Boston, 1999.

4

Fundamentals of Color

The presence or absence of light is what causes the sensation of color. Light is a physical phenomenon, but color depends on the interaction of light with our visual apparatus and is therefore a psychophysical phenomenon.

When one wants to describe an object, one often mentions its color. But what does it mean to say that an object is red? Does "red" mean the same to different people? These questions make it necessary to specify precisely the meaning of colors, to give a quantitative definition that will allow the creation of a common language based on an appropriate model. There are several models, and the most appropriate one can only be chosen given the context: a model that is valid from the perceptual point of view, for example, can yield inaccurate results when used as a computational model.

In this chapter we study physical and psychophysical aspects of color theory. The problem of displaying a color in a graphics device will be deferred to Chapter 16, on image systems, since it is directly connected with graphics output devices.

4.1 Paradigms in the Study of Color

Color is a perceptual manifestation of light, which in turn is an electromagnetic signal. Last chapter's paradigm for the study of signals, therefore, applies perfectly well to this chapter (Figure 4.1).

We must first understand the process of color formation in the physical universe. Next we must study mathematical models for color (the elements of the mathematical universe) and the relationship between these models and the physical universe of color: this study is based on the psychophysics of color, which investigates the various aspects of color perception by our visual apparatus. After that we must study models of color representation (the elements of the representation universe) and problems of conversion between various such representations. The last step, color encoding on the computer,

L. Velho et al., *Image Processing for Computer Graphics and Vision*,
Texts in Computer Science, DOI 10.1007/978-1-84800-193-0_4,
© Springer-Verlag London Limited 2009

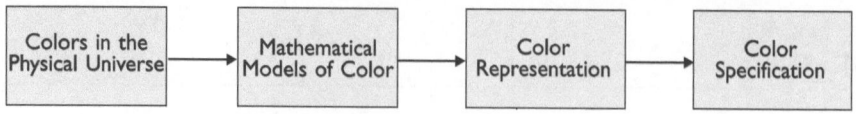

Fig. 4.1. Abstraction levels in the study of color.

is directly related to image encoding and will be deferred to Chapter 6, on digital images.

4.2 The Physical Universe of Color

To understand color one must first understand the nature of light. Physics has definitively established the dual (wave/particle) nature of light: a light ray is made up of particles, called *photons*, and photons in motion determine a wave whose intensity at each point represents the probability of finding a photon at that point. This dual model explains perfectly the physical phenomena where light behaves sometimes as a particle and sometimes as an electromagnetic wave.

A photon moves at a velocity c, which depends on the medium; the associated wave has a certain *frequency* f. The frequency and velocity of a photon determine its *wavelength* λ, the length of a full cycle in the wave:

$$\lambda f = c. \tag{4.1}$$

The energy E of each photon is related to the frequency by means of *Planck's equation*

$$E = hf, \tag{4.2}$$

where $h \approx 6.626 \times 10^{-34}$ Joules · sec is *Planck's constant*.

When photons hit the retina, they give rise to electrical impulses, which, on reaching the brain, are translated into color. Different wavelengths are perceived as different colors. However, not every wavelength can be perceived by the human eye: only those between 380 nm and 780 nm, approximately, where 1 nm, or nanometer, is one billionth of a meter (10^{-9} m; one can also write mμ, where μ is the micron, one millionth of a meter). This interval is the *visible range* of the spectrum, or simply *visible spectrum*. Table 3.1 shows the correspondence between the main color words in English and subranges of the visible spectrum.

4.2.1 Color Formation

The colors that we perceive in our everyday experience arise from very diverse physicochemical processes. Some objects are light sources, while others

color	range
violet	380–440 nm
blue	440–490 nm
green	490–565 nm
yellow	565–590 nm
orange	590–630 nm
red	630–780 nm

Table 3.1. Wavelengths corresponding to different colors.

merely reflect incident light. In order to search for mathematical models for the various processes of color formation in the physical universe, we must first understand and classify these processes. That is the goal of this section.

In general, we can talk about three main color formation processes: *additive*, *subtractive*, and *by pigmentation*.

Additive Color Formation

It is well known that white light is decomposed into spectral colors when it goes through a prism. It was believed that the colors were created by the prism, until the British physicist and mathematician Isaac Newton verified experimentally that white light is made up of colored components and that the prism simply separates these components. Newton's experiment consisted in using another prism to show that the spectral colors that came out of the first prism could be combined again, yielding white light.

In additive color formation, the spectral distributions corresponding to two or more light rays that are being combined are added. Thus, in the resulting color, the number of photons within a given range of wavelengths is the sum of the number of photons in the same range present in the component colors.

Subtractive Color Formation

Subtractive color formation occurs when the light we see is transmitted through a *light filter* or a *dye*.

A light filter is a partly transparent solid material, that is, a solid object that absorbs part of the light that reaches it and transmits the rest, the transmitted fraction depending on the wavelength. Thus, a red filter lets through radiation in the red part of the spectrum, while radiation with other wavelengths is blocked. The color may be intrinsic, but more commonly it is the result of colored particles (pigments) dispersed in a solid medium; see the next section. Several filters can be used in series, the resulting color being made up of those wavelengths that can go through all of them.

A dye is a colored liquid: either a liquid colored compound, or a solution or dispersion of a colored compound. Like a filter, it blocks radiation selectively.

Even when the coloring is due to dispersed particles (pigments), the effect of reflection in these particles is negligible.

Color Formation by Pigmentation

A *pigment* consists of colored particles in suspension in a liquid, or spread over a surface (after the liquid in which they were applied to the surface evaporates). These particles can absorb, reflect, or transmit the light that reaches them. When a light ray reaches a surface covered with pigment, it is scattered by the particles, with successive and simultaneous events of reflection, transmission, and absorption. These events determine what light is returned by the surface.

Opaque paints and opaque inks used in painting work by pigmentation. Pigmentation would be the most appropriate process to emulate on the computer, for use in so-called paint programs.

All three color formation processes are common. Additive color is used in TV monitors. Subtractive color formation occurs, for example, when we project color slides onto a screen. Color formation by pigmentation allows us to see the colors in a painted masterwork. In some industrial applications, such as offset printing (Chapter 16), two or more of the basic processes can be combined.

We will not study mathematical models of color arising from subtractive or pigmentation color formation processes. Our perception of color is triggered by the radiant energy that reaches the eye, and, as already discussed, radiant energy behaves additively in combination. Therefore, additive color formation is the process taken as the base for colorimetric studies.

4.2.2 Photometry and Colorimetry

Photometry is the branch of science that deals with the psychophysical aspects of radiant energy (without regard to color), and specifically with the measurement of light, or the comparison of different light source intensities. *Colorimetry* is the theory that deals with the psychophysical aspects of color, and especially with the measurement, specification, and determination of colors. The color formation method used as a basis for colorimetry is the additive process: The number of photons corresponding to two or more light rays is added, which corresponds to adding the spectral distributions of the rays.

We will not go into a detailed study of the nature of light. Such a study, which would lead us immediately to Maxwell's equations, the foundation of electromagnetism, would be necessary in order to understand the interaction of light with matter and other physical phenomena involving light. However, in this chapter and the next, we are primarily interested in colorimetry, which uses mathematical models to measure color information. Light–matter interaction is directly related with lighting models and will not be studied in this book. For a better understanding of this chapter, the reader may want to look first at Chapter 12, on radiometry and photometry.

4.3 The Mathematical Universe of Color

With very few exceptions, light rays are not made up of photons all of the same wavelength; rather, a light source radiates photons in a range of wavelengths, and one may quantify the rate of emission for photons in each subrange of wavelengths, giving rise to a *spectral distribution* for the number of photons.

If $n(\lambda)\,d\lambda$ photons are emitted per unit time in the wavelength range $(\lambda, \lambda + d\lambda)$, we conclude from (4.2) and (4.1) that the energy per unit time radiated in the same range of wavelengths is $hc\lambda^{-1}n(\lambda)\,d\lambda$. We thus get a spectral distribution for the energy, that is, a function that associates to each wavelength its energy. Likewise we can consider a spectral distribution for other magnitudes of colorimetric interest (see Chapter 12). We will return to this point in the beginning of Section 4.3 (see Figure 4.2).

In physical terms, what we mean by a *color* will usually be a spectral distribution, that is, a mix of wavelengths in certain proportions. A source of radiation that emits photons all of the same wavelength—that has an impulse function as its distribution—is called *monochromatic*. The color is called a *spectral color*, or a *pure color*, if its wavelength is in the visible range.

In sum, the spatial model for color signals is the spectral distribution function. This function associates to each wavelength a measurement representing one of the physical magnitudes associated with radiant energy. This is illustrated in Figure 4.2. For more details, see Chapter 12, on radiometry and photometry.

A *spectrophotometer* is an instrument that gives the spectral distribution of a given color. Such instruments are commonly used in experiments and in applications.

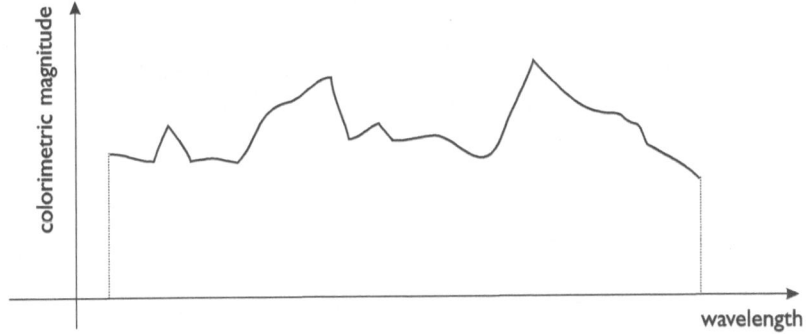

Fig. 4.2. Possible spectral distribution of a color signal.

The spatial model for color (spectral distribution) is of paramount importance in the study of color. The color space associated with this model is traditionally known as *spectral color space* and will be denoted in this chapter by $\mathcal{E}$.

4.4 The Representation Universe of Color

The representation process seeks to replace a continuous model of a signal by a discrete one. In the case of color, we want to replace the spectral space, which is infinite-dimensional, by a finite-dimensional space. What representation to use? To find the answer we must keep in mind that, in practice, we are only interested in colors directly associated with physical systems, so that the color space that interests us is the color space associated with such a system. We must seek a representation of spectral color space $\mathcal{E}$ in this space.

As we discussed in the previous chapter, there are two operations associated with the representation of a color signal: sampling, which approximates the infinite-dimensional color space by some finite-dimensional space, and color reconstruction, which obtains a spectral color from given samples.

In the physical universe of physical color systems, the operation of sampling is associated with physical receptors, and the operation of reconstruction is associated with physical emitters.

4.4.1 Color Sampling

A *physical color receptor* consists of a finite number of sensors $s_1, s_2, \ldots, s_n$, each with a certain *spectral response function* $s_i(\lambda)$. This function gives, for each wavelength, the weight with which light of this wavelength contributes to the sensor's output. Thus, if the receptor is exposed to light with spectral color distribution $C(\lambda)$, the resulting signal is given by the n numbers

$$C_i = \int_{\mathbb{R}} C(\lambda) s_i(\lambda) \, d\lambda. \tag{4.3}$$

In an *ideal receptor*, the spectral response function of each sensor s_i is a Dirac impulse at a given wavelength λ_i, so (4.3) reduces to

$$C_i = \int_{\mathbb{R}} C(\lambda) \, \delta(\lambda - \lambda_i) = C(\lambda_i).$$

Thus, an ideal receptor performs a point sampling of spectral color space (Figure 4.3) at n different values of wavelength.

In general, the spectral response of each sensor in a receptor is not an impulse function, but it approximates a pulse function as defined in the preceding chapter. More precisely, it is usually a unimodal positive function whose mode (maximum value) is achieved at some wavelength λ_0 (Figure 4.4). The smaller

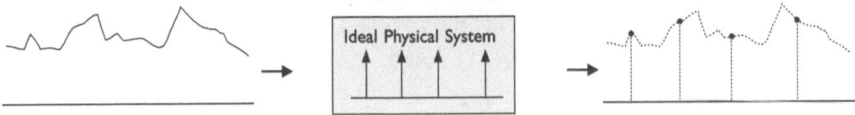

Fig. 4.3. Point sampling of color in an ideal receptor.

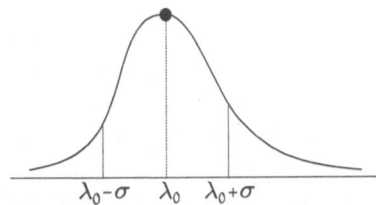

Fig. 4.4. Typical spectral response function of a sensor in a physical receptor.

the standard deviation of the pulse function, the more localized the response of the receptor, and the higher the sensitivity to that spectral band.

Nonetheless, it is useful to think of the action of a receptor as a sampling of the visible spectrum.

A receptor establishes a linear transformation $\mathcal{R} : \mathcal{E} \to \mathbb{R}^n$, defined by

$$\mathcal{R}(C) = (C_1, \ldots, C_n), \tag{4.4}$$

where each component C_i is given by (4.3). The transformation $\mathcal{R}$ is a *representation transformation*, as defined in Chapter 2. The representation transformation therefore establishes the relationship between spectral color space and its finite representation $\mathbb{R}^n$.

Note that, in the representation process, colors with distinct spectral distributions can be represented by the same color vector in $\mathbb{R}^n$. Two colors $C(\lambda), C'(\lambda) \in \mathcal{E}$ are said to be *metamerous* with respect to the receptor with representation transformation $\mathcal{R}$ if $\mathcal{R}(C(\lambda)) = \mathcal{R}(C'(\lambda))$. The occurrence of metamerism means that physically distinct colors (colors with different spectral distributions) can be the same from the point of view of the receptor. A *metamerism relation* $\simeq$ can be defined in spectral color space $\mathcal{E}$ by setting

$$C(\lambda) \simeq C'(\lambda) \iff \mathcal{R}(C) = \mathcal{R}(C').$$

Since the representation transformation is linear, this is an equivalence relation. The space $\mathbb{R}^n$ is identified in a natural way with the quotient space $\mathcal{E}/\simeq$ of spectral color space $\mathcal{E}$ by the metamerism relation, as indicated by this commutative diagram:

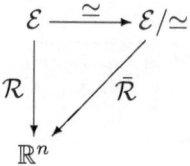

In sum, we see that a physical color receptor performs a natural representation of spectral color space $\mathcal{E}$ in a finite-dimensional vector space, which we call the *color space* of the physical receptor.

Example 4.1 (Color space of the human eye). Isaac Newton believed that the human eye possessed infinitely many photosensitive molecules, corresponding to the different frequencies of the visible spectrum. In the early nineteenth century, the physicist Thomas Young proposed, based on experiments, a trichromatic model for the human eye. In Young's model, the eye has only three types of photosensitive molecules, one for low frequencies of the visible spectrum, one for middle frequencies, and one for high frequencies. In the terminology introduced above, Young's model says that the eye is a physical color receptor with three sensors. Figure 4.5 shows a rough sketch of the spectral response of these three sensors; more precise information can be found in the literature on color theory.

We observe that the spectral response curves of the eye have maxima at the wavelengths corresponding to the colors red, green, and blue. As discussed above, we can consider that the eye samples spectral space at these three wavelengths. Thus, according to Young's theory, the color space of the human eye is a three-dimensional vector space. Much color information is, of course, lost in the sampling process.

In the beginning, there was a lot of resistance within the scientific community to Young's three-color theory. Several perceptual experiments carried out by Maxwell and Helmholtz made Helmholtz support Young's model, which

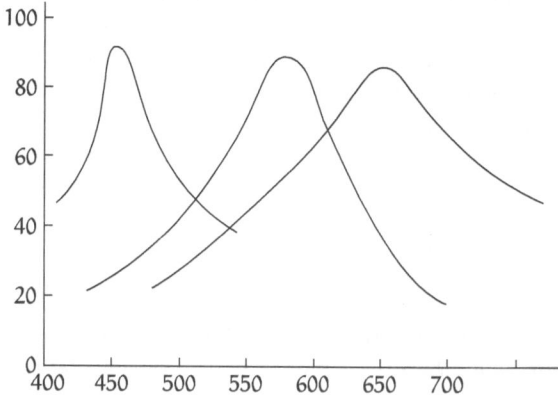

Fig. 4.5. Spectral response curves for the three sensors present in the human eye.

became know as Young–Helmholtz's theory. In the early 1960s it was verified that the eye does, in fact, possess three types of color-sensitive cells, with spectral responses roughly as in Figure 4.5.

The metamerism phenomenon that occurs in the color sampling process of the human vision can be easily illustrated: a pure green light of wavelength (say) 530 nm and a pure red light of wavelength 700 nm, when combined in certain proportions, yield a color that is perceived by the eye as being the same as a pure yellow of wavelength 570 nm mixed with a bit of white. The spectral distributions are very different, but the color, in the everyday sense of the word, is the same.

4.4.2 Color Reconstruction

In the preceding section we studied color sampling systems, where the color signal is processed (filtered) through n sensors $s_1, s_2, \ldots, s_n$, which gives a representation of the spectral color space as an n-dimensional space.

In order to perceive color from a physical color system, the eye must receive visible radiation with an appropriate spectral distribution. This continuous (analog) signal must have been previously reconstructed by the physical system. The reconstruction of colors is performed by physical emitters. In such systems, the sensors emit light when they receive an external stimulus.

From the mathematical point of view, an emitter has a finite number of sensors $s_1, s_2, \ldots, s_n$, and the basic emission of each s_i is a color with spectral distribution $P_i(\lambda)$. The set $\mathcal{B} = \{P_k(\lambda)\}$, for $k = 1, \ldots, n$, is a basis of the emitter's *color space*. Each color $P_k(\lambda)$ is called a *primary color*. Thus, every reconstructed color that can be produced by the emitter is a linear combination

$$C(\lambda) = \sum_{k=1}^{n} \beta_k P_k(\lambda) \tag{4.5}$$

of the emitter's primary colors. The basis $\mathcal{B}$ is called the *primary basis* of the emitter. The vector $(\beta_1, \ldots, \beta_n)$ defines the components of the color C in the emitter, called the *primary components* of C. Physically, C can be obtained as an additive combination of the primary colors.

The vector space $\langle P_1(\lambda), \ldots, P_n(\lambda) \rangle$ generated by the primary colors $P_k(\lambda)$, for $k = 1, \ldots, n$, is a finite-dimensional subspace of spectral color space $\mathcal{E}$. Thus, we can hardly expect that an arbitrary color $C(\lambda)$ will be exactly reconstructed using the primary basis of a physical emitter. In fact, this is not the correct way to set the problem of color reconstruction. The process of reconstruction must be considered together with that of representation in a receptor. More precisely, there are three elements involved:

- a color $C(\lambda)$, defined by its spectral distribution;
- color sampling, performed by some receptor $\mathcal{S}_r$ with sensors $s_1, \ldots, s_n$;
- color reconstruction, performed by some emitter $\mathcal{S}_e$ with primary basis $P_1(\lambda), \ldots, P_n(\lambda)$.

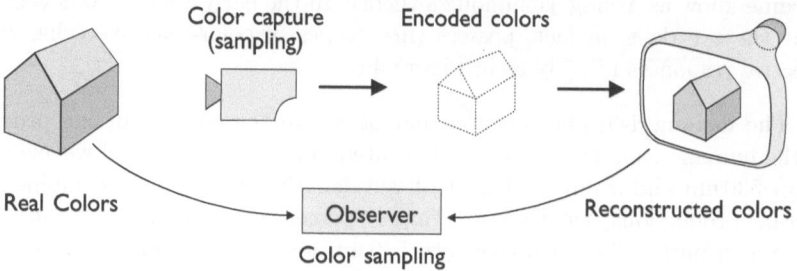

Color capture (sampling)   Encoded colors

Real Colors   Observer   Reconstructed colors

Color sampling

Fig. 4.6. Metamerism and reconstruction.

To make things clearer, we illustrate the situation with a concrete example (see Figure 4.6): the receptor system is the eye of the observer. The continuous spectral distribution $C(\lambda)$ comes from a real scene, and the reconstructed colors come from a television set. There is another sampling color system in the figure (the camera that captures the image), but it is not important for our problem.

In this arrangement, the eye samples the colors from the real scene and from the television set. The reconstruction system of the TV set uses a basis with three primaries, red, green, and blue, and it is certainly not able to reconstruct perfectly the spectral distribution of the color from the real scene. But, in fact, it does not have to do that in order to reconstruct a good image; it is enough that the reconstructed image should look the same when processed by the sampling system of the eye.

Going back to our generic setting, the receptor $\mathcal{S}_r$ yields a discrete representation $(c_1, \ldots, c_n)$ of the color $C(\lambda)$, where each c_i is defined by (4.3). The reconstruction problem consists in obtaining a color

$$C_r(\lambda) = \sum_{k=1}^{n} \beta_k P_k(\lambda), \qquad (4.6)$$

in the color space of the of the emitter $\mathcal{S}_e$, such that the representation of the reconstructed color $C_r(\lambda)$ in the receptor $\mathcal{S}_r$ is the same vector $(c_1, \ldots, c_n)$ representing the original spectral color $C(\lambda)$. In other words, the problem consists in reconstructing a spectral color metameric to the spectral color $C(\lambda)$ with respect to the receptor $\mathcal{S}_r$. This is expressed by the diagram

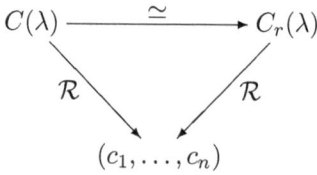

where $\simeq$ is the metamerism equivalence relation and $\mathcal{R}$ is the color representation functional.

In general, the reconstruction problem is posed for the case where the receptor is the human eye. As a concrete example, consider the case of a real image, captured by a video camera and displayed (that is, reconstructed) on a monitor screen (Figure 4.6). To the human eye (the receptor), the reconstructed image is identical to the original image from the point of view of color.

Once the color reconstruction problem has been correctly posed, we will from now on adopt the common convention of omitting the index r in (4.6). We obtain then the initial equation (4.5), and we say that the scalars β_k are the *primary components* of the spectral color C in the system with basis $\{P_k(\lambda)\}$. This doesn't cause any problems, so long as one keeps in mind that the equality in (4.5) is, in fact, equality up to metamerism. In the next section we discuss a method for computing the primary components β_k in (4.5).

4.4.3 Computation of Primary Components

Consider a color, with spectral distribution $C(\lambda)$, and its reconstruction

$$C(\lambda) = \sum_{k=1}^{n} \beta_k P_k(\lambda)$$

in the color space of a physical emitter with primary basis $\{P_k(\lambda)\}$, for $k = 1, \ldots, n$.

If $s_1(\lambda), \ldots, s_n(\lambda)$ are the spectral response curves of a receptor, the representation of the color $C(\lambda)$ in this receptor is given by the vector $(\alpha_1(C), \ldots, \alpha_n(C))$, where, by (4.3), we have

$$\alpha_i(C) = \int_{\mathbb{R}} \left(\sum_{k=1}^{n} \beta_k P_k(\lambda) \right) s_i(\lambda)\, d\lambda$$

$$= \sum_{k=1}^{n} \beta_k \int_{\mathbb{R}} P_k(\lambda) s_i(\lambda)\, d\lambda, \quad \text{for } i = 1, 2, \ldots, n. \tag{4.7}$$

Set

$$a_{ik} = \alpha_i(P_k) = \int_{\mathbb{R}} P_k(\lambda) s_i(\lambda)\, d\lambda,$$

so that a_{ik} is the color response of the i-th sensor of the receptor to the k-th color of the primary basis. We can then write (4.7) in the form

$$\sum_{k=1}^{n} \beta_k a_{ik} = \alpha_i(C) = \int_{\mathbb{R}} C(\lambda) s_i(\lambda). \tag{4.8}$$

In this equation, if we know a_{ik} and $s_i(\lambda)$, which are characteristics of the two physical systems (receptor and emitter), together with the spectral distribution $C(\lambda)$, we can compute the primary components $\beta_1, \ldots, \beta_n$ of the color $C(\lambda)$.

Normalized Coordinates

In practice, we calibrate a primary basis $P_k(\lambda)$ with respect to a reference color, in order to adjust the components β_k correctly. To do this, we take a standard light source with spectral distribution $W(\lambda)$ and determine the primary components w_i:

$$W(\lambda) = \sum_{k=1}^{n} w_k P_k(\lambda).$$

The components β_k of the color $C(\lambda)$ are normalized by

$$T_k(C) = \frac{\beta_k}{w_k}. \tag{4.9}$$

The purpose of normalizing coordinates by means of a reference color is to adjust the sensors or the emitters of the physical system so as to create a correct color scale for each color coordinate. For the case of a video camera, this calibration is known as the *white balance*. Its purpose is to adapt the sensors of the camera to the prevailing lighting conditions. The operator places a white card in the camera's field of view and pushes the white-equalization button. This causes a special circuit to adjust the gain of the camera's sensors to such values that the output signal, when reconstructed, does indeed look white. This also guarantees the correct reproduction of neutral colors (grays), a necessary step toward the faithful rendering of the scene's colors.

Color Reconstruction Functions

The spectral response curves $S_i(\lambda)$ of a receptor are, in general, difficult to find experimentally, and this complicates the calculation of the normalized primary coordinates $T_k(\lambda)$ through Equation (4.8). For this reason we use *color reconstruction functions* $C_k(\lambda)$, defined in such a way that

$$\delta(\lambda - \lambda_0) = \sum_{k=1}^{n} C_k(\lambda_0) P_k(\lambda).$$

Understanding this equation is very important: for each wavelength λ_0 of the spectrum, the values $C_k(\lambda_0)$, for $k = 1, \ldots, n$, are the primary coordinates of the pure color of wavelength λ_0 (that is, having spectrum $\delta(\lambda - \lambda_0)$) in the primary basis $\{P_k(\lambda)\}$ of the system.

Substituting $C(\lambda) = \delta(\lambda - \lambda_0)$ in (4.8), we obtain

$$\sum_{k=1}^{n} \beta_k a_{ik} = \int_{\mathbb{R}} s_i(\lambda) \delta(\lambda - \lambda_0) \, d\lambda = s_i(\lambda_0).$$

From (4.9), we get $\beta_k = T_k(C) w_k = C_k(\lambda_0) w_k$, so

$$\sum_{k=1}^{n} w_k a_{ik} C_k(\lambda_0) = s_i(\lambda_0) \tag{4.10}$$

for $i = 1, 2, \ldots, n$ and λ_0 arbitrary.

When the receptor is the human eye, the color reconstruction functions $C_k(\lambda)$, for $k = 1, 2, 3$, are called *color matching functions*. We spell out their physical meaning in this context: for each wavelength λ_0, the color matching functions encode the intensities with which we must combine each of the three primary colors $P_k(\lambda)$ to obtain a color metameric to the pure color of wavelength λ_0.

The following theorem shows that, if one knows the color reconstruction functions, one can easily obtain the primary coordinates of any color.

Theorem 4.2. *Let $C(\lambda)$ be a spectral distribution, and let $C_k(\lambda)$ be the color reconstruction functions associated with the primary basis of a color reconstruction system (emitter). The normalized components $T_k(C)$ of the color C in this basis are*

$$T_k(C) = \int_{\mathbb{R}} C(\lambda) C_k(\lambda) \, d\lambda. \tag{4.11}$$

Proof. Applying (4.10) to an arbitrary wavelength λ_0 and multiplying both sides by $C(\lambda)$, we can write

$$\sum_{k=1}^{n} w_k a_{ik} C(\lambda) C_k(\lambda) = C(\lambda) s_i(\lambda).$$

Integrating over λ, we get

$$\sum_{k=1}^{n} w_k a_{ik} \int_{\mathbb{R}} C(\lambda) C_k(\lambda) \, d\lambda = \int_{\mathbb{R}} C(\lambda) s_i(\lambda) \, d\lambda.$$

This, together with (4.8), gives

$$\sum_{k=1}^{n} a_{ik} \left(w_k \int_{\mathbb{R}} C(\lambda) C_k(\lambda) \, d\lambda - \beta_k \right) = 0.$$

Physical considerations, based on the choice of the sensors and the basis of primary colors, show that the matrix (a_{ik}) is invertible. We conclude that

$$w_k \int_{\mathbb{R}} C(\lambda) C_k(\lambda) = \beta_k,$$

that is,

$$\int_{\mathbb{R}} C(\lambda) C_k(\lambda) = \frac{\beta_k}{w_k} = T_k(C),$$

as we wished to show.

We observe that, in general, the color reconstruction functions $C_k(\lambda)$ have compact support, so the integral in (4.11) is computed in an interval $[\lambda_a, \lambda_b]$, for $\lambda_a, \lambda_b \in \mathbb{R}$, of the visible spectrum. Thus the integral in (4.11) can be easily solved numerically, yielding the normalized components of a color with spectral distribution $C(\lambda)$. A simple method is to take a partition

$$\lambda_a = \lambda_0 < \lambda_1 < \cdots < \lambda_n = \lambda_b$$

on n subintervals of the interval $[\lambda_a, \lambda_b]$, and approximate each integral by the *Riemann sum*

$$c_k = \sum_{i=1}^{n} C(\lambda_i^*) c_k(\lambda_i^*)(\lambda_i - \lambda_{i-1}),$$

where λ_i^* is an element of the interval $[\lambda_i, \lambda_{i+1}]$. This method is the rectangle rule for numerical integration. If necessary, one can integrate to a higher degree of approximation, using better numerical methods such as Simpson's rule or Gaussian quadrature.

Equation (4.11) in Theorem 3.1 allows the calculation of the components of a spectral color using the color reconstruction functions, instead of requiring the spectral response functions of the system as (4.8) does. As we shall see, its advantage is that the color matching functions can be computed experimentally with great precision.

4.5 CIE-RGB Representation

In this section we study in detail an application of the preceding section's theory, to a color representation system that has great importance in many color processes, and in particular in computer graphics.

We have seen that color representation consists in a reduction of the spectral color space to a finite-dimensional space. Young–Helmholtz's trichromatic model (Example 4.1) states that, for the purposes of color perception, the human eye works as a receptor of dimension three, with sensors that sample the spectrum at the red, green, and blue regions of the visible spectrum. Thus, it is natural to seek a three-dimensional color reconstruction system whose primary basis has as elements colors in these same three regions.

Such a representation system was adopted by the International Commission on Illumination (CIE, for Commission Internationale de l'Éclairage) in 1931. The basis chosen is

$$P_1(\lambda) = \delta(\lambda - \lambda_1) \quad \text{for } \lambda_1 = 700\,\text{nm (red)},$$
$$P_2(\lambda) = \delta(\lambda - \lambda_2) \quad \text{for } \lambda_2 = 546\,\text{nm (green)},$$
$$P_3(\lambda) = \delta(\lambda - \lambda_3) \quad \text{for } \lambda_3 = 435.8\,\text{nm (blue)},$$

where δ is a Dirac delta (impulse function).

From (4.8) we now get

$$a_{ik} = \int_{\mathbb{R}} s_i(\lambda)\delta(\lambda - \lambda_k)\, d\lambda = s_i(\lambda_k),$$

for $i, k = 1, 2, 3$. The reference white used by the CIE has, by definition, a uniform-energy spectral distribution: $W(\lambda) \equiv 1$. Thus, using (4.3), the w_i component of $W(\lambda)$ is given by

$$w_i = \int_{\mathbb{R}} s_i(\lambda)\, d\lambda,$$

for $i = 1, 2, 3$. Using the values computed above for a_{ik} and w_i in (4.8), and taking into account that $C(\lambda) = W(\lambda) \equiv 1$, we get

$$\sum_{k=1}^{3} w_k a_{ik} = \sum_{k=1}^{3} w_k s_i(\lambda_k) = \int_{\mathbb{R}} s_i(\lambda)\, d\lambda = w_i,$$

for $i = 1, 2, 3$, which allows us to compute the coefficients w_k of the CIE reference white with respect to the primary basis

$$\{P_1(\lambda), P_2(\lambda), P_3(\lambda)\}.$$

Substituting the values of w_k thus computed in (4.10), we get the values of the color matching function $C_k(\lambda)$ for any λ. The graphs of these functions are shown in Figure 4.7, each in a different style for clarity.

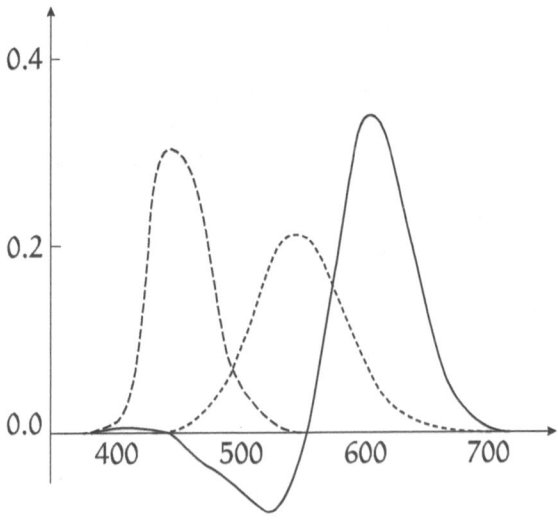

Fig. 4.7. Color reconstruction functions in the CIE-RGB system.

4.5.1 Color Matching Experiments

Note that, in the calculations above, knowledge of the spectral response curves $s_i(\lambda)$ of each of the three types of sensors in the human eye is of paramount importance. The measurement of these values is only possible through indirect methods, and the results are not very precise. As already remarked, more precise results are obtained by taking another route, namely, establishing the color matching functions directly, by means of experiments that we now describe.

An observation panel, Figure 4.8, is set up, consisting of a surface that reflects light uniformly in all directions and in all wavelengths (diffuse reflector). The panel is divided into two halves. One half is lit with light having spectral distribution $C(\lambda)$, called the *test light*. The other half is lit with a white reference light, having spectral distribution $W(\lambda)$, together with three primary light sources with spectral distributions $P_1(\lambda)$, $P_2(\lambda)$, and $P_3(\lambda)$. The goal of the experiment is to find the components of the color of the test light $C(\lambda)$ with respect to the primary colors P_1, P_2, and P_3.

The three primary lights are shone on the same area, and their intensities are adjusted until their mix is perceptually indistinguishable from the reference white W. The intensity values w_1, w_2, and w_3 are recorded. The next step is to obtain the coordinates of the test light; to do this, one again adjusts the intensities of the primary lights until the mix has perceptually the same color as the test light. If these new intensities are denoted by β_1, β_2, and β_3, the normalized coordinates $c_k(\lambda)$ of the test light are then $c_k = \beta_k/w_k$, for $k = 1, 2, 3$.

Experiments of this type allow one to obtain with great precision the values of the color matching functions defined by the primary colors P_1, P_2, and P_3. One starts by taking a partition $\lambda_a = \lambda_0 < \lambda_1 < \cdots < \lambda_{n-1} < \lambda_n = \lambda_b$ of the visible spectrum. For each λ_i one takes as a test light a pure color of wavelength λ_i and carries out the matching experiment. The components c_k obtained are the values $c_k(\lambda_i)$ of the color reconstruction functions at λ_i. The graphs of color matching functions of the CIE-RGB system shown in

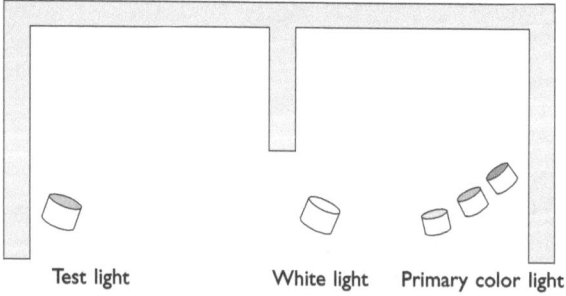

Test light   White light   Primary color light

Fig. 4.8. Color matching experiment.

Figure 4.7 were obtained from CIE data compiled using experiments of the type just described.

Note that one of these graphs takes negative values. Physically, this means that, in the experiment above, it is impossible to combine the primary colors P_1, P_2, and P_3 in such a way as to match the test light at that wavelength. In this case the experiment must be modified as follows: one mixes the primary color P_1 (for example) with the test color C and tries to adjust the other two primaries, P_2 and P_3, to get a match. If this is possible, the coordinates of C are

$$T_1(C) = -\frac{\beta_1}{w_1}, \quad T_2(C) = \frac{\beta_2}{w_2}, \quad T_3(C) = \frac{\beta_3}{w_3}.$$

If this still doesn't work, one mixes P_2 with the test color and compares with a combination of P_1 and P_3, and so on.

4.6 Luminance and Chrominance

As a way to motivate the main topic of this section, we make some remarks about the evolution of color perception theory.

According to the Young–Helmholtz model (Example 4.1), the eye has three types of molecules, sensitive to electromagnetic radiation in overlapping ranges that we loosely correspond to red (R), green (G), and blue (B). In the late nineteenth century, the physiologist Hering carried out several experiments in color perception and concluded that the Young–Helmholtz theory did not explain in a convincing way the results of all of them. Hering then formulated a new theory of color perception, according to which the color apparatus of the eye possesses three double channels. One channel carries information about black and white, which when combined yield light/dark, or *luminance*, information. The other two channels carry the red/green and blue/yellow information, and in each of them there is no color combination.

Hering's model could explain several experiments that the Young–Helmholtz model could not cope with. However, as we have already seen, the fundamental ideas of the Young–Helmholtz theory were confirmed in the mid-1960s, with the discovery of three types of photosensitive molecules in the eye, just as the theory predicted. In order to account for the experimental results that led Hering to postulate his theory, modern color perception theory has elaborated on the Young–Helmholtz model, as follows.

The process of color perception is believed to take place in two steps. In a first step, the signals R, G, and B are generated by cells containing the corresponding types of photosensitive molecules, just as in the classical Young–Helmholtz theory. In a second step, these signals are combined, and what is sent to the brain are the combined signals $R+G$, $R-G$, and $B-(R+G)$. It is an empirical fact that the B component has very little influence on whether a color is perceived as light or dark, so it can be said that the composite signal $R+G$ corresponds roughly to the luminance. The other two, $R-G$

and $B - (R + G)$, encode the remainder of the color information, technically known as *chrominance*: intuitively speaking, this information corresponds to the hue. Thus, the eye sends to the brain the light signal decomposed into a (two-dimensional) *chrominance* component and a (one-dimensional) *luminance* component. This section studies the geometry of the decomposition of the color space into these two components.

Given a color sampling system with sensors $s_1, \ldots, s_n$ having spectral response functions $s_i(\lambda)$, for $i = 1, \ldots, n$, we define the *average spectral response curve* by setting

$$V(\lambda) = \sum_{i=1}^{n} a_i s_i(\lambda), \tag{4.12}$$

where the positive real constants a_i depend on the characteristics of the color system.

For the human eye's three-color receptor, the curve $V(\lambda)$ is called the *relative light-efficiency function*. Its values can be obtained experimentally and have been tabulated. See Figure 4.9 for a graph of $V(\lambda)$.

Let $C(\lambda)$ be the spectral distribution of radiance of a color C in spectral color space $\mathcal{E}$. (See Chapter 12 for the meaning of radiance and a discussion of units of measurement such as lumens.) The *luminance* of C with respect to a physical color receptor having spectral response curve $V(\lambda)$ is defined by

$$L(C(\lambda)) = K \int_{\mathbb{R}} C(\lambda) V(\lambda) \, d\lambda, \tag{4.13}$$

where the constant K equals approximately 680 lumens/watt.

It is important to interpret Equation 4.13, which defines the luminance. It represents a weighted average of the values of the color spectral distribution function, the weights being given by the spectral response curve of the receptor color system. The modulation $C(\lambda) V(\lambda)$ inside the integral shows that the luminance depends on the wavelength. This means that light sources with the same energy power, but with different spectral distributions, might have completely different luminance values.

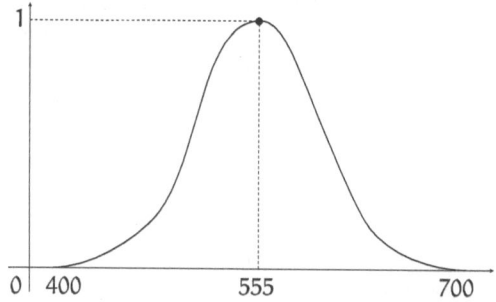

Fig. 4.9. The human eye's light efficiency function.

The reader should consult Chapter 12 for more details about luminance.

Luminance defines a linear functional $L : \mathcal{E} \to \mathbb{R}$ from spectral color space into the real numbers. Let $\mathcal{R} : \mathcal{E} \to \mathbb{R}^n$ be a representation of spectral color space in the color space of an n-sensor receptor. If $C(\lambda)$ and $C'(\lambda)$ are metameric spectral distributions in $\mathcal{E}$, so that $\mathcal{R}(C(\lambda)) = \mathcal{R}(C'(\lambda))$, we have $L(C) = L(C')$. Indeed, metamerism implies that

$$\int_{-\infty}^{+\infty} C(\lambda)s_i(\lambda)\,d\lambda = \int_{-\infty}^{+\infty} C'(\lambda)s_i(\lambda)\,d\lambda.$$

This, together with (4.13) and (4.12), gives

$$L(C) = K \int_{-\infty}^{\infty} C(\lambda)V(\lambda)\,d\lambda = K \int_{-\infty}^{\infty} C(\lambda)\left(\sum_{i=1}^{n} a_i s_i(\lambda)\right) d\lambda$$

$$= K \sum_{i=1}^{n} \int_{-\infty}^{\infty} C(\lambda)a_i s_i(\lambda)\,d\lambda = K \sum_{i=1}^{n} \int_{-\infty}^{\infty} C'(\lambda)a_i s_i(\lambda)\,d\lambda$$

$$= K \int_{-\infty}^{\infty} C'(\lambda)\left(\sum_{i=1}^{n} a_i s_i(\lambda)\right) d\lambda = K \int_{-\infty}^{\infty} C'(\lambda)V(\lambda)\,d\lambda$$

$$= L(C').$$

This shows that the luminance functional L induces naturally, via the metamerism equivalence relation, a linear functional $\bar{L} : \mathbb{R}^n \to \mathbb{R}$ in the color space of the color physical system, as follows:

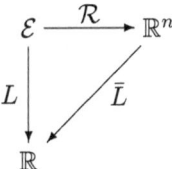

In other words, $\bar{L}(\mathcal{R}(C(\lambda))) = L(C(\lambda))$. The functional $\bar{L}$ is called the *luminance* of the physical system. In what follows we will simplify the notation, denoting the luminance functional by L rather than $\bar{L}$.

Consider a physical color reconstruction system (emitter) with primary basis $\{P_1(\lambda), \ldots, P_n(\lambda)\}$. The primary colors $\{P_i(\lambda)\}$ generate an n-dimensional vector subspace V of the color space. The restriction $L|V$ of the luminance functional to V defines the luminance of the color system. Thus, the luminance of a color

$$C(\lambda) = \sum_{k=1}^{n} c_i P_i(\lambda)$$

in V is

$$L(C(\lambda)) = \sum_{k=1}^{n} c_i L(P_i(\lambda)).$$

If we denote the luminance of each primary color $L(P_i(\lambda))$ by ℓ_i, we deduce that

$$L(C) = \sum_{k=1}^{n} c_i\ell_i.$$

The kernel of the luminance functional $L : V \to \mathbb{R}$ is an $(n-1)$-dimensional vector subspace of V. If C_1 and C_2 are colors with the same luminance, we have $L(C_1) = L(C_2)$ and $L(C_1 - C_2) = 0$, so $C_1 - C_2 \in \ker L$; see Figure 4.10(a). Geometrically, this shows that the color vectors of C_1 and C_2 lie in an affine hyperplane $C_v = \ker L + v$ (where $v \in \mathbb{R}^n$), parallel to the kernel of the luminance functional L. What varies, then, in each hyperplane C_v, is only the color information. This information is called *chrominance*, or *chroma*. Each hyperplane C_v is therefore known as a *chrominance hyperplane*.

We can write $V = \ker L \oplus \pounds$, where $\pounds$ is a subspace complementary to $\ker L$, that is, a one-dimensional subspace generated by a vector $v \notin \ker L$. Thus, every color vector C can be written in a unique way as

$$C = C_c + C_l, \tag{4.14}$$

where $C_c \in \ker L$ and $C_l \in \pounds$; see Figure 4.10(b). The vector C_c is called the *chrominance component* of C, and C_l is called its *luminance component*. The decomposition in (4.14) is called the *chrominance–luminance decomposition* of the system's color space.

The chrominance–luminance decomposition is not unique. It is, nonetheless, of the greatest importance in the definition of various coordinate systems in the color space. Some of these systems will be studied in the next chapter, on color systems.

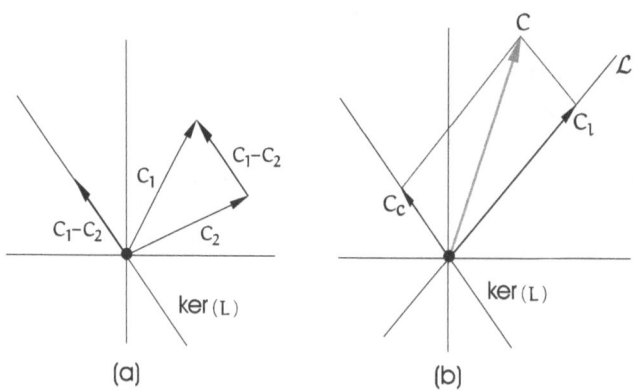

(a)   (b)

Fig. 4.10. Chrominance-luminance decomposition.

4.7 The Color Solid

Recall that the spectral color space $\mathcal{E}$ is the space of functions consisting of all spectral distributions. We say that a color is *visible* if its spectral distribution has nonzero energy somewhere in the visible range.

If $C(\lambda)$ is a visible color and $t > 0$ is a positive real number, $tC(\lambda)$ is also a visible color. Therefore, the set of all visible colors is a cone in spectral color space.

It is also immediate to check that, if C_1 and C_2 are visible colors and $t \in [0, 1]$ is a real number, the color $C = (1 - t)C_1 + tC_2$ is also visible. Thus the set of visible colors is convex.

We will denote by $\mathcal{V}_\mathcal{E}$ the subset of the spectral color space consisting of all visible colors. Let $R : \mathcal{E} \to \mathbb{R}^n$ be a representation of the spectral color space. Since R is linear, the image $R(\mathcal{V}_\mathcal{E})$ of the set of visible colors is also a convex cone in $\mathbb{R}^n$. This cone is called the *color solid*. Only vectors in the color solid of $\mathbb{R}^n$ represent colors from the physical world. The other vectors of $\mathbb{R}^n$ can be interpreted as being associated to nonvisible colors.

A *color map* is a curve $\varphi : I \to \mathcal{S}$ from a closed interval $I \subset \mathbb{R}$ into the color solid $\mathcal{S}$ of a physical system. A color map, therefore, parametrizes a one-dimensional set of colors in the system's color space. The set of spectral colors (pure colors) defines a color map that has a natural parametrization φ, where $\varphi(\lambda)$ is the pure color with wavelength λ.

4.7.1 Chromaticity Space

We saw earlier that the luminance operator gives rise to a decomposition of color space into a chrominance subspace and a luminance subspace. Classically, the set of chrominances in a color space is represented by a *chromaticity diagram*, which we explain in this section.

By linearity, when we multiply a color C by a real number $t > 0$, we also multiply its luminance by the same number: $L(tC) = tL(C)$. Clearly, the color information itself does not vary with multiplication by t, since we are simply multiplying by the same factor the energy corresponding to each wavelength in the color's spectral distribution. Thus, given a color C, the set of colors

$$C' = \{tC : t \in \mathbb{R}, t \neq 0\}$$

has the same color (chroma) information. We can define the color information of C as the set C' itself. Geometrically, this set is a straight line going through the origin of color space, excluding the origin itself. This is called the chrominance line. Thus the set of possible values for chroma information corresponds to the set of "lines" of $\mathbb{R}^3 - \{0\}$ passing through the origin—in other words, the projective plane. How can we obtain a representation of this space? Among the various possible models of the projective plane, the representation using homogeneous coordinates seems natural. However, we must emphasize two important facts:

- We are interested only in the visible colors, that is, in a subset of the projective plane.
- We would like to obtain a representation of chroma space in such a way as to parametrize the subset of perceptible colors using a single coordinate system.

The classical solution is to introduce coordinates in chroma space using the Maxwell plane, as described below.

The Chromaticity Diagram

Our goal is to understand the geometry of the set of chromaticities of the color solid of a system. To simplify the notation, we will carry out our study for three-dimensional color representation spaces, but the results extend easily to n-dimensional representations.

Consider, then, a three-dimensional color space associated with some physical system. After fixing a basis, we can identify this space in a natural way with Euclidean space $\mathbb{R}^3$. The primary colors (elements of the chosen basis) will be denoted P_1, P_2, and P_3, and the coordinates of a color C in this basis by C_1, C_2, and C_3.

Let the *chrominance plane* or *Maxwell plane* be the plane $\mathcal{M}$ having equation $x + y + z = 1$ (see Figure 4.11). The triangle formed by the intersection of this plane with the axes of color space is called the *Maxwell triangle*. Note that, in general, the Maxwell plane intersects the plane of zero luminance (the kernel of the luminance operator) in a line of space, which we call the *line of zero luminance*.

Each chrominance line is entirely determined by its intersection point with the Maxwell plane. This gives a parametrization of the chromaticity space by associating to each chroma line its intersection with the Maxwell plane. What we are doing, geometrically speaking, is introducing local coordinates in the

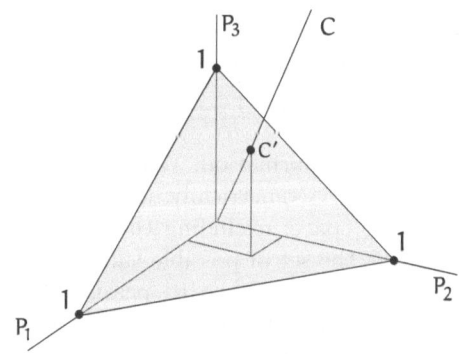

Fig. 4.11. The Maxwell triangle.

projective plane, which, as we have seen, is the model for chroma space. This local system is sufficient because we are interested only in the set of visible colors. The coordinates thus obtained are called *chromaticity coordinates*.

Calculation of chromaticity coordinates is immediate. Indeed, given a color vector C with nonzero luminance, there exists a unique positive real number t_0 such that $t_0 C$ lies in the Maxwell plane, that is,

$$t_0 C = C', \tag{4.15}$$

where C' is in the Maxwell plane. Geometrically, C' is the radial projection of C in the Maxwell plane (Figure 4.11).

The Cartesian coordinates of the vector C' are the chromaticity coordinates of the color C. Let the coordinates of C and C' be denoted by C_1, C_2, C_3, and C'_1, C'_2, C'_3, respectively. From (4.15) we obtain

$$t_0 C_1 = C'_1, \quad t_0 C_2 = C'_2, \quad \text{and} \quad t_0 C_3 = C'_3.$$

Since C' is in the Maxwell plane, we have $C'_1 + C'_2 + C'_3 = 1$. Combining this with the preceding equation, we obtain

$$t_0 = \frac{1}{C_1 + C_2 + C_3}.$$

Thus the chroma coordinates of C, obtained by substituting the value of t_0 in (4.15), are

$$C'_1 = \frac{C_1}{C_1 + C_2 + C_3},$$
$$C'_2 = \frac{C_2}{C_1 + C_2 + C_3},$$
$$C'_3 = \frac{C_3}{C_1 + C_2 + C_3}. \tag{4.16}$$

The radial projection of the color solid in the Maxwell plane is a subset of this plane called the *chromaticity diagram*. It represents the set of all visible colors in color space, up to changes in luminance.

Visualization of Color Space

Visualization of the color space solid, and in particular of the chromaticity diagram, is very important in order to have a perceptual knowledge of the distribution of colors on space. We will postpone a detailed study of this problem until the next chapter, but we make some comments about it here.

Since the color solid is a convex cone, it follows that the chromaticity diagram is a convex subset of the Maxwell triangle.

In order for the chromaticity diagram to be contained in the Maxwell triangle, we must take a primary basis $\{P_1, P_2, P_3\}$ such that, if C is a visible color, the components C_1, C_2, C_3 of

$$C = C_1 P_1 + C_2 P_2 + C_3 P_3$$

are all positive. Geometrically this means that all the visible colors are in the first octant of color space.

As seen earlier, the CIE-RGB system (Section 4.5) does not have the property just discussed. The next chapter, about color systems, will return to this issue. We will introduce there a primary color basis that possesses the property in question, and we will be able to visualize the chromaticity diagram and the color space cone.

The achromatic point

The color white is represented by a point in the chroma diagram called the *achromatic point*. The line connecting the origin (which represents black) to the achromatic point is called the *achromatic line*. The colors along this line correspond to black, white, and all shades of gray in between.

4.8 Grassmann's Laws

As discussed above, the representation linear transformation of $\mathcal{R}$ defined in (4.4) is usually not injective. Thus, if C_1 and C_2 are spectral distributions with $\mathcal{R}(C_1) = \mathcal{R}(C_2)$, this does not imply that $C_1(\lambda) = C_2(\lambda)$, but simply that C_1 and C_2 are metameric, or indistinguishable from the point of view of the sampling color system. This loss of information is a consequence of the sampling of the spectral space performed by the system. For the human eye, this means that colors with very different spectral distributions may be perceptually indistinguishable.

A clear example of this occurs in TV monitors: as shown by Newton, the white light produced by the sun consists of a mix of electromagnetic waves with wavelengths throughout the visible range of the spectrum; but on a TV monitor, white is obtained by mixing only three primary colors, produced by the red, green, and blue phosphors present on each screen dot.

Many experiments with primary light sources, similar to the one described in Section 4.5.1, were carried out in the nineteenth century. The results were summarized by H. Grassmann, in what is now known as the five *Grassmann laws* of colorimetry. As we should expect, Grassmann's laws are a natural consequence of the theory of color representation developed earlier. They apply not only to the human eye, but to any color representation system. Grassmann's laws simply restate properties of the metamerism relation between two colors,

$$C_1 \simeq C_2 \iff \mathcal{R}(C_1(\lambda)) = \mathcal{R}(C_2(\lambda)).$$

These laws were obtained empirically, through color perception experiments. We state some of them now:

Equivalence

The relation $\simeq$ is an equivalence relation. As already seen, this is an immediate consequence of the linearity of the color representation transformation $\mathcal{R}$.

Additivity

If C_1, C_2, and C_3 are elements of $\mathcal{E}$, we have

$$C_1 \simeq C_2 \iff C_1 + C_3 \simeq C_2 + C_3.$$

To prove this, note that $C_1 \simeq C_2$ means $\mathcal{R}(C_1) = \mathcal{R}(C_2)$. Then $\mathcal{R}(C_1 + C_3) = \mathcal{R}(C_1) + \mathcal{R}(C_3) = \mathcal{R}(C_2) + \mathcal{R}(C_3) = \mathcal{R}(C_2 + C_3)$, as we wished to show.

Multiplication by a scalar

If $C_1, C_2 \in \mathcal{E}$ and $t > 0$ is a real number, we have

$$C_1 \simeq C_2 \implies tC_1 \simeq tC_2.$$

Dimensionality

There exists a set of n colors $P_1, P_2, \ldots, P_n$ such that every color C is metamerous to a linear combination of these n colors:

$$C \simeq \sum_{i=1}^{n} \beta_i P_i,$$

where the β_i are scalars. This follows from the discussion on color sampling and reconstruction given earlier in this chapter. The colors P_i form a primary basis for the color emitter.

Linearity of luminance

If L is the luminance operator, we have

$$C = \sum_{i=1}^{n} \beta_i P_i \implies L(C) = \sum_{i=1}^{n} \beta_i L(P_i).$$

This is an immediate consequence of the definition of luminance given earlier (4.13).

Grassmann's laws follow from the definition of a representation of spectral color space. Formally, however, we can take them as axioms and develop from them all of classical colorimetry. This is the point of view adopted in many books on color theory.

4.9 Comments and References

This chapter is an introduction to the study of color. Since there is no concise and comprehensive treatment of the subject, geared toward computer graphics applications, in the literature, we opted for an exposition as self-contained as possible, assuming minimal prerequisites. This chapter is complemented by Chapter 12 on radiometry and photometry, which gives necessary background, and by Chapter 5, which covers the various color systems used in computer graphics.

An elementary, yet reasonably complete, introduction can be found in (Padgham and Saunders 1975). This book also has a concise discussion of the evolution of color perception theories, starting with the Young–Helmholtz model. In particular, the reader can find in this book a more precise sketch of the spectral response curves of the human eye than the one seen in Figure 4.5.

For a comprehensive coverage of many aspects of color and light science, see (Wyszecki and Stiles 1982). This work is particularly useful for its wealth of tables, equations, and other quantitative information on the standards of colorimetry and photometry. It includes tables for the values of the light intensity function $V(\lambda)$ and a discussion of experimental methods for determining these values. It also has a good survey of the various perceptual models of color vision and discusses the anatomy of the human eye.

Though outdated, (Walsh 1958) is still a good reference on techniques for photometric and radiometric measurements. It includes a discussion of the instruments used in radiometry, colorimetry, and photometry.

We mention also (Fishkin 1982), an introductory but comprehensive study of the use of color in computer graphics. This reference contains a more complete discussion of mathematical models for color formation by subtraction and by pigmentation.

The Grassmann laws can be regarded as axioms of a system of additive color formation. Therefore, it is common to encounter different formulations for them in the literature, tied to one or another way of introducing the various properties of a trichromatic color space. The reader interested in their original formulation should consult (Grassmann 1854), a translation of the original German article. An exposition of the laws using a rigorous axiomatic formalism, together with a discussion of the algebraic consequences of this formalism, can be found in (Krantz 1975).

This section is far from being a survey of the large bibliography on color science. It is restricted to the main references consulted by the authors while writing the chapter; the reader can turn to these references' bibliographies for a more complete picture.

References

[Fishkin 1982]Fishkin, K. P. (1982). *Applying color science to computer graphics.* Master thesis, UC Berkeley.

[Grassmann 1854]Grassmann, H. (1854). On the theory of compound colours. *Philosophical Magazine*, 7:254–264.

[Krantz 1975]Krantz, D. H. (1975). Color measurement and color theory: I. Representation theorems for Grassmann structures. *Journal of Mathematical Psychology*, 12:283–303.

[Padgham and Saunders 1975]Padgham, C. A. and Saunders, J. E. (1975). *The Perception of Light and Color*. Academic Press, New York.

[Walsh 1958]Walsh, J. T. (1958). *Photometry*. Dover, New York.

[Wyszecki and Stiles 1982]Wyszecki, G. and Stiles, W. S. (1982). *Color Science*. John Wiley & Sons, New York.

5

Color Systems

We have seen that the mathematical model appropriate for the representation
of spectral color space is a finite-dimensional vector space. A representation
space is associated with every physical color system, be it a receptor or an
emitter. A receptor samples the spectral distribution function of the incident
light, while an emitter performs color reconstruction by combining the elements of its basis of primary colors, which generate the emitter's color space.

In this chapter we shall see how different choices of primary colors for a
representation space allow great flexibility in defining coordinate systems in
a color space and how those choices can be geared to the different needs of
various industrial color processes, particularly in computer graphics.

5.1 Preliminary Notions

A *color system* is a color solid in which we have defined a coordinate system.
A common example of a color system comes from taking a basis $\{P_1, \ldots, P_n\}$
of primary colors in a color representation space. Then a color vector C can
be written as

$$C = \sum_{i=1}^{n} c_i P_i$$

and therefore has coordinates $(c_1, \ldots, c_n)$.

Industrial processes involving color rely on different color systems, each
geared toward the relevant applications. This is a particular case of a general
principle: for each problem we seek the coordinate system that is best suited
to that problem's statement and solution.

The use of *standard color systems* is of great importance for the comparison
of colors between different systems and is of great interest to the industry. Such
a standard system should be defined independently of the application. The
goal of these systems is to establish paradigms that can be used as a base for
defining color systems suited to particular applications.

L. Velho et al., *Image Processing for Computer Graphics and Vision*,
Texts in Computer Science, DOI 10.1007/978-1-84800-193-0_5,
© Springer-Verlag London Limited 2009

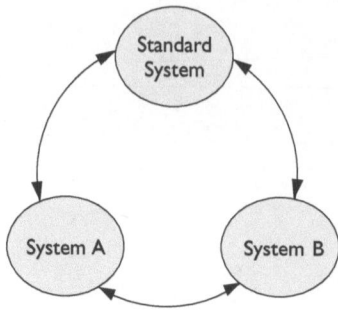

Fig. 5.1. A standard color system and changes between color systems.

An important aspect of the existence of different color systems is the possibility of converting between color coordinates in two or more systems. Mathematically, this amounts to a change in the coordinate system. This is illustrated in Figure 5.1: two systems A and B are specified with respect to a standard system, and we need ways to perform a color change between the three systems.

5.2 Changing Between Color Systems

In general, color conversion between systems can be a complicated and laborious process. A particular case is when the systems are defined through a basis; then the problem reduces to a change of basis in a vector space.

The transformation of color coordinates from one basis of primaries, say $\{P_1, P_2, \ldots, P_n\}$, to another, say $\{Q_1, Q_2, \ldots, Q_n\}$, is called a *change of primaries*. If we denote by (a_{ij}) the change of basis matrix, we have

$$Q_i = \sum_{i=1}^{n} a_{ij} P_j. \tag{5.1}$$

A color C in the relevant color space can be written in the bases $\{P_i\}$ and $\{Q_i\}$ as

$$C = \sum_{i=1}^{n} q_i Q_i \quad \text{and} \quad C = \sum_{i=1}^{n} p_j P_j. \tag{5.2}$$

Thus, the vectors $(q_1, \ldots, q_n)$ and $(p_1, \ldots, p_n)$ represent C in the bases $\{Q_i\}$ and $\{P_i\}$.

Using (5.1), we can write

$$C = \sum_{i=1}^{n} q_i \left(\sum_{j=1}^{n} a_{ij} P_j \right) = \sum_{j=1}^{n} \left(\sum_{i=1}^{n} a_{ij} q_i \right) P_j.$$

Comparing this with (5.2) gives

$$p_j = \sum_{i=1}^{n} a_{ij} q_i,$$

which expresses the relation between the coordinates of C in the two coordinate systems.

Now let $P_i(\lambda)$ and $Q_i(\lambda)$ be the color reconstruction functions of each of the bases $\{P_i\}$ and $\{Q_i\}$. The coefficients q_i and p_i in (5.2) can be computed as follows (see Theorem 4.2 in Chapter 4):

$$q_i = \int_{\mathbb{R}} C(\lambda) Q_i(\lambda) \, d\lambda \quad \text{and} \quad p_j = \int_{\mathbb{R}} C(\lambda) P_j(\lambda) \, d\lambda. \tag{5.3}$$

Therefore, we can write

$$C = \sum_{i=1}^{n} q_i \left(\sum_{j=1}^{n} a_{ij} P_j \right) = \sum_{j=1}^{n} \sum_{i=1}^{n} q_i a_{ij} P_j$$

$$= \sum_{j=1}^{n} \sum_{i=1}^{n} \left(\int_{\mathbb{R}} C(\lambda) Q_i(\lambda) d\lambda \right) a_{ij} P_j$$

$$= \sum_{j=1}^{n} \left(\int_{\mathbb{R}} C(\lambda) \left(\sum_{i=1}^{n} a_{ij} Q_i(\lambda) \right) d\lambda \right) P_j.$$

Using the expression for C in the basis P_j, given in (5.2), we obtain

$$p_j = \int_{\mathbb{R}} C(\lambda) \left(\sum_{i=1}^{n} a_{ij} Q_i(\lambda) \right) d\lambda.$$

Comparing this with the value of p_j in (5.3), we get

$$P_j(\lambda) = \sum_{i=1}^{n} a_{ij} Q_i(\lambda). \tag{5.4}$$

Thus we see that the same matrix that performs the change of coordinates from the basis $\{Q_i\}$ to the basis $\{P_i\}$ can be used to transform the color reconstruction functions $Q_i(\lambda)$ for the basis $\{Q_i\}$ into the corresponding functions $P_j(\lambda)$ for the basis $\{P_j\}$.

In practice, changing coordinates from one color representation space to another is a bit more complicated. This is because the specification of a color—even a primary color—is generally done in terms of its chromaticity coordinates. As mentioned in the preceding chapter, chroma space is the projective plane, so the change of coordinates between two systems defined by primary color bases is naturally expressed by a projective transformation. We will see concrete examples later.

5.3 Color Systems and Computer Graphics

In computer graphics we can single out four important types of color systems: standard systems, device systems, interface systems, and computational systems.

Standard Color Systems

Standard systems were established to allow the specification of colors independently of the particularities of a device or application. Thus, standard systems are extremely useful in problems of color comparison, in the definition of new systems, and in the storage of color information.

Device Color Systems

Device systems are color systems associated with input, processing, and output devices. Such systems are very important because it is through them that we provide or receive color information.

Interface Color Systems

The purpose of interface systems is to allow the user to specify color information easily. Such systems usually serve as a bridge between graphics devices and the user.

Computational Color Systems

Color representation systems associated with physical color systems are not always the most suitable for calculations. Such processing is of great importance, for example, in the calculation of the light energy in the area of image synthesis.

We now study these systems in more detail, including examples.

5.4 Standard Color Systems

The International Lighting Commission (CIE) is the organization responsible for establishing standards in photometry and colorimetry. It has established several color standards, for the most part in the 1930s; no new standards have been introduced specifically for computer graphics. In this section we study the two basic CIE color standards.

5.4.1 The CIE-RGB Standard

The CIE-RGB standard system, already mentioned in Section 4.5, was defined in 1931. It defines a trichromatic (three-color) color space, whose basis of primaries are pure colors in the low, middle, and high portions of the visible spectrum, or red, green, and blue, respectively; hence the abbreviation RGB. The wavelengths of the primary colors are

$$\lambda_R = 700\,\text{nm (red)},$$
$$\lambda_G = 546\,\text{nm (green)},$$
$$\lambda_B = 435.8\,\text{nm (blue)}.$$

The graphs of the color reconstruction functions $R(\lambda)$, $G(\lambda)$, and $B(\lambda)$ of this system are shown in Figure 4.7. They reflect values obtained experimentally by means of color perception experiments of the type described in Section 4.5.1.

These tabulated values allow one to obtain the image of the map of spectral colors

$$\varphi(\lambda) = (R(\lambda), G(\lambda), B(\lambda))$$

in color space. The two endpoints of the map represent the frequencies at the limits of the visible spectrum: the longest-wavelength red and the shortest-wavelength blue-violet. The segment joining these points contains then the various hues of purple, obtained by interpolating red and blue; it is therefore called the *purple line*. Since the image of the color map φ is comprised of pure spectral colors, it is contained in the boundary of the color solid.

The chromaticity diagram of the CIE-RGB color system is obtained by projecting this solid radially onto the Maxwell plane $x + y + z = 1$. The boundary of the diagram is given by the radial projection of the map $\varphi(\lambda)$, together with the purple line. The projection of the color map is expressed by

$$C(\lambda) = \frac{1}{R(\lambda) + G(\lambda) + G(\lambda)}(R(\lambda), G(\lambda), G(\lambda)). \tag{5.5}$$

To obtain a two-dimensional representation of the diagram, we project orthogonally from the chroma plane onto the RG plane in color space; that is, we discard the last coordinate of the map. The result is shown in Figure 5.2.

Note that, in the literature, the expression *chromaticity diagram* is used for the object in Figure 5.2, that is, the orthogonal projection of the chromaticity diagram defined in the preceding chapter. The segment connecting blue to red in Figure 5.2, which is the projection of the purple line as we have defined it, is also called the *purple line* in the literature.

In the CIE-RGB representation, the color reconstruction functions take on negative values. The physical meaning of this was explained in Section 4.5.1. The existence of these negative values is reflected in the chromaticity diagram

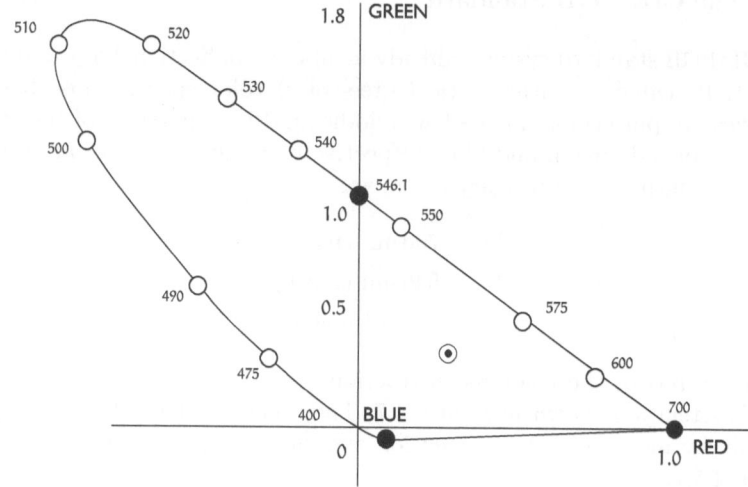

Fig. 5.2. Chromaticity diagram of the CIE-RGB system.

(Figure 5.2): the diagram is not contained in the first quadrant of the plane. Colors falling outside the first quadrant cannot be reproduced by means of a positive linear combination of primary colors in this model.

The units of measurement along the axes of the CIE-RGB system are adjusted in such a way that the chromaticity coordinates of the equal-energy white are $(\frac{1}{3}, \frac{1}{3}, \frac{1}{3})$. It can be shown that the luminance of a color C with coordinates (R, G, B) in the CIE-RGB system is given by

$$L(C) = 0.176R + 0.81G + 0.011B. \tag{5.6}$$

5.4.2 The CIE-XYZ Standard

As a standard, the CIE-RGB model studied in the preceding section has several drawbacks:

- The primary basis does not span the color solid.
- The color reconstruction functions take on negative values, which complicates the calculation of a color's coordinates starting from its spectral distribution: One must compute the integral for the negative part separately, and then subtract (see Theorem 4.2 in the preceding chapter).
- In order to obtain an achromatic point with chromaticity coordinates $(\frac{1}{3}, \frac{1}{3}, \frac{1}{3})$, it is necessary to change the scales of the primary colors, so the region under the graph of each color reconstruction function does not have the same area.

- Photometric magnitudes are not obtained directly from the three-color coordinates. For example, to compute the luminance, one must perform an integration using (4.13) or compute the linear combination in (5.6).

Due to these problems, the CIE established in 1931 a new standard, with primary colors X, Y, and Z, that is designed to simplify, as much as possible, calculations involving colorimetric magnitudes. For this purpose, the primaries must satisfy the following conditions:

1. All XYZ components for all visible colors should be nonnegative.
2. Two of the primaries should have zero luminance.
3. As many spectral colors as possible should have at least one zero XYZ component.

Let's analyze what these conditions mean in terms of the CIE-RGB chromaticity diagram in Figure 5.2.

1. The color reconstruction functions for the CIE-RGB system take on negative values because the primaries are visible colors. By choosing the three primary colors R, G, B to lie in the visible spectrum, we obtain a triangle in the chromaticity plane; any color outside this triangle necessarily has at least one negative RGB coordinate. To avoid this, then, we must choose primary colors such that the triangle they form entirely encloses the set of visible colors, that is, the chromaticity diagram in Figure 5.2. Such primaries do not correspond to physical color stimuli (actual visible colors), but the computational convenience gained outweighs this drawback.

2. To simplify the calculation of photometric magnitudes, the CIE declared that two of the primaries, X and Z, should have zero luminance, while the color reconstruction function of the primary color Y should be the light-efficiency function $V(\lambda)$ itself. This condition is consistent with the preceding one, since the primary colors need not be visible colors, but just vectors in our three-dimensional color space.

To get an XYZ basis satisfying condition 2, then, it is enough to perform a chrominance–luminance decomposition of the CIE-RGB space, choosing vectors X and Z in the chrominance component of the space and the vector Y along the luminance component.

In other words, the vertices X and Z of the triangle of primaries should lie on the zero-luminance line (see Section 4.7.1). By (5.6), the zero-luminance plane in RGB color space is given by

$$0.176R + 0.81G + 0.011B = 0.$$

The zero-luminance line on the RG-plane of the chromaticity diagram is obtained by substituting $B = 1 - R - G$ in the equation just given, thus obtaining

$$0.066R + 0.70G + 0.11 = 0.$$

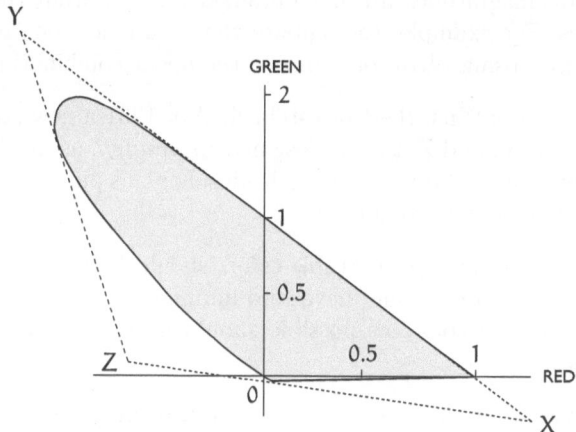

Fig. 5.3. Choosing the chroma coordinates of the primaries X, Y, and Z.

It is on this line of the RG-plane that we must place the primary colors X and Z (see Figure 5.3).

3. Finally, to get some spectral colors to have at least one zero XYZ component, one should make the triangle of primaries touch the spectral color map (the curved boundary of the shaded region in Figure 5.3) as much as possible. The XY-side of the triangle can be taken to be the red-to-yellow portion of the spectral color map, which is almost straight. The YZ-side of the triangle should touch the boundary somewhere and is arbitrarily chosen so that the area of the XYZ triangle is minimized in terms of the RGB coordinates. The resulting triangle is shown in Figure 5.3.

5.4.3 Changing Between the CIE-RGB and CIE-XYZ Systems

Not all colors in the CIE-XYZ system are physically realizable. We can obtain the colorimetric magnitudes of colors in this system in terms of those in the CIE-RGB system. If we knew the vectors corresponding to the primary colors X, Y, and Z, the problem would boil down to a change of basis in a vector space, as we saw in Section 5.2. However, we know only the chromaticity coordinates of the primary colors, so the transformation is a bit more complicated; we study it in this section.

These calculations are important, not only because they allow us to obtain the values of a color in the XYZ system in terms of its values in the RGB system, but also because they serve as a foundation for change-of-system calculations among other color systems.

To begin with, careful choices and computations based on comments 1, 2, and 3 of the previous section allow us to obtain the chromaticity coordinates in the CIE-XYZ system of the three primaries R, G, B of the CIE-RGB system. They are as follows:

primary	XYZ chromaticity		
	x	y	z
R	0.73467	0.26533	0.0
G	0.27376	0.71741	0.00883
B	0.16658	0.00886	0.82456

These chromaticity coordinates determine three colors of the CIE-XYZ color space, namely

$$c_1 = \rho \times (0.737467, 0.26533, 0),$$
$$c_2 = \gamma \times (0.27376, 0.71741, 0.00883),$$
$$c_3 = \beta \times (0.16658, 0.00886, 0.82456),$$

where ρ, γ, and β are positive real numbers. These three colors constitute a basis of the space; therefore the change of coordinate matrix should map the vectors $R = (1, 0, 0)$, $G = (0, 1, 0)$, and $B = (0, 0, 1)$ to c_1, c_2, and c_3, respectively. The transformation is given by

$$\begin{pmatrix} X \\ Y \\ Z \end{pmatrix} = \begin{pmatrix} 0.73467\,\rho & 0.27376\,\gamma & 0.16658\,\beta \\ 0.26533\,\rho & 0.71741\,\gamma & 0.00886\,\beta \\ 0.0 & \rho\ 0.00883\,\gamma & 0.82456\,\beta \end{pmatrix} \begin{pmatrix} R \\ G \\ B \end{pmatrix}. \tag{5.7}$$

The constants ρ, γ, and β must be determined. In order to do this, we require that the white color have coordinates $\left(\frac{1}{3}, \frac{1}{3}, \frac{1}{3}\right)$ in both systems, that is,

$$T\left(\tfrac{1}{3}, \tfrac{1}{3}, \tfrac{1}{3}\right) = \left(\tfrac{1}{3}, \tfrac{1}{3}, \tfrac{1}{3}\right).$$

Substituting this into (5.7), we obtain the system

$$\begin{pmatrix} 1 \\ 1 \\ 1 \end{pmatrix} = \begin{pmatrix} 0.73467 & 0.27376 & 0.16658 \\ 0.26533 & 0.71741 & 0.00886 \\ 0.0 & 0.00883 & 0.82456 \end{pmatrix} \begin{pmatrix} \rho \\ \gamma \\ \beta \end{pmatrix},$$

whose solution is

$$\rho = 0.666952, \quad \gamma = 1.132407, \quad \beta = 1.200641.$$

Substituting these values in (5.7), we finally obtain the desired transformation:

$$\begin{pmatrix} X \\ Y \\ Z \end{pmatrix} = \begin{pmatrix} 0.489989 & 0.310008 & 0.2 \\ 0.176962 & 0.81240 & 0.010 \\ 0.0 & 0.01 & 0.99 \end{pmatrix} \begin{pmatrix} R \\ G \\ B \end{pmatrix}. \tag{5.8}$$

The preceding computations can be better understood by observing that the chromaticity space is the projective plane. Therefore, we would like to obtain a projective transformation between the chromaticity spaces of these two systems. By the fundamental theorem of projective geometry, a transformation between projective spaces of dimension n is specified by what it does to $n+2$ projectively independent points; thus we need to specify the action of the transformation on a fourth vector in XYZ space. A natural choice is to make the equal-energy white have coordinates $\left(\frac{1}{3}, \frac{1}{3}, \frac{1}{3}\right)$ as we did before. Thus the projective transformation from the RGB system to the XYZ system is defined by setting $T(R, G, B) = (X, Y, Z)$, where the linear transformation T is given by Equation 5.7, and the constants ρ, γ, and β must be determined using the condition

$$T\left(\tfrac{1}{3}, \tfrac{1}{3}, \tfrac{1}{3}\right) = \left(\tfrac{1}{3}, \tfrac{1}{3}, \tfrac{1}{3}\right).$$

The inverse transformation, going from the XYZ system to the RGB system, is given by

$$\begin{pmatrix} R \\ G \\ B \end{pmatrix} = \begin{pmatrix} 2.3647 & -0.89658 & -0.468083 \\ -0.515155 & 1.426409 & 0.088746 \\ 0.005203 & -0.014407 & 1.0092 \end{pmatrix} \begin{pmatrix} X \\ Y \\ Z \end{pmatrix}. \tag{5.9}$$

As we saw in the preceding chapter, the chromaticity coordinates x, y, z of a color with coordinates X, Y, Z in the CIE-XYZ system are given by

$$x = \frac{X}{X+Y+Z}, \qquad y = \frac{Y}{X+Y+Z}, \qquad z = \frac{Z}{X+Y+Z}.$$

Substituting the values of X, Y, and Z given by (5.8), we immediately obtain the expression of x, y, z in terms of the chromaticity coordinates r, g, b in the CIE-RGB system:

$$x = \frac{0.49000\,r + 0.31000\,g + 0.20000\,b}{0.66697\,r + 1.13240\,g + 1.20063\,b},$$

$$y = \frac{0.17697\,r + 0.81240\,g + 0.01063\,b}{0.66697\,r + 1.13240\,g + 1.20063\,b}, \tag{5.10}$$

$$z = \frac{0.01000\,g + 0.99000\,b}{0.66697\,r + 1.13240\,g + 1.20063\,b}.$$

Analogously, using (5.9), we obtain r, g, b as a function of x, y, z. In particular, the chromaticity coordinates of the XYZ primaries in the RGB system are given by:

primary	RGB chromaticity		
	r	g	b
X	1.2750	−0.2779	0.0029
Y	−1.7395	2.7675	−0.0280
Z	−0.7431	0.1409	1.6022

If x, y, and z are the chromaticity coordinates of a color with coordinates X, Y, Z, we have $y = Y/(X+Y+Z)$. It follows that $X+Y+Z = Y/y$, and, since $Y = V$, the light-efficiency function, we have $X + Y + Z = V/y$. Therefore,

$$X = x(X + Y + Z) = \frac{x}{y}V,$$

$$Z = z(X + Y + Z) = \frac{z}{y}V.$$

(5.11)

One can derive from this an expression for the color reconstruction functions of the XYZ system in terms of the color matching functions of the RGB system. Indeed, using (5.10), we get the functions $x(\lambda)$, $y(\lambda)$, and $z(\lambda)$ in terms of $r(\lambda)$, $g(\lambda)$, and $b(\lambda)$. Using (5.11), we get

$$X(\lambda) = \frac{x(\lambda)}{y(\lambda)}V(\lambda),$$

$$Y(\lambda) = V(\lambda),$$

$$Z(\lambda) = \frac{z(\lambda)}{y(\lambda)}V(\lambda).$$

These calculations allow one to obtain quantitative information about the CIE-XYZ system from information about the CIE-RGB system, which, in turn, is obtained by means of experiments, as described in the preceding chapter.

We are now in a position to sketch the graphs of the color reconstruction functions and the chromaticity diagram of the CIE-XYZ system. They are shown in Figures 5.4 and 5.5.

Clearly, we can obtain the color reconstruction functions using the result established in (5.4), together with Equation (5.8), which provides the transformation matrix (a_{ij}) from the XYZ system to the RGB system.

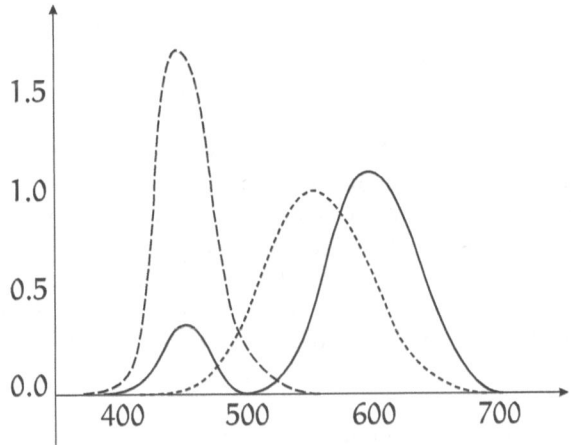

Fig. 5.4. Color reconstruction functions of the CIE-XYZ system.

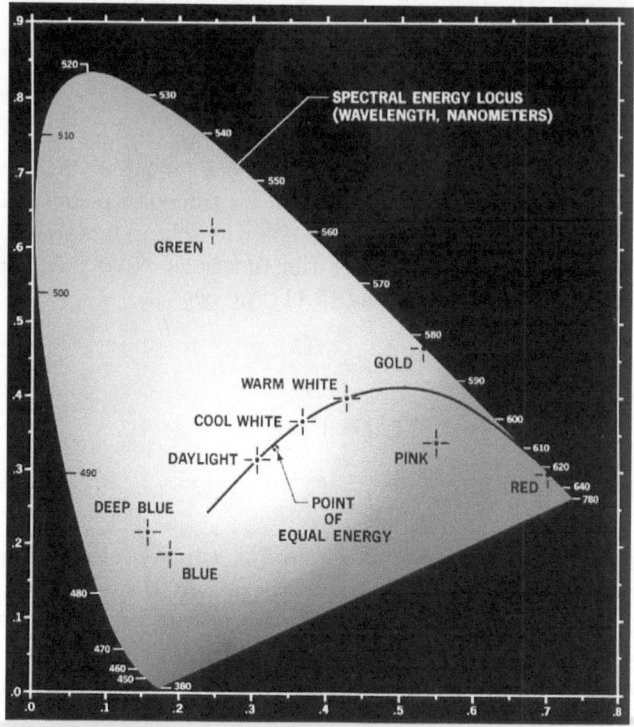

Fig. 5.5. Chromaticity diagram of the CIE-XYZ system. See Plate 1 in color insert.

In 1964 the CIE changed the CIE-RGB and CIE-XYZ standards, based on new color perception experiments. An important parameter in these experiments is the angle of view of the standard observer, which in the 1931 experiments was 2° and in the 1964 experiments was increased to 10°. The 1931 standards are still used in computer graphics, on the grounds that the angle of view of images seen on a monitor is small.

As mentioned earlier, the XYZ primaries were chosen so the resulting system would allow the representation of all visible colors with positive coordinates. Therefore, the chromaticity diagram of the CIE-XYZ system, shown in Figure 5.5, contains a representation of all the colors, up to changes in luminance. (However, the printed figure cannot exactly reproduce the colors corresponding to many points of the diagram, because of the reconstruction process used in the offset printing. We will return to this point in Chapter 16, on image systems. Nonetheless, for pedagogical reasons, it is worth printing this diagram in color.)

Finally, we stress that in applications we work with physical color systems, generally associated with graphics devices. Thus the chromaticity diagram of these systems is a subset of the chromaticity diagram of the CIE-XYZ system shown in Figure 5.5.

5.4.4 Complementary Color Systems

Two colors c_1 and c_2 are called *complementary* if, when additively combined in the appropriate proportion, they form an achromatic color (pure gray or white). This can be illustrated geometrically using the chromaticity diagram (Figure 5.6). Given a color c in the chromaticity diagram, its complement c' can be obtained as follows. Take the line going through c and the achromatic point O in the diagram. The complementary color c' lies on that same line.

An important case is that of pairs of complementary spectral colors. In the construction just given, such colors are given by the intersections of a line going through O with the boundary of the chromaticity diagram. Some spectral colors don't possess a complement; they are indicated in Figure 5.6 by the dashed part of the curved boundary.

Color systems based on complementarity start from three primary colors, C_1, C_2, and C_3, but use as a basis for color space the three complementary colors C_1', C_2', and C_3'. For example, starting from the CIE-RGB standard, we get the CMY system, whose primaries are cyan (complementary to red), magenta (complementary to green), and yellow (complementary to blue).

In the RGB system, the combination of the three primary colors in equal amounts yields white, as shown in Figure 5.7 (left). In Figure 5.7 (right) we see the combination of the primary colors in the complementary CMY system. Notice that when the three complementary basis colors combine, we obtain the color black. This system is quite appropriate for the imitation of the subtractive process of color formation: as we add colors from the complementary basis to the paper, the white reflected color is subtracted, in such a way that when all three colors are added we get no reflection at all (black). For this reason,

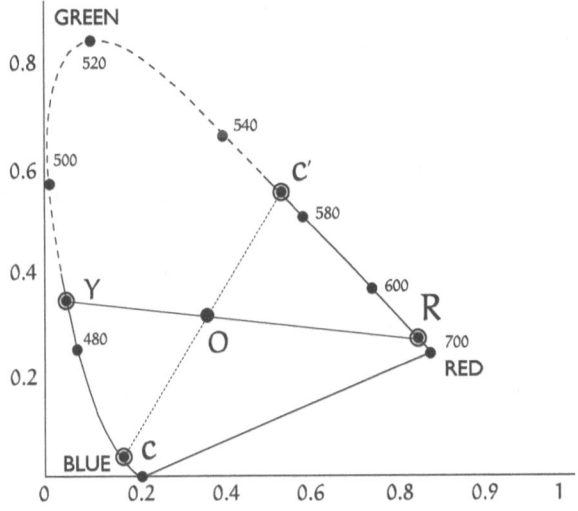

Fig. 5.6. Finding the complement of a color c.

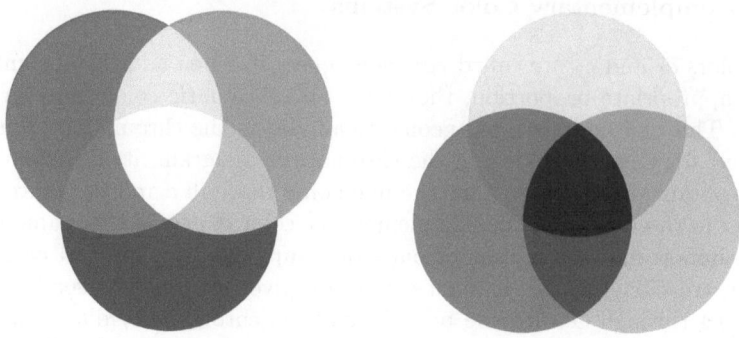

Fig. 5.7. Colors in the additive RGB system (left) and their complements (right). See Plate 2 in color insert.

in the literature the CMY model is sometimes called "subtractive"; however, this terminology is incorrect and confusing, because the CMY system *does not* uses additive complementary colors to reconstruct colors.

The CMY system is useful when one considers a process that involves subtractive color formation, such as color printing. In printing we start with white (the color of the paper), and as we add inks there is a decrease in luminance. Zero luminance (black) is obtained by superimposing all three primary colors. From the additive point of view, cyan is a mixture of green and blue RGB primaries, so putting a layer of cyan ink on paper eliminates the red component from the light scattered by the paper. We will return to color printing in Chapter 16, on image systems.

5.4.5 Uniform Color Systems

Visual sensitivity to small differences between colors is of fundamental importance in color perception experiments. Let c_0 be a color indicated by a point in the chromaticity diagram, as in Figure 5.8. We consider the problem of *finding the set of colors c in the diagram that are perceptually at the same distance from the color c_0.*

Experiments carried out by MacAdam in the early 1940s showed that this set is an ellipse centered at c_0. The eccentricity and the axes of this ellipse vary as we move c_0 around the chromaticity diagram. This shows that perceptual distances in color space cannot be measured with the Euclidean metric; if they were, the set in the figure would be a circle. The perceptual metric is known in the literature as the *jnd metric* ("just noticeable difference metric").

We say that a color system is *perceptually uniform* if the jnd metric is the Euclidean metric on the color solid. Several standard color systems that attempt to be uniform have been introduced: for example, the CIE-Luv and the CIE-Lab systems. These two systems use a chrominance–luminance

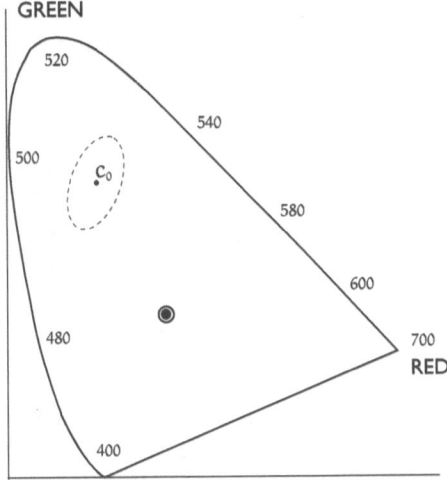

Fig. 5.8. MacAdam ellipse.

decomposition of color space. However, luminance is replaced by a related measure called *lightness*, which records how light intensity is perceived by the human eye. The Luv system performs a transformation of the CIE-XYZ system to establish perceptual uniformity in the chromaticity plane, whereas the CIE-Lab system establishes perceptually uniform coordinates in three-dimensional color space.

5.5 Device Color Systems

We now turn to color systems used in certain display devices, such as CRT (cathode-ray tube) monitors and other video equipment.

5.5.1 The Monitor RGB System

The color space of a color CRT monitor is indicated by mRGB. It has a primary color basis of the three color vectors (red, green, and blue) defined by the phosphors used in the monitor. The color solid of this system is a bounded subset of the space generated by the primaries, because each primary has a maximum possible intensity. Using an appropriate scale along each primary axis, we can normalize these coordinates so that the maximum is 1. Therefore, the color solid is a cube, as shown in Figure 5.9. The figure also shows the corresponding Maxwell triangle.

This color solid is called the *RGB cube*. The origin $(0, 0, 0)$ of the cube corresponds to black, and the point with coordinates $(1, 1, 1)$ corresponds to the monitor's brightest white.

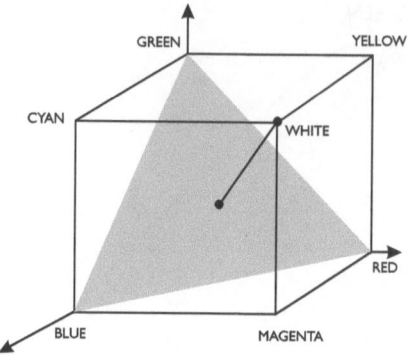

Fig. 5.9. Chromaticity triangle of the mRGB space.

The primary colors of the mRGB system can be represented by three points in the CIE-XYZ chromaticity diagram. The chromaticity diagram of the color solid of the mRGB system is then a triangle having these points as vertices, as shown in Figure 5.10, left.

The basis of primary colors of the mRGB system, as already mentioned, is determined by the phosphors used on the monitor screen. In general, monitors from different manufacturers have different color spaces. Figure 5.10, right, shows the chromaticity diagrams of two distinct mRGB systems. Only the colors lying in the polygonal region formed by the intersection of the two triangles can be reconstructed by both monitors.

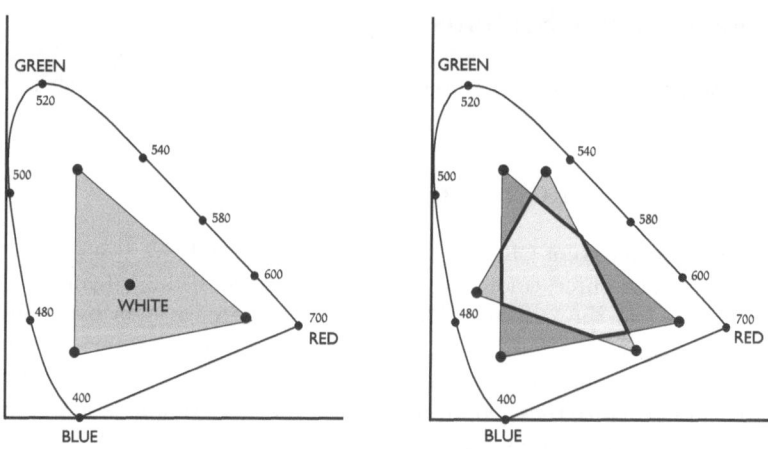

Fig. 5.10. Left: chromaticity triangle of an mRGB space, inside the CIE-XYZ chromaticity diagram. Right: comparison of chromaticity triangles for different mRGB spaces.

A common task is to perform a color transformation between an mRGB space and the CIE-XYZ or CIE-RGB standard. In general, we know the chromaticity coordinates of the primaries of the mRGB system, because the monitor's manufacturer provides them. The problem is then entirely analogous to one we solved earlier, namely transforming three-color coordinates between the CIE-RGB and CIE-XYZ standards.

5.5.2 Monitor-Complementary Systems

Associated with the mRGB system of a monitor, we have the complementary mCMY system for the same monitor, which is of great importance in electronic publishing. Normalizing the color coordinates in the mRGB cube so that they lie in the interval $[0, 1]$, we have

$$C = B + G,$$
$$M = R + B,$$
$$Y = R + G,$$

where C represents cyan, M magenta, and Y yellow (see Figure 5.11).

An easy way to obtain a change of coordinates into the CMY system is to consider a coordinate system whose origin is at the white point $W = (1, 1, 1)$ and whose coordinate axes at this point are given by $\{-R, -G, -B\}$. In this case, the change of coordinates can be performed by changing to the basis $\{-R, -G, -B\}$ and then translating by the vector W. Thus, given a color with coordinates (r, g, b) in the mRGB system, the change-of-coordinates transformation T acts in the following way:

$$T(r, g, b) = (1, 1, 1) - rR - gG - bB$$
$$= (1 - r)R + (1 - g)G + (1 - b)B.$$

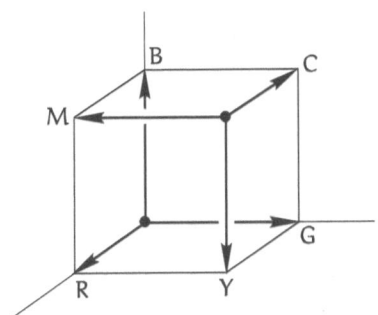

Fig. 5.11. The mRGB and mCMY coordinate systems.

We therefore get

$$(c, m, y) = T(r, g, b) = (1 - r, 1 - g, 1 - b).$$

We have thus shown that the components (c, m, y) of a color in the mCMY system can be obtained from its components (r, g, b) in the mRGB system using the relations

$$c = 1 - r, \qquad m = 1 - g, \qquad y = 1 - b.$$

The colors cyan, magenta, and yellow, together with the primaries red, green, and blue, make up six vertices of the color solid of the mRGB system (the unit cube). The two remaining vertices are the endpoints of the gray line, a diagonal of the cube. The orthogonal projection of these six vertices of the cube onto the plane with equation $x + y + z - 3 = 0$, which is perpendicular to the gray line and goes through the white point $(1, 1, 1)$, forms a hexagon, as shown in Figure 5.12. This geometric construction is important in understanding the color solid of some color systems that we will study later.

5.5.3 Component Video Systems

Color systems using the chrominance–luminance decomposition of color space are very common in the video and television industries. The recent advances

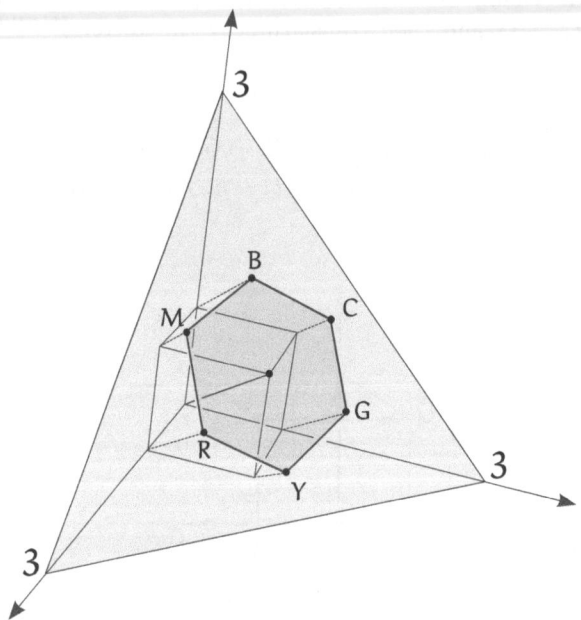

Fig. 5.12. The hexagon formed by RGB and their complements CMY.

in multimedia technology and its various applications have led to the dissemination of the joint use of video and computer graphics. This has contributed to the increasing use of these systems in computer graphics.

An important observation, often not encountered in textbooks, is that the primary color components used in the video and television industries incorporate a correction to compensate for the nonlinearity of the video monitors. (This *gamma correction* will be studied in Chapter 16.) Despite this, we will use the same notation for primary colors in the CIE-RGB system and in the various video systems.

The standard luminance used by the television industry is computed by

$$Y = 0.299R + 0.587G + 0.114B. \tag{5.12}$$

This corresponds approximately to the average luminance of the reconstruction color space of a television monitor; it is called NTSC luminance. By observing the luminance equation (5.12), we conclude that the R component contributes 30%, the G component 60%, and the B component 10% toward the luminance of a color.

As we stressed in the preceding chapter, the human visual system has less sensitivity toward variations of color than toward variations of luminance. This can be exploited in color encoding processes that use color systems based in a chrominance–luminance decomposition of color space.

A simple way to obtain a chrominance–luminance decomposition is to calculate the luminance Y, using (5.12), and make this one of the components of the system. Next, one subtracts Y from the other color components, obtaining the combinations $R - Y$, $G - Y$, and $B - Y$, which do not carry luminance information. Any two of these combinations, together with the luminance component Y, define a coordinate system in color space. Since the green contribution to luminance is the highest, a compact encoding can be obtained using the components $R - Y$ and $B - Y$.

Using (5.12), we see that the change of coordinates between the Y, $R - Y$, $B - Y$ system and the RGB system is given by

$$
\begin{aligned}
Y &= 0.299R + 0.587G + 0.114B, \\
R - Y &= 0.711R - 0.587G - 0.114B, \\
B - Y &= -0.299R - 0.587G + 0.990B,
\end{aligned}
$$

that is,

$$
\begin{pmatrix} Y \\ R - Y \\ B - Y \end{pmatrix} = \begin{pmatrix} 0.299 & 0.587 & 0.11 \\ 0.711 & -0.587 & -0.11 \\ -0.299 & -0.587 & 0.99 \end{pmatrix} \begin{pmatrix} R \\ G \\ B \end{pmatrix}.
$$

Figure 5.13(a) shows the result of the transformation of the cube RGB of a monitor color system, under the matrix in the preceding equation. The parallelogram shown in this figure is the color solid in the Y, $R - Y$, $B - Y$ space.

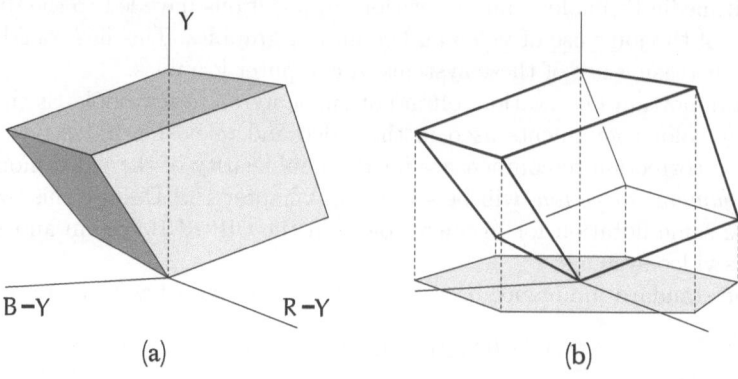

Fig. 5.13. (a) Color solid in the Y, R − Y, B − Y space. (b) Projection of the Y, R − Y, B − Y solid in the R − Y, G − Y plane.

Color systems based on the Y, R − Y, B − Y decomposition are called *video component systems* since they are predominantly used in the video and television industries. Recently some of these systems have started to be widely used in computer graphics.

The chromaticity diagram of the Y, R − Y, B − Y system is obtained by an orthogonal projection of the color solid of Figure 5.13(a) onto the R − Y, B − Y plane. This projection yields a hexagon, as displayed in Figure 5.13(b). In Figure 5.14 we show this hexagon in the R − Y, B − Y plane.

The transformation and projection of the RGB cube just discussed are important in understanding the differences among the various color systems used by the industry, based on the Y, R − Y, B − Y decomposition of color space.

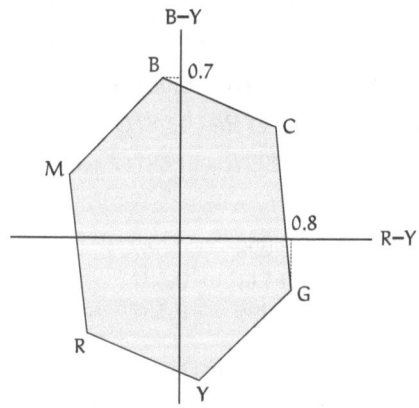

Fig. 5.14. Chrominance hexagon of the Y, R − Y, B − Y system.

These systems differ by changes of scale or by changes in the chrominance axes $R - Y$ and $B - Y$. We give two examples.

The Betacam System

The $Y p_b p_r$ system used by SONY in its Betacam line of video equipment is characterized by a scale change. Geometrically, the purpose of this adjustment is to obtain a regularized chromaticity diagram and to normalize the maximum and minimum values of the chrominance components $R - Y, B - Y$. The scaling is as follows:

$$p_b = \frac{0.5}{1 - 0.114} (B - Y), \qquad p_r = \frac{0.5}{1 - 0.299} (R - Y).$$

This results in the chrominance hexagon shown in Figure 5.15.

Digital Video System

The $Y C_b C_r$ system is the international standard for digital video signals. It is obtained from the Y, $R - Y$, $B - Y$ system by means of the following transformation:

$$Y = 16 + 235Y,$$

$$C_b = 128 + 112\left(\frac{0.5}{1 - 0.114} (B - Y)\right),$$

$$C_r = 128 + 112\left(\frac{0.5}{1 - 0.299} (R - Y)\right).$$

This is the color system used in the JPEG image compression standard and in the MPEG video compression standard.

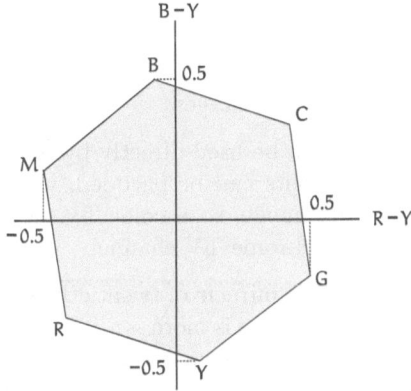

Fig. 5.15. Chrominance hexagon of SONY's Betacam system.

5.5.4 Composite Video Systems

In the preceding section we saw a chrominance–luminance decomposition of color space that is the basis for the definition of several color systems used by the video and television industries. The broadcast video systems used (mainly) by the television industry—NTSC, PAL, SECAM, etc.—are also based on the Y, $R-Y$, $B-Y$ decomposition of color space. As in the case of video component systems, these systems are defined by means of changes of coordinates in the chrominance plane. However, in order to be broadcast, all components of the system must be combined into a single signal; hence the establishment of *composite video systems.*

The YUV System

The YUV system is used as the basis for the encoding of a composite video signal. It is obtained from the Y, $B-Y$, $R-Y$ system by the change of coordinates

$$U = 0.493(B - Y), \qquad V = 0.877(R - Y).$$

The YIQ System

In the YUV encoding, the UV components occupy the same bandwidth of the spectrum. By rotating the UV by 33°, we obtain the color components IQ:

$$\begin{pmatrix} I \\ Q \end{pmatrix} = \begin{pmatrix} \cos 33° & -\sin 33° \\ \sin 33° & \cos 33° \end{pmatrix} \begin{pmatrix} U \\ V \end{pmatrix}.$$

In the YIQ system thus obtained, the Q component occupies a narrower band of the spectrum than the I component. This can be seen geometrically in Figure 5.14.

Video broadcasting uses either the YUV or YIQ system. The use of these color systems has at least two advantages:

- The luminance signal Y can be used directly by monochrome TV sets.
- The chrominance components can be encoded, with minimal perceptual loss in the quality of the image, so a color TV signal can use the same spectrum band as a monochrome TV channel.

This latter fact is a veritable miracle of twentieth-century electronics. It is possible only because the human eye is more sensitive to changes in luminance than to changes in chrominance information. Color information represents only about 5% of the bandwidth in the standard TV signal received and decoded by a home television set.

5.6 Color Interface Systems

Color specification systems based on a vector space model, such as the RGB system, are computationally practical, but they present a human interface problem in that a user cannot easily and intuitively specify a desired color in them. For example, suppose we're given a color in the RGB system and we want to make it "lighter" or "darker" (change the amount of white in it) without changing the hue. To achieve this, it is necessary to change all three RGB components, and the changes have no direct intuitive connection with the perceptual feature concerned, which is known as *saturation*.

Features such as luminance, saturation, and hue are much more directly linked to the way humans respond to color. Therefore, a model in which these features can be directly controlled is preferable from the viewpoint of color specification. The study of such models requires some basic terms from color perception theory; we introduce these terms here, relating them to the corresponding colorimetric concepts.

Dominant Wavelength and Hue

In colorimetry the *dominant wavelength* of a color is the wavelength of the pure color that, when (additively) combined with white, yields the given color. The dominant wavelength of a color can be easily read from the chromaticity diagram, as follows (see Figure 5.16): draw a line from the achromatic point (white) to the point representing the given color; the intersection of this line with the boundary of the diagram gives the dominant wavelength.

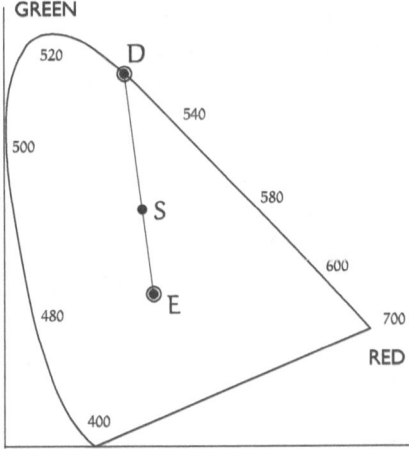

Fig. 5.16. Point D represents the dominant wavelength of color S.

From the perceptual point of view, the dominant wavelength corresponds to the color's *hue;* intuitively, it represents the color in its pure form (green, yellow, blue, and so on).

Note that the dominant wavelength of some colors lies on the purple line of the diagram. We have seen, however, that in practice we work with a color representation model whose chromaticity diagram is a subset of the chromaticity diagram of the visible colors—for example, the chromaticity triangle of the mRGB system. The problem, therefore, does not arise in practice.

Purity and Saturation

Once we have determined the hue of a color, we can change the color by mixing it with white. The less white, the more saturated the color. This perceptual idea of saturation corresponds to the magnitude called *purity* in colorimetry. In the notation of Figure 5.16, the purity of a color is geometrically defined by $p = \frac{ES}{ED}$. As the point representing the color approaches the achromatic point E, the segment ES gets shorter. Since ED has constant length (we're keeping the hue fixed), purity decreases until it reaches 0 at E. Thus, saturation grows monotonically with purity.

Luminance and Brightness

The parameters of hue and saturation determine the chroma characteristics of the color. To fully determine a color, we need a third parameter related to the color luminance, which is connected with the perceptual notion of color intensity. In practice, interface models use some magnitude that varies monotonically with luminance. The name of this magnitude depends on the model, but we will refer to it generically as *brightness*.

Many color representation models are based on the perceptual notions of hue, saturation, and brightness. From a purely descriptive point of view, the color solid of these models is a three-dimensional set parametrized by cylindrical coordinates, as indicated in Figure 5.17: hue changes along horizontal circles, saturation increases radially, and brightness increases as one goes up, orthogonally to the hue–saturation plane.

There are two methods for defining color specification systems: by coordinates and by samples.

Specification by Coordinates

In this method the user specifies a color by choosing its coordinates in a given system. One can use the mRGB or mCMY system, or any other, but in general one tries to use color systems involving the perceptual parameters just discussed: hue, saturation, and brightness.

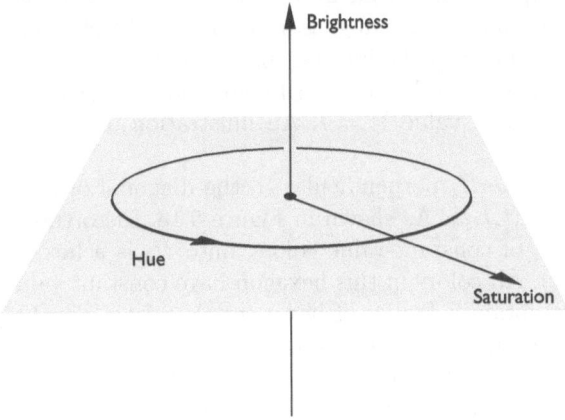

Fig. 5.17. Cylindrical coordinates representing hue, saturation, and brightness.

Specification by Samples

This method performs a discretization of the color solid of the system. This discretization leads to a finite number of color samples, which are generally collected in a *color atlas*, grouped into *color charts*.

Recently there has been great progress in the area of color specification by computer. Color specification is closely related to the more general problem of *color management*. Specialized software has been developed to cope with it; such programs, called *color management systems*, not only offer the user an interface for color specification but also attempt to store color information in a way that is independent of the various device-dependent color systems in use, converting to the appropriate device color system at display time.

In the rest of this section, we give a geometric description of some color specification systems that use either the method of specification by coordinates or the method by samples. These systems are widely used in computer graphics and in the color industry in general.

5.6.1 The HSV System

The HSV model derives its name from the parameters hue, saturation, and *value*, the name of the magnitude corresponding to brightness in this system. It is defined in terms of the mRGB system. By definition, the value of a color C with mRGB coordinates (C_R, C_G, C_B) is

$$V(C) = \max\{C_R, C_G, C_B\}.$$

As we have seen, the color solid in the mRGB is a unit cube I^3, called the color cube, whose main diagonal goes from $(0,0,0)$ (black) to $(1,1,1)$ (white). For each real number t in the interval $[0,1]$, we get a cube C_t parallel to the unit cube I^3, having side length t. All colors on the sides $R = t$, $G = t$, or $B = t$ have the same value $V = t$. An illustration of this fact is shown in Figure 5.18.

Consider a plane Π_t perpendicular to the diagonal of the cube and containing the point (t,t,t). As shown in Figure 5.18, the orthogonal projection of each cube C_t, of constant-value colors, onto Π_t is a hexagon centered at the point (t,t,t). All colors in this hexagon have constant value t. As t varies from 1 to 0 we obtain a family of hexagons that form a right pyramid with a hexagonal base and vertex at the origin. The base of this pyramid is the hexagon corresponding to the unit cube C. In this hexagon each vertex corresponds to one of the RGB primary colors or to one of the complementary colors CMY, as can be seen in Figure 5.19, left.

Note that each hexagonal cross section parallel to the base of the pyramid represents a set of colors of the unit cube having the following characteristics: The hue of each color is on the border of the hexagon, and the saturation decreases as we approach the center along a radius; the value of the colors in each hexagon is constant and varies in proportion to the distance from the plane of the hexagon to the vertex of the pyramid. The axis of the pyramid, formed by the centers of the hexagons, corresponds to the diagonal of the cube and therefore to the achromatic colors (black, white, and grays). Figure 5.19, right, shows the variation of these parameters in the HSV pyramid.

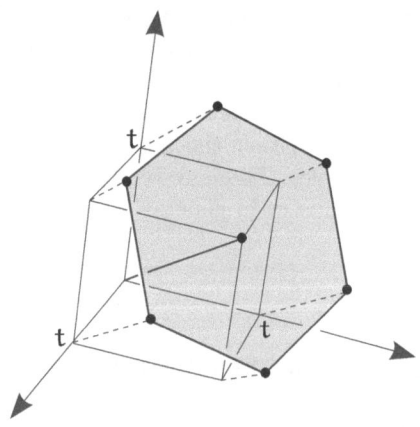

Fig. 5.18. Cube of colors with constant value in the HSV system.

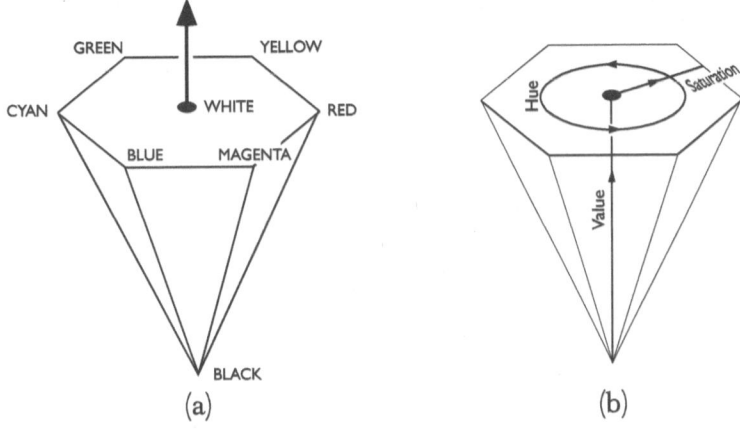

Fig. 5.19. Color solid of the HSV system.

5.6.2 The HSL System

The HSL model (hue, saturation, and lightness) is also defined in terms of an mRGB system. The magnitude corresponding to brightness, called *lightness*, is defined for a color C with mRGB coordinates (C_R, C_G, C_B) as

$$L = \tfrac{1}{2}(\max(C_R, C_G, C_B) + \min(C_R, C_G, C_B)).$$

Thus, the color $(0, 0, 0)$, or black, has lightness zero, and the color $(1, 1, 1)$, or white, has lightness one. The corresponding color solid is made up of two congruent right cones of altitude 1, whose bases coincide (Figure 5.20).

The link between the perceptual parameters and the geometry of the two cones is the following: lightness varies from 0 to 1 along the common axis; the hue is determined by the points on the outer circle of the common base; and saturation varies with distance to the axis.

In the literature, the HSV color solid appears with other geometries. See Section 5.9 at the end of this chapter.

The HSV and HSL systems have certain drawbacks as interface systems, from the perceptual point of view. For instance, colors with the same value in the HSV system, or the same lightness in the HSL system, do not necessarily have the same luminance.

5.6.3 The Munsell System

Whereas the HSV and HSL models define a color solid parametrized by hue, saturation, and brightness, the Munsell model uses the color atlas method:

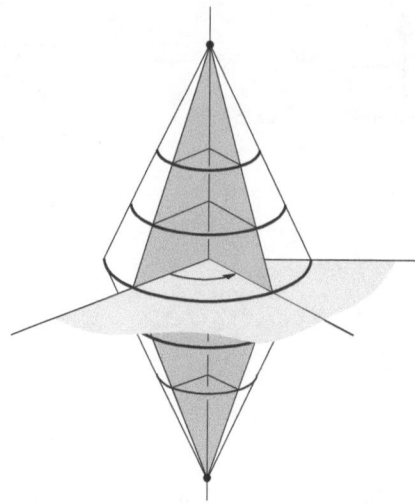

Fig. 5.20. Color solid for the HSL system.

hue, saturation, and brightness are used for sampling. In this system, brightness is called *value*, and saturation is called *chroma*. Each chart of the color atlas is defined by fixing a hue, and each color in the chart is obtained by varying chroma and value. All the charts together make up the atlas, which is the color solid of the system. Figure 5.21 illustrates part of the color solid (four charts).

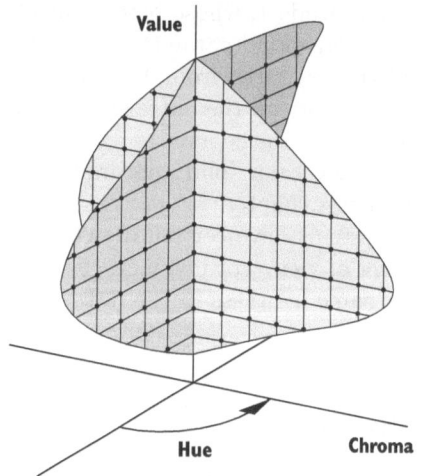

Fig. 5.21. Color solid of the Munsell system.

We mention two important facts. The Munsell system was conceived in 1915 and significantly predates the CIE color standards, which, as we mentioned, were established in 1931. The choice of color samples in the Munsell system is based on a criterion of perceptual uniformity, as discussed in Section 5.4.5. The uniform standard CIE-Lab of Section 5.4.5 is, in fact, obtained by applying cubic interpolation to the samples of the Munsell system.

There are other classical color systems constructed in a way similar to Munsell's: for example, the Ostwald system.

The method used by the Munsell and Ostwald systems for constructing color charts establishes a paradigm for the creation of user interface systems geared toward computer color selection, based on hue, saturation, and brightness parameters. In either system, we obtain a color atlas by means of the following sampling process: we initially take a finite number of hues; each hue t_0 determines a section of the color solid, defined by cutting the solid with the half-plane going through the axis of the color solid and the chosen hue (Figure 5.22). The chart for hue t_0 is constructed by choosing a finite number of colors in this section, as shown in Figure 5.22 for the HSV system. When the color atlas is defined in this way, the determination of the color coordinates corresponding to a given sample is immediate.

5.6.4 The Pantone System

Another color interface system that uses the sampling method is the *Pantone Color Matching System*. Introduced in the early 1960s, this system starts

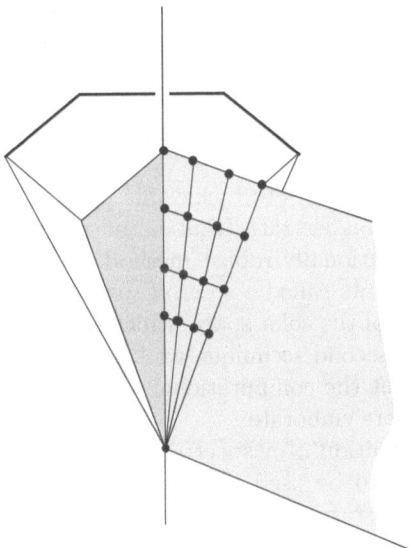

Fig. 5.22. Definition of a color atlas in the HSV system.

from a certain number of basic colors, including black and white, and uses them to determine the color samples in the various charts of the color atlas. The Pantone system was developed to aid in the specification of color for printing on paper and is still widely used for this purpose. For this reason, besides charts with colors defined from basic colors, it includes certain special charts with metallic colors, phosphorescent colors, and others. With the developments in electronic publishing, this system has gradually been making its way into sample-based color selection systems in computer graphics.

There are other atlas-based color systems like Pantone: we mention Truematch and Focoltone.

5.7 Computational Color Systems

Two problems must be considered when we discuss computational aspects of color: color resolution, and computational color systems.

Color resolution is the number of bits needed to represent each color component in the computer. This number is of great importance in calculations involving color, and it will be discussed in detail in the next chapter.

We know from the preceding chapter that any physical color space is a representation of spectral color space $\mathcal{E}$; in other words, color spaces are obtained essentially by sampling of $\mathcal{E}$. Although this sampling is perceptually acceptable, it can lead to computational problems, as we discussed in Chapter 2. One specific source of problems is that the interaction of light with the environment takes place over the whole visible spectrum, so two objects whose colors have metameric spectral distributions, and are therefore indistinguishable in white light, may have very different colors under colored light. The effect of this interaction leads to erroneous results if we restrict our calculations to three-color models.

There are two possible ways to tackle this problem. One can either sample the spectrum at a higher rate (that is, at more than three points) or sample using a computationally robust method. In the first case the system obtained is inappropriately called a *spectral system*. In the first technique we increase the dimension of the color space, which raises the computational cost of calculations. In the second technique we keep the dimension of the representation space low, but the computational cost may still be higher because the calculations are more elaborate.

In general, the definition of a spectral system involves the spectral distribution of the colors to be used in the system. Thus one must use color reconstruction methods starting from the color specified by the user. The use of spectral systems for computational color problems is of great importance in image synthesis, intervening in the correct calculation of the color intensity function.

5.8 Color Transformations

Given a subset U of a color space E_1, and a subset V of a color space E_2, a *color transformation* is a map $T : U \to V$.

One important color transformation is the change of coordinates from one color system to another. Such a transformation associates to each color in one system a color in the other and is therefore called a *change of system*. Change-of-system transformations are important because they allow the use of different color systems in an application.

Many important changes of system cannot be performed simply by a change of basis in color space—usually because one or both coordinate systems are not defined by means of a basis of primary colors. Such is the case, for instance, when going from an mRGB system to the corresponding mHSV or mHSL system. Such change-of-system transformations are nonlinear. One must then, in each case, take into account the geometry of the color solid, so as to obtain the formulas for the change of coordinates. The algorithms involved in the determination of these transformations will not be discussed here.

Since the color representation model we are considering is a vector space, any transformation that can be defined on this space can be used as a color transformation. Many color transformations are used frequently in computer graphics, usually in connection with the color space of an image; we will discuss them later. In particular, in the next chapter we will study the quantization transformation, which is crucially involved in the display of an image on a graphics device.

5.9 Comments and References

Quantitative information, such as tables, equations, and so on, about the standards used in colorimetry and photometry can be found in (Wyszecki and Stiles 1982). In particular, the book tabulates approximate values for the color reconstruction functions for the CIE-RGB and CIE-XYZ models and discusses in detail the 1931 CIE representation models, as well as the modifications introduced in 1964.

The HSV interface model was first published in (Smith 1978). The HSL model was introduced in the specification of the CORE graphics system produced by an ACM SIGGRAPH committee in 1977, and refined in 1979 (Michener and Van Dam 1979). The geometry of the color solid in this specification is the same one we used here (two right cones glued at the base), but other geometries occur in the literature. In particular, (Rogers 1985) and (Foley et al. 1990) replace the cones by hexagonal prisms, while (Joblove and Greenberg 1978) use a right cylinder.

A comparative discussion of the HSV and HSL color systems, leaning toward the HSV system, can be found in (Smith 1981). Algorithms for converting between several color systems are given in (Rogers 1985).

The Munsell model is discussed in detail in (Wyszecki and Stiles 1982), where the criteria for the choice of samples are explained.

Uniform color systems are covered well in (Padgham and Saunders 1975). The transformation equations between these systems and the CIE-XYZ system are given in (Wyszecki and Stiles 1982). These equations are complex, and their application is computationally expensive.

Spectral color models in computer graphics were introduced in (Hall and Greenberg 1983); see also (Hall 1989) for details. A method for robust sampling of spectral space using few samples is discussed in (Meyer 1988). Several methods to obtain the spectral distribution of a color starting from its three-color coordinates can be found in (Wyszecki and Stiles 1982).

Further details on color systems for the video and television industries can be found in the immense literature on television technology.

References

[Foley et al. 1990]Foley, J. D., van Dam, A., Feiner, S. K., and Hughes, J. F. (1990). *Fundamentals of Interactive Computer Graphics*, second ed. Addison-Wesley, Reading, MA.

[Hall 1989]Hall, R. A. (1989). *Illumination and Color in Computer Generated Imagery*. Springer-Verlag, New York.

[Hall and Greenberg 1983]Hall, R. A. and Greenberg, D. P. (1983). A testbed for realistic image synthesis. *IEEE Computer Graphics and Applications*, 3:10–20.

[Joblove and Greenberg 1978]Joblove, G. H. and Greenberg, D. (1978). Color spaces for computer graphics. *Computer Graphics (SIGGRAPH '78 Proceedings)*, 12(3):20–25.

[Meyer 1988]Meyer, G. W. (1988). Wavelength selection for synthetic image generation. *Computer Vision, Graphics and Image Processing*, 41:57–79.

[Michener and Van Dam 1979]Michener, J. C. and Van Dam, A. (1978). A functional overview of the Core System with glossary. *ACM Computing Surveys*, 10:381–387.

[Padgham and Saunders 1975]Padgham, C. A. and Saunders, J. E. (1975). *The Perception of Light and Color*. Academic Press, New York.

[Rogers 1985]Rogers, D. F. (1985). *Procedural Elements for Computer Graphics*. McGraw-Hill, New York.

[Smith 1978]Smith, A. R. (1978). Color gamut transform pairs. *Computer Graphics (SIGGRAPH '78 Proceedings)*, 12(3):12–19.

[Smith 1981]Smith, A. R. (1981). Color tutorial notes. Technical Report No. 37, Lucasfilm.

[Wyszecki and Stiles 1982]Wyszecki, G. and Stiles, W. S. (1982). *Color Science*. John Wiley & Sons, New York.

Digital Images

Digital images are the focus of many computer graphics processes. They are links between the user and these processes, revealing the results of the latter. We may even say that all areas of computer graphics involve digital images, whether as a final product, as in the case of visualization, or as an essential intermediate step in the interaction process, as in the case of modeling. Thus, an understanding of the meaning of images in this context is essential. A rigorous formulation of the various notions associated with digital images is necessary in order to allow an analysis of the data structures used in image representation and of the algorithms used in image creation and manipulation.

This chapter is devoted to the conceptual underpinnings of digital images, to abstract models for images, and to the various ways of representing and encoding images on the computer.

6.1 Abstraction Paradigms for Images

In order to represent and manipulate images on the computer, we must define appropriate mathematical models. Once more, the four-universe paradigm of Chapter 1 is helpful in understanding the various image models we are about to study.

An image is the result of light stimuli produced by a two-dimensional support. This is the case whether the image arises through an intermediate step, as in the case of a photograph, or ultimately through the projection of our three-dimensional world onto our retina.

We must establish a mathematical universe in which we can define abstract models for images. Then we must create a representation universe, where we try to find schemes to allow the discrete representation of these models, with the purpose of obtaining an encoding of the image on the computer.

An image is a two-dimensional signal; therefore, we can use the conceptual framework introduced in Chapter 2, on signal theory. We will now specialize those concepts to the case of images. We have three abstraction levels, corresponding to continuous models, discrete representations, and symbolic encoding of images, as illustrated in Figure 6.1.

L. Velho et al., *Image Processing for Computer Graphics and Vision*,
Texts in Computer Science, DOI 10.1007/978-1-84800-193-0_6,
© Springer-Verlag London Limited 2009

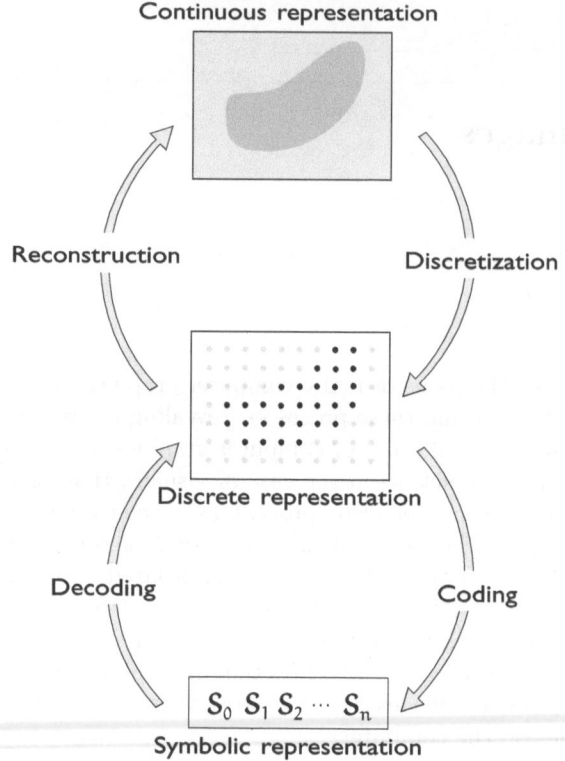

Fig. 6.1. Abstraction levels in the representation of an image.

Note that these levels will be realized concretely in different ways in an image processing system. For this reason, in order to obtain a unified scheme for image processing, we must use transformations to pass from one level to another, and we must also be able to manipulate descriptions on a single level (see Chapter 2).

6.2 The Spatial Model

Although there are several mathematical models appropriate for the description of images, we will stress in this book the so-called spatial model, which is the one best suited for computer graphics applications.

6.2.1 Continuous Images

When we look at a photograph or a real-life scene, we receive from each point in space a light impulse, which associates color information to that point.

Thus, a natural mathematical model for describing an image is a function defined on a two-dimensional surface and taking values in a color space.

A *continuous image* is a map $f : U \to C$, where $U \subset \mathbb{R}^2$ is a subset of the plane, and C is a vector space. (We stress that *continuous* here means *nondiscrete*; it doesn't mean that the map f is continuous in the topological sense.) In most applications, U is a rectangle of the plane, and C is a color space. However, it is convenient to allow C to be any vector space, in general containing color space as a subspace. The function f is called the *image function*. The set U is called the *support* of the image, and the set of values of f (a subset of C) is called the set of *values* of the image, or the *image color gamut*.

When C is a one-dimensional color space, we talk of a *monochrome* or *grayscale image*. The image can then be regarded geometrically as the graph $G(f)$ of the image function f:

$$G(f) = \{(x, y, z) : (x, y) \in U \text{ and } z = f(x, y)\},$$

where we consider the intensity values as the height $z = f(x, y)$ of the graph at each point (x, y) of the domain. Figure 6.2 shows a grayscale image, together with a sketch of the graph of the corresponding image function $f(x, y)$. This geometric interpretation allows a more intuitive visualization of certain aspects of the image. In Figure 6.2, for example, it is easy to identify the discontinuities of the function, corresponding to abrupt variations in the image's intensity. This approach of manipulating images as geometric models and vice versa makes clear the connection between image processing, geometric modeling, and, further down the line, computer vision.

In agreement with current image processing literature, an image as defined above should more appropriately be called a *two-dimensional image*. This is because modern computer graphics, especially in the area of scientific visualization, deals also with *three-dimensional images*, or *volumes*. We could extend the notion of an image to cover the three-dimensional case as well (simply by not requiring that the support set be two-dimensional); however, we prefer not to do so, because two-dimensional images are the ones that

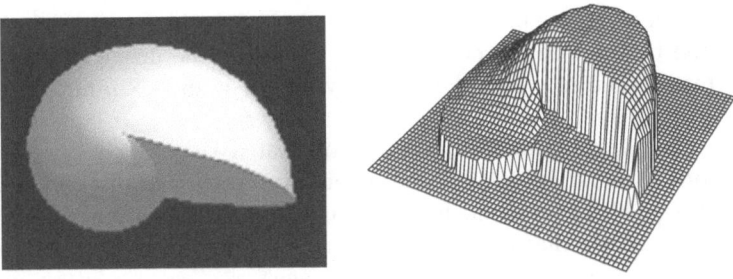

Fig. 6.2. A halftone and the graph of its image function.

can be exhibited directly in today's graphical output devices. Still, many of the techniques described in this book generalize to volume images. From the viewpoint of computer graphics, the study of volume images is related to the area of geometric modeling.

6.2.2 Image Representation

The most common representation of an image $f : U \subset \mathbb{R}^2 \to C$ in computer graphics consists in taking a discrete subset $U' \subset U$ of the image's domain and sampling the image function f in the set U'. In this case the image $f(x, y)$ will be spatially continuous or discrete, depending on whether the coordinates (x, y) of each point vary in the set U or U', respectively. Each point (x_i, y_i) of the discrete subset U' is called a *pixel*. We stress that "continuous" and "discrete" here refer to the discreteness of the domain of the image function, not to topological continuity.

In order to encode the image in the computer, we must also work with image models where the image function f takes on values in a discrete subset of the color space C. This discretization of an image's color space is called *quantization*. Although the floating-point representation of real numbers on the computer is itself a discretization, in image processing we consider a color space parametrized by floating-point coordinates as a continuum. This is reasonable, because only when we use a very small number of bits in representing the colors of an image does the error introduced lead to perceptual or computational difficulties.

Matrix Representation

The most common case of spatial discretization of an image consists in taking as the domain a rectangle

$$U = [a, b] \times [c, d] = \{(x, y) \in \mathbb{R}^2 : a \le x \le b \text{ and } c \le y \le d\},$$

choosing positive real numbers Δx and Δy, and discretizing the rectangle U using the two-dimensional orthogonal lattice

$$\{(x_j, y_k) \in U : x_j = j \cdot \Delta x, \ y_k = k \cdot \Delta y \text{ with } j, k \in \mathbb{Z}\}.$$

This is shown in Figure 6.3. Each pixel (x_j, y_k) of the image can therefore be represented by the integer coordinates (j, k). Thus, the image can be conveniently represented in *matrix form*, say by means of an $m \times n$ matrix A with entries $a_{jk} = f(x_j, y_k)$.

Each entry a_{jk} represents the value of f at a lattice point (x_j, y_k) and is therefore a vector in color space, expressing the color of the pixel with coordinates (j, k). For a monochrome image, $A = (a_{jk})$ is a real matrix, each entry being a scalar that expresses the corresponding pixel's luminance.

The number m of rows in A is the image's *vertical resolution*, and the number n of columns is the *horizontal resolution*. The *spatial resolution* or

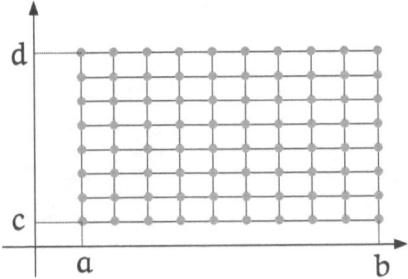

Fig. 6.3. A uniform lattice gives rise to a matrix representation for the image.

geometric resolution of the representation is the product $m \times n$. The spatial resolution establishes the final sampling rate for the image. Thus, the higher the resolution, the more detail (high frequencies) the matrix representation captures. Each row of the matrix is usually called a *scanline* of the image.

In absolute terms, the spatial resolution does not tell us much about the actual fineness of the image as realized on a physical device, since the device pixel size can vary. Usually, the more appropriate measurement is the *resolution density*, which gives the number of pixels per unit length. The most common unit for the resolution density is *pixels per inch* (ppi), also known as *dots per inch* (dpi). The resolution and resolution density of the matrix representation of an image enable us to obtain the dimensions—width and length—of the image.

Figure 6.4 shows the same image at four different spatial resolutions. The pixel sizes are chosen so that all images have the same dimensions (width and length). This clearly illustrates the effect of low resolutions.

The *color resolution* or *color depth* of an image is the number of bits used in storing the color vector a_{jk} associated with each pixel. We will return to this topic in Chapter 11.

The natural isomorphism between the space of $m \times n$ matrices and $\mathbb{R}^{mn}$ is commonly used in image processing in order to identify the matrix representation of a digital image with a vector whose coordinates are the rows of the image's matrix representation. The matrix representation, or the equivalent representation by means of a vector with mn coordinates, allows the use of linear algebra techniques in image processing. An example of this fact appeared in Chapter 2, where the calculation of the discrete Fourier transform was reduced to matrix multiplications.

6.2.3 Digital Images

Although on the computer we must work with discrete representations, conceptually it is important to be able to idealize an image in any of the possible combinations described in Chapter 2, depending on the nature of the domain

Fig. 6.4. The same image sampled at different spatial resolutions.

and range: *continuous-continuous, continuous-quantized, discrete-continuous,* and *discrete-quantized.*

In practice, on the one hand continuous-continuous images serve as a concept used in the development of mathematical methods for image processing; on the other hand, the discrete-quantized image is the representation used by many graphics devices. A discrete-continuous image is convenient for most image operations, for in it the image function takes on floating-point values, which (although represented by a finite number of bits) approximate real values. A discrete-quantized image is also called a *digital image.*

Elements of a Digital Image

The elements of a digital image are, essentially, the pixel coordinates and the color information at each pixel. These two elements define the spatial resolution and color resolution of the image. The *number of components* of the pixel is the dimension of the color space in question. Thus, each pixel in a monochrome image has a single component. The *gamut* of a digital image is the set of colors of the quantized color space. A monochrome image whose gamut has only two colors is called a *bilevel image,* or *bitmap.* A monochrome image whose gamut has more than two levels is a *grayscale* image.

If the color space of an image has dimension k, for most processes we can consider each color component separately. The image can therefore be decomposed into k grayscale images, each of which is a *component* of the original image. It is very common in image processing to work separately with each component. This simplifies certain operations considerably. However, processing by components does not take advantage of the correlation in color information that exists between the components of an image.

In addition to the color information, other components of the vector space C that forms the range of a digital image can be used to carry additional information. A scalar value, called *opacity*, can be used to define a mask for the purposes of image compositing (see Chapter 14). For synthetic images, it is very common to store the *pixel depth* (distance from the observer to the point in space represented by the pixel). Such information is generated and used by graphics algorithms for various purposes.

6.2.4 Digital Topology

When we use geometrical and topological methods to work with discrete-domain images, we need to use results from *digital topology*, which studies topological concepts associated with a space decomposition. We will not go into a detailed study of digital topology, but it is important to have an understanding of its fundamentals.

Several graphs can be associated to any space partition. The graph nodes and edges are defined by the geometric subsets of the partition and by the adjacency relationship between them. The combinatorial topology of this graph induces a combinatorial topology of the decomposed space. When the space decomposition is defined by some lattice, this induced topology is called the *digital topology* of the lattice.

Different graphs induce different topologies for the same underlying space decomposition. This can be illustrated with the concept of connectedness, associated with the matrix representation of a digital image.

The regular lattice used for the matrix representation of an image defines a decomposition of the plane into rectangles. We can associate a graph to this decomposition by defining the nodes to be the rectangles and by connecting two nodes if the associated rectangles have a common edge (see Figure 6.5(a)).

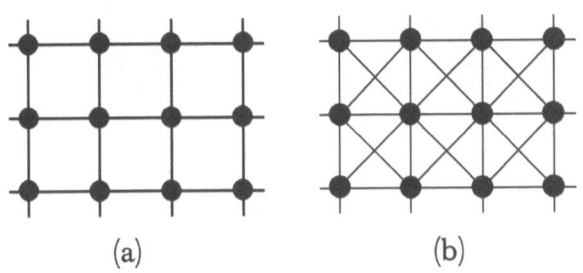

(a)   (b)

Fig. 6.5. Graphs associated to the matrix representation.

Another graph can be associated to the same rectangular decomposition of the image domain by connecting two rectangles (graph nodes) if they have either an edge or a vertex in common (see Figure 6.5(b)).

Now we recall that the neighborhood of a graph node is the set of all nodes connected to it. The two graphs defined above induce two common types of discrete neighborhoods used in image processing.

The graph in Figure 6.5(a) induces the 4-connected neighborhood. Given an element $a_{i,j}$, its *4-connected neighborhood* is the set of elements $a_{i-1,j}$, $a_{i+1,j}$, $a_{i,j-1}$, and $a_{i,j+1}$, as shown in Figure 6.6(a).

The graph in Figure 6.5(b) induces the *8-connected neighborhood*. In this case, the neighborhood of an element a_{ij} consists of the elements of its 4-connected neighborhood, plus the elements $a_{i-1,j-1}$, $a_{i+1,j-1}$, $a_{i-1,j+1}$, and $a_{i+1,j+1}$, as shown in Figure 6.6(b).

An important remark is that sometimes it is possible to use some metric on the plane such that the neighborhood defined by the digital topology of the lattice can be obtained by using metric relations between the subsets of the space partition. We illustrate this for the 4-connected and 8-connected neighborhoods defined above. An 8-connected neighborhood is defined by the norm $|(x,y)| = \max\{|x|, |y|\}$, the *maximum norm*. A 4-connected neighborhood is defined by the norm $|(x,y)| = |x| + |y|$, called the *sum norm*.

In Figure 6.7 we show a curve going from pixel A to pixel B. This curve is 8-connected but not 4-connected.

6.2.5 Pixel Shape

Consider the uniform polygonal mesh defined by the orthogonal lattice of the matrix representation of an image (see Figure 6.3).

The continuous image gives rise to two discrete images. One is the image defined in the initial pixel lattice, and the other is the image defined in the *dual*

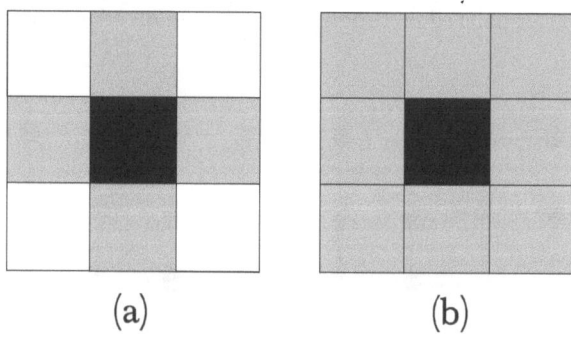

(a)   (b)

Fig. 6.6. The 4-connected neighborhood (a) and 8-connected neighborhood (b) of a pixel.

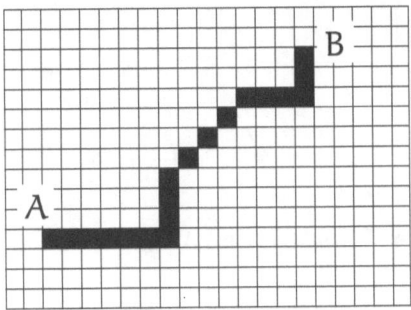

Fig. 6.7. A curve that is 8-connected but not 4-connected.

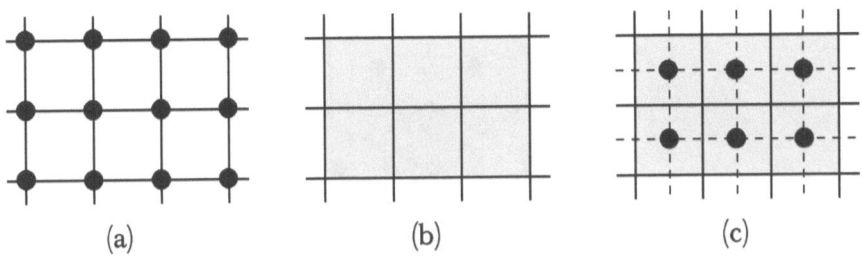

Fig. 6.8. A pixel lattice (a), the corresponding mesh (b), and the dual lattice (c).

lattice. In the dual lattice, each pixel is located in the center of the rectangle of the polygonal mesh. See Figure 6.8.

The matrix representation of an image is a particular case of a family of image representations based on regular lattices of the plane, not necessarily orthogonal. Let v_1 and v_2 be linearly independent vectors of the plane $\mathbb{R}^2$. The corresponding lattice in $\mathbb{R}^2$ is the set of all linear combinations of v_1 and v_2 with integer coefficients,

$$\Delta_{(v_1,v_2)} = \{jv_1 + kv_2 : j, k \in \mathbb{Z}\}.$$

This is illustrated in Figure 6.9. The lattice used for the matrix representation is a particular case, where the vectors v_1 and v_2 are orthogonal.

The fundamental parallelogram of the lattice is defined by

$$R_{(v_1,v_2)} = \{xv_1 + yv_2 : |x| < 1, \ |y| < 1, \ x, y \in \mathbb{R}\}.$$

This is illustrated by the shaded parallelogram in Figure 6.9. Notice that by translating the fundamental parallelogram to the lattice vertices, we cover the whole plane.

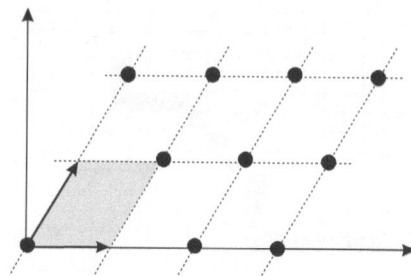

Fig. 6.9. Nonorthogonal, regular lattice of the plane.

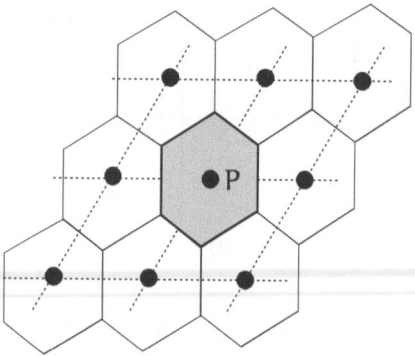

Fig. 6.10. Shape of a pixel in a nonorthogonal lattice.

The shape of a pixel P in a lattice $\Delta_{(v_1,v_2)}$ is the set of all points in $\mathbb{R}^2$ closer to P than to any other pixel of the lattice (for those familiar with computational geometry, this is the Voronoi cell of the pixel P). The pixel shape is illustrated in Figure 6.10.

By appropriately choosing the lattice generators v_1 and v_2, it is possible to obtain a lattice whose pixel shape is a regular hexagon. This is illustrated in Figure 6.11(a). In this case the digital topology induced by the adjacency graph of the lattice has a very symmetric neighborhood, as shown in Figure 6.11(b), and it is defined by the usual Euclidean metric of the plane. This hexagonal image representation is very useful when we are interested in using topological methods for image processing and analysis in the discrete domain.

One problem that arises when we work with representations in which the pixel topology is not rectangular is that graphical devices for digital im-

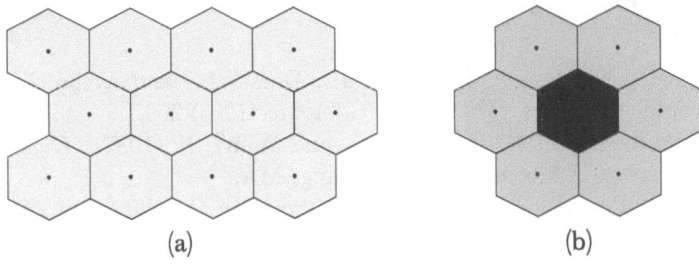

Fig. 6.11. (a) Hexagonal discretization. (b) Neighborhood of a pixel in the hexagonal discretization.

ages tend to use a representation matrix, which presupposes a rectangular discretization. This is a typical case where we must reconstruct the image and then resample it for display.

When we work with geometric pixels in the continuous domain, the color information can change greatly within each polygon that defines the geometric pixel. Thus, in order to compute the digital image in the dual lattice, we must take into account this variation, so as to minimize aliasing. Two commonly used methods are area sampling and supersampling, which we mentioned in Chapter 2. We will return to this question in Chapter 8.

6.3 Comments and References

This chapter introduced what we can call the *functional model* of an image. In this model, an image is a function of two variables, taking values in a color space. There are other possible models; we concentrated on the functional model because it is the most commonly used, and the most appropriate, in computer graphics. Other models, especially stochastic ones, are discussed in (Rosenfeld 1993). Chapter 10 will be devoted to Probabilistic Image Models.

The concepts of image and of volumetric image constitute examples of graphical objects. For a generic definition of a graphical object, several examples, and applications, the reader should consult (Gomes et al. 1996).

Among the many books devoted exclusively to image processing, we mention the two classics (Pratt 1978) and (Rosenfeld and Kak 1976), which cover this chapter's topics very well. For a more advanced approach and a fuller discussion of recent developments in digital image processing, see (Jain 1989), which also contains many references. We also mention (Gonzalez e Wintz 1987), a comprehensive introduction appropriate for a first course.

An elementary approach covering certain topics related to computer graphics and image processing can be found in (Pavlidis 1982).

References

[Gomes et al. 1996]Gomes, J., Costa, B., Darsa, L., and Velho, L. (1996). Graphical objects. *The Visual Computer* 12(6):269.

[Gonzalez and Wintz 1987]Gonzalez, R. and Wintz, P. (1987). *Digital Image Processing*. Addison-Wesley, Reading, MA.

[Pavlidis 1982]Pavlidis, T. (1982). *Algorithms for Graphics and Image Processing*. Computer Science Press, Rockville, MD.

[Pratt 1978]Pratt, W. (1978). *Digital Image Processing*. Wiley–Interscience, New York.

[Rosenfeld 1993]Rosenfeld, A. (1993). Image modelling during the 1980's: A brief overview. In *Markov Random Fields, Theory and Applications*, 1–10.

[Rosenfeld and Kak 1976]Rosenfeld, A. and Kak, A. C. (1976). *Digital Picture Processing*. Academic Press, New York.

7

Operations on Images

Image operations play an important role in computer graphics. Unless we explicitly say otherwise, in this chapter we will suppose that a digital image is given by its matrix representation. We'll illustrate certain operations using one-dimensional signals instead of images; this allows a better understanding of the two-dimensional case. You can always think of a one-dimensional signal as the restriction of an image to a single scanline (row of its matrix representation).

Chapter 2, on signals, is a prerequisite for this chapter. Nonetheless, there is some overlap between the two chapters; this is for the benefit of readers who have already studied signal processing and who may want to read this chapter independently.

7.1 Arithmetic Operations

The space of images $\mathcal{I} = \{f : U \subset \mathbb{R}^2 \to C\}$ has a vector space structure, with operations of addition of functions and multiplication of a function by a scalar:

$$(f + g)(x, y) = f(x, y) + g(x, y) \quad \text{and} \quad (\lambda f)(x, y) = \lambda f(x, y).$$

In matrix representation, these operations reduce to the standard operations of addition and scalar multiplication on matrices with vector entries.

These operations have applications in several contexts, where they acquire a concrete meaning. For example, the difference of two images, which we define by

$$f - g = f + (-1)g,$$

can be used to detect the motion of some object present in the images.

We can define other operations in $\mathcal{I}$. Generally speaking, an *operation* between elements of $\mathcal{I}$ is a map

L. Velho et al., *Image Processing for Computer Graphics and Vision*,
Texts in Computer Science, DOI 10.1007/978-1-84800-193-0_7,
© Springer-Verlag London Limited 2009

$$T : \underbrace{\mathcal{I} \times \mathcal{I} \times \cdots \times \mathcal{I}}_{m \text{ times}} \times \underbrace{\mathbb{R} \times \mathbb{R} \times \cdots \times \mathbb{R}}_{n \text{ times}} \to \mathcal{I},$$

where $\times$ denotes the Cartesian product. Thus, an operation associates to an m-tuple $(f_1, \ldots, f_m)$ of images and to the vector $(\lambda_1, \ldots, \lambda_n) \in \mathbb{R}^n$ an image in $\mathcal{I}$, the result of the operation. The addition operation, defined above, is a map $\mathcal{I} \times \mathcal{I} \to \mathcal{I}$, and multiplication by a scalar is a map $\mathcal{I} \times \mathbb{R} \to \mathcal{I}$.

As a matter of terminology, when $m = 1$ and $n = 0$, we say that the operation is *unary*. Thus, a unary operation transforms one image into another. Otherwise, when $m > 1$ and $n \geq 0$, we say that the operation is *m-ary*; such operations combine two or more images to produce a new image. One important case occurs for $m = 2$, when the operation is called *binary*. In general, it's possible to reduce an m-ary operation to a sequence of binary operations.

In the monochrome case, the image is given by a real-valued function $f : U \subset \mathbb{R}^2 \to \mathbb{R}$, so we can define the product of two images f and g as

$$(f \cdot g)(x, y) = f(x, y) \, g(x, y),$$

where on the right-hand side we have the product of two real numbers. For color images, we can define the product by multiplying each color component separately.

As we saw in Chapter 2, a *transform* is a linear operator, generally invertible, that allows us to pass from one functional image space, using a certain model, to another space, using a different model.

Operations can be classified according to the scope of their action into *local operations* and *point operations*. For a local operation T, the value of the result at a pixel p depends on the values of the image's pixels in a neighborhood of p. For a point operation, the value of T at each pixel p does not depend on the behavior of the image in a neighborhood of p.

Example 7.1 (Luminance). Given an image $f(x, y)$, we can obtain a monochrome image g by calculating, for each pixel, the brightness (luminance) at that point. That is, we set $g(x, y) = L(f(x, y))$, where L is the luminance operator. For example, if $f(x, y) = (R(x, y), G(x, y), B(x, y))$ is given by its components in the RGB system, we can compute g using the NTSC luminance operator, defined by

$$g(x, y) = 0.176 R(x, y) + 0.81 G(x, y) + 0.011 B(x, y).$$

This is the standard method to obtain a grayscale image from a color image. The operator L is a unary point operator.

Example 7.2. As an example of a local unary operation on an image f, we define, for each pixel (i, j), the result $T(i, j)$ as the average

$$T(i, j) = \tfrac{1}{3}(f(i - 1, \, j) + f(i, j) + f(i + 1, \, j))$$

of the values of f at the pixel itself and at its two neighbors on the same row. (At the beginning and end of the row we have only one neighbor, so we take the average of two pixels.)

The Fourier Transform

In Chapter 2 we defined the Fourier transform of an n-dimensional signal. In particular, that definition applies to images. It is straightforward to extend the discrete Fourier transform, given by Equation (2.25) in Chapter 2, to two-dimensional discrete signals:

$$F(k_1, k_2) = \frac{1}{N_1 N_2} \sum_{n_1=1}^{N_1} \sum_{n_2=1}^{N_2} f(n_1, n_2) e^{-2\pi i k_1 n_1 / N_1} e^{-2\pi i k_2 n_2 / N_2}.$$

The inverse transform is given by

$$f(n_1, n_2) = \sum_{k_1=1}^{N_1} \sum_{k_2=1}^{N_2} F(k_1, k_2) e^{2\pi i k_1 n_1 / N_1} e^{2\pi i k_2 n_2 / N_2}.$$

Observe that, in order to compute $F(k_1, k_2)$ for each pair (k_1, k_2), we must carry out $N_1 N_2 - 1$ additions and $N_1 N_2$ multiplications. Since there exist $N_1 N_2$ different pairs (k_1, k_2), we need a total of $N_1^2 N_2^2$ multiplications and $N_1 N_2 (N_1 N_2 - 1)$ additions to compute $F(k_1, k_2)$. It is easy to see that we can compute the transform of an image in a two-step procedure: first we transform the rows and then we apply the DFT again to the columns of the resulting image. This reduces the computation of the two-dimensional DFT to the computation of two one-dimensional ones. For this, we can apply the fast Fourier transform discussed in Chapter 2.

7.2 Filters

Unary operations on images are also called *filters*. Figure 7.1 illustrates the action of a filter.

Filtering and Computer Graphics

Filtering is of paramount importance in several stages of the process of image synthesis. We mention its uses at three stages: visualization, mapping, and postprocessing.

Fig. 7.1. Action of a filter on an image.

Visualization

During visualization, filters are used to attenuate the high-frequency components in the sampled image. This helps ensure that the reconstructed image has as little aliasing as possible and is free of other defects.

Mapping

The various mappings used in computer graphics (texture mapping, reflection mapping, and so on) work essentially by applying a deformation transform to the texture to be mapped. In this context, filters play an important role in resampling (reconstruction with later sampling) of the mapped image. We will study deformation filters in a later chapter.

Postprocessing

Filters are used for several purposes in the postprocessing of the synthesized image. As an example, we mention the resampling of an image to adjust it to a different geometry (an image might be generated at video resolution, say 512×512, and then might have to be transformed into a 35-mm slide at a higher resolution and a different aspect ratio). Another example is the use of filters to obtain special effects: thus, the use of a lowpass filter gives the feeling of an out-of-focus image. This application is quite useful, because virtual camera models generally do not take into account depth of field.

7.2.1 Classification

Filters can be classified according to the linear structure of the space of images (linear versus nonlinear), according to the computational method used (statistic versus deterministic), and according to their domain of action (topological versus amplitude filters).

Linear Filters

A filter is *linear* if the corresponding operator T is linear, that is, if

$$T(\lambda f) = \lambda T(f) \quad \text{and} \quad T(f + g) = T(f) + T(g),$$

where λ is a real number and f, g are images.

From the mathematical point of view, linear filters preserve the vector-space structure of the space of images. The first equation above says that the response of the filter is invariant under a scaling transformation of the image values. Thus, applying a linear filter to an image f and then making a constant scaling of its values is equivalent to scaling the values of the original image and then applying the filter; the order does not matter. The second equation says that the response of a linear filter when applied to two images that are added together can be obtained by adding together the response for each of the images.

The filters in Examples 7.1 and 7.2 are linear. Examples of nonlinear filters are very easy to obtain. An obvious example is the filter defined on the space of grayscale images by $T(f) = f^2$.

Statistical Filters

A *statistical filter* uses statistic properties of the image to determine the result at each pixel. Two important examples are the *median filter* and the *mode filter*, used in noise elimination.

Median Filter of Order n

In this filter, we take for each pixel p an 8-connected neighborhood of p with n pixels $p_1, p_2, \ldots, p_n$. The response of the filter at p is defined as the median of the pixel values at $p_1, \ldots, p_n$, these values being ordered according to their intensity. Note that, if the intensity of a pixel is very different from that of neighboring pixels, that pixel will certainly not be the median, so its value after filtering will change to the original value of a more representative pixel in the neighborhood. Therefore, this filter eliminates intensity values that are "outliers" in a neighborhood ("speckles"); such values are often due to noise. The resulting image intensities are more uniform.

Mode Filter of Order n

The mode filter is defined in the same way as the median filter, but instead of taking the median of the pixels in a neighborhood, we take their mode, that is, the intensity value that occurs most frequently in that neighborhood. It is clear that, as for the median filter, the resulting image intensities are more uniform.

The mode and median filters are local and nonlinear.

Amplitude Filters and Topological Filters

Regarding the domain of action, we can have amplitude filters and topological filters. *Amplitude filters* act directly on the color space of the image, while *topological filters* act on the support set of the image. Amplitude filters directly change the color of the pixels, while topological filters change the topology, or structure, of the objects present in the image.

The filters of Examples 7.1 and 7.2, as well as the median and mode filters, are amplitude filters. An important class of amplitude filters in computer graphics consists of those that change the color system of the image or perform color adjustments such as gamma correction, gamut transformation, and color clipping. Details on these operations will be given in Chapter 16.

An important class of topological filters consists of *warping filters*, which apply a deformation to the domain of the image, resulting in a change in the geometric structure of the image objects. Such filters will be studied in Chapter 10, together with *morphing transformations*, which combine warp and amplitude filters.

Topological filters are also important in mathematical morphology, which we now discuss briefly.

7.2.2 Morphological Filters

Mathematical morphology is the field that studies topological and structural properties of objects based on their images. The techniques used in this field, especially in the case of continuous-domain images, are similar to some techniques used in geometric modeling.

The filters used in mathematical morphology are defined based on a *structural element*, that is, a subset of the plane that interacts with the image of the object to determine topological properties of the object and reveal structural information about it. Two basic examples are *erosion filters* and *dilation filters*. We define them first in the continuous domain.

Given a subset B of the plane, we denote by $B(x,y)$ the translate of B by the vector (x,y). The *erosion* of a set $X \subset \mathbb{R}^2$ by B is the set $X \ominus B$ consisting of points $(x,y) \in X$ such that $B(x,y) \subset X$:

$$X \ominus B = \{(x,y) \in X : B(x,y) \subset X\}.$$

The set B is the structural element. The gray area of Figure 7.2(b) shows the erosion of the set of Figure 7.2(a) by the disc of radius r centered at the origin.

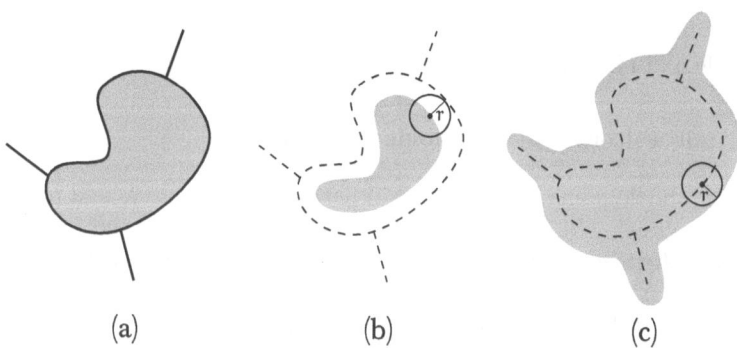

(a)   (b)   (c)

Fig. 7.2. Parts (b) and (c) show the erosion and the dilation of the set shown in (a).

The *dilation* $X \oplus B(x, y)$ of a set X by the structural element $B(x, y)$ is defined as $X \oplus B = (X^c \ominus B(x, y))^c$, where the superscript ^{c} denotes the set complement operation. It is easy to see that this definition is equivalent to

$$X \oplus B = \{(x, y) \in \mathbb{R}^2 : B(x, y) \cap X \neq \varnothing\}; \tag{7.1}$$

that is, $X \oplus B$ is the set of points in the plane for which the structural element B intersects X. The gray area of Figure 7.2(c) shows the dilation of the set X of Figure 7.2(a) by a disc of radius r centered at the origin. The dilation operation can be considered as a *geometric convolution*.

Intuitively, the erosion operation takes away points from X, while dilation adds points to the set. This intuition is true if the structural set is convex, but otherwise it is not always accurate. Figure 7.3 shows an example of a dilation with a disconnected structural set. The resulting set is disconnected and disjoint from the original set.

On a binary image, the image function f assumes values 0 or 1 only. Thus, the image defines a subset of the image plane where the point membership classification function (characteristic function) of the set is defined by $\chi(x, y) = f(x, y)$. Therefore, the operations just defined of dilation and erosion of subsets of the plane can be immediately extended to binary images. Extending them to grayscale and color images needs an additional effort.

In the discrete domain the erosion and dilation filters are defined analogously, but using discrete sets. In Figure 7.4(c) we show the dilation of the set of black pixels in (b), using as structural element B the set of pixels in (a), the basepoint being the center pixel. The pixels added by this process are shown in gray in (c).

As an application of morphological filters in obtaining topological information about the objects of an image, we consider the problem of determining the boundary of a given object. We define

$$\partial X = X - (X \ominus B), \tag{7.2}$$

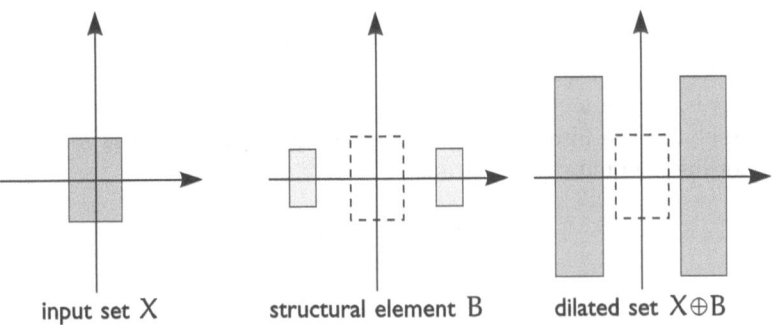

input set X \qquad structural element B \qquad dilated set $X \oplus B$

Fig. 7.3. Dilation can yield a set disjoint from the original.

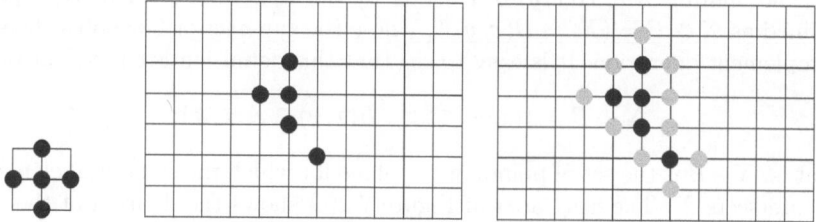

Fig. 7.4. Part (c) shows the dilation of the set in (b) by a disc of radius r (a).

Fig. 7.5. Computation of the boundary of an object using an erosion filter.

for an appropriately chosen structural element B. Figure 7.5 illustrates the action of this filter ∂. Part (a) shows the set X; part (b) shows the structural element, whose basepoint is the center pixel; and part (c) shows the set $X \ominus B$ in gray, while the boundary pixels, computed according to (7.2), are shown in black.

Besides the boundary operator, several other operations in mathematical morphology can be derived from the erosion and dilation filters. The literature on the subject is fairly large; see Section 7.6.

7.2.3 Spatially Invariant Filters

Given an operation T on the space of images, a filter F is *invariant with respect to* T if, for every image $g \in \mathcal{I}$, we have $(F \circ T)(g) = (T \circ F)(g)$. In other words, T commutes with the filter. This means that we get the same result applying the filter before or after the operation. This is indicated by the following commutative diagram:

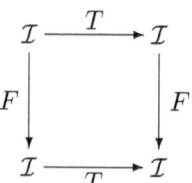

An important case of invariance occurs when T is a translation operator on the plane. This means that there exists a vector with coordinates (x_0, y_0) such that $T(g(x, y)) = g(x - x_0, y - y_0)$, where g is the image function. Thus, T has the effect of translating the image by the vector (x_0, y_0). When a filter F is invariant with respect to any translation, we say that F is *spatially invariant*, and we have

$$(Fg)(x - x_0, y - y_0) = F(g(x - x_0, y - y_0)).$$

Geometrically, spatial invariance means that the behavior of the filter is the same in all points of the domain of the image.

Example 7.3. Consider a filter T defined by $T(f) = af + b$, where a and b are real numbers, with $a \neq 0$. It is easy to see that T is spatially invariant. Indeed,

$$(Tf)(X - X_0) = (af + b)(X - X_0) = a(X - X_0) + b = T(f(X - X_0)).$$

Moreover, T is linear if and only if $b = 0$.

On the other hand, consider the filter T defined by $Tf(x, y) = f(y, x)$, so that T reflects the image f in the diagonal line $x = y$ of the plane. T is clearly linear, but it is easily seen not to be spatially invariant.

Example 7.4. One can easily check that the dilation and erosion filters of Section 7.2.2 are spatially invariant.

An important class of filters that are not spatially invariant consists of filters that change their behavior according to the properties of the image in a neighborhood of the pixel in question. Such filters are called *adaptive*.

7.3 Spatially Invariant Linear Filters

The *impulse response* of a filter T is the image $h(x, y)$ obtained by applying the filter to a unit impulse input, or Dirac delta function:

$$h(x, y) = T(\delta(x, y)).$$

For images, the impulse function $\delta(x, y)$ is an image with zero luminance everywhere except at the origin, where the luminance is maximal (pure white). Thus, the impulse response function $h(x, y)$ tells how the filter spreads a point of light; for this reason it is also called the *point spread function*. If a spatially invariant filter has impulse response function h, its response to a point source $\delta(x - x_0, y - y_0)$ is the image $h(x - x_0, y - y_0)$ obtained by translating the filter's point spread function.

A filter has *finite impulse response* if its impulse response has compact support. We also say in this case that we have an FIR filter. Otherwise we say that the filter has infinite impulse response (IIR).

For spatially invariant linear filters, we have the following result:

Theorem 7.5. *A spatially invariant linear filter is completely characterized by its point spread function h (impulse response function).*

Proof (Informal proof). Indeed, if f is an image, we can write f as an infinite sum of point sources (Dirac deltas):

$$f(x) = \int_{-\infty}^{+\infty} f(u, v)\delta(u - x, \, v - y) \, du \, dv.$$

Since the filter is linear,

$$Tf(x, y) = T\left(\int_{-\infty}^{+\infty} f(u, v)\delta(u - x, v - y) \, du \, dv \right)$$

$$= \int_{-\infty}^{+\infty} f(u, v)T(\delta(u - x, \, v - y)) \, du \, dv.$$

Because the filter is spatially invariant, its response to the point signal $\delta(u - x, v - y)$ is $h(u - x, v - y)$. The previous equation therefore becomes

$$Tf(x, y) = \int_{-\infty}^{+\infty} f(x, y)h(u - x, \, v - y) \, du \, dv. \tag{7.3}$$

Thus, the filtered image is an infinite average of translations of the point spread function, weighted by the input image values. This proves the theorem.

The impulse response function of a spatially invariant linear filter is called its *kernel*. The integral in (7.3) is called the *convolution* of f and h and is denoted by $f * h$. The proof of the theorem, therefore, shows that processing an image f through a spatially invariant linear filter is the same as convolving f with the filter's kernel. As we saw in Chapter 2, this is equivalent to multiplying the Fourier transform of f by the *transfer function* of the filter, which is, by definition, the Fourier transform of the filter's impulse response.

Separable Filters

A filter is *separable* if its kernel satisfies

$$h(x, y) = h_1(x)h_2(y)$$

for some functions h_1 and h_2. Separable filters are very important because, when applying them to images, it is possible to use a matrix representation for the image, and the filtering operation can be carried out independently on each line (or column) of the image. This generally reduces the filtering of the image to one-dimensional operations over the lines and columns of the image, which is advantageous in terms of computational efficiency.

There is another property of a separable kernel that will be used later in this chapter: *the Fourier transform of a separable kernel* $h = h_1 \cdot h_2$ *is the product of the Fourier transform of* h_1 *and* h_2. *That is,*

$$\hat{h} = \hat{h}_1 \cdot \hat{h}_2. \tag{7.4}$$

The proof is immediate:

$$\hat{h}(u,v) = \int_{\mathbb{R}}^2 h(x,y)e^{-2\pi i(ux+vy)}\,dx\,dy$$

$$= \int_{\mathbb{R}}^2 h_1(x)h_2(y)e^{-2\pi iux}e^{-2\pi ivy}\,dx\,dy$$

$$= \int_{\mathbb{R}} h_1(x)e^{-2\pi iux}\,dx \int_{\mathbb{R}} e^{-2\pi ivy}\,dy$$

$$= \hat{h}_1(u) \cdot \hat{h}_2(v).$$

7.3.1 Discrete Filters

In practice, we work with discrete images, so we must introduce a discrete version of the filters described in the previous section. We will limit ourselves to spatially invariant linear filters.

Discretization of the Kernel

In general, if h is the convolution kernel of a linear and spatially invariant filter, we assume that

$$\int_{\mathbb{R}^2} h(x,y)\,dx\,dy = 1. \tag{7.5}$$

This equality guarantees that the filtering process does not change the average intensity of the image pixels.

A simple and commonly used discretization method consists in fixing resolutions $\Delta x, \Delta y \in \mathbb{R}$, taking the plane lattice Δ of points $(x_j, y_k) \in U$ with $x_j = j\,\Delta x$ and $y_k = k\,\Delta y$, where $j, k \in \mathbb{Z}$, and sampling the kernel at the vertices of Δ. In this way we obtain a matrix $h_{ij} = h(i,j)$. If the kernel is defined in a rectangular region of the plane, this discretization defines a matrix representation (h_{ij}) of h, analogous to the matrix representation of an image.

The entries h_{ij} of the kernel matrix representation (where $i = 1, \ldots, m$ and $j = 1, \ldots, n$) should satisfy

$$\frac{1}{mn} \sum_{i=1}^{m} \sum_{j=1}^{n} h_{ij} = 1. \tag{7.6}$$

This says that the mean of discretized kernel entries is 1. It is the analog, in the discrete domain, of Equation (7.5), and it guarantees that the filtering process does not change the average intensity of the pixels of the image.

In certain exceptional cases, it is not desirable to preserve the average intensity of the pixels of the filtered image; one then chooses a kernel that does not satisfy Equation (7.6).

We know from the foregoing discussion that applying a linear and spatially invariant filter to an image is equivalent to taking the convolution of the image with the filter's kernel. Thus, the study of filtering in the discrete domain reduces to that of the convolution of two functions in the discrete domain. In what follows, we discuss this operation in detail, treating separately the one- and two-dimensional cases for ease of comprehension.

One-dimensional Filters

If f and h are two discrete signals, defined on the set $\mathbb{Z}$ of integers, we define the discrete convolution product $f * h$ of f and h by analogy with Equation (7.3):

$$(f * h)(n) = \sum_{k=-\infty}^{+\infty} f(k)h(n - k). \tag{7.7}$$

Naturally, we are supposing that the sum converges. This occurs, for example, if one or both of f and h have finite support: indeed, in this case f or h are discretized into a finite number of samples, and, for each value of n, the sum has only finitely many nonzero summands.

When the impulse response function h of the filter has compact support (FIR filters), the kernel $h(k)$ (where $k \in \mathbb{Z}$) has only a finite number of nonzero samples. In this case we generally call h the *mask of the filter*. The process of filtering then amounts to replacing each pixel $f(n)$ by the weighted average, with weights $h(i)$, of the pixels $f(i)$ (where $-k_1 \leq i \leq k_2$). This follows immediately from (7.7).

Figure 7.6 illustrates the calculation of the convolution for an FIR filter: we show the sequence $h(k)$ defining the kernel, the sequence $h(-k)$, and the sequence $h(n - k)$. The part of the signal $f(k)$ corresponding to the nonzero elements $h(n - k)$ is indicated by a darker shading. To obtain the convolution $f * h$ at the point n, we multiply corresponding elements in the sequences $h(n - k)$ and $f(k)$ and add the results.

A common case is that of symmetric FIR filters h whose kernel is discretized into a finite number of samples. If h has $2k + 1$ samples, we can index them from $-k$ to k, so the filter is determined by the values

$$h(-k), h(-k + 1), \ldots, h(0), \ldots, h(k - 1), h(k), \tag{7.8}$$

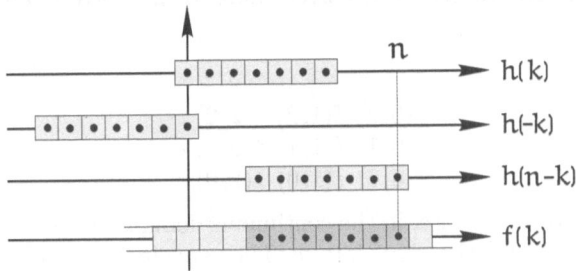

Fig. 7.6. Computation of the one-dimensional discrete convolution.

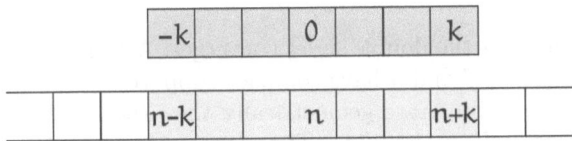

Fig. 7.7. Convolution with a symmetric FIR filter of odd order.

where $h(-j) = h(j)$ for $j = 0, \ldots, k$. The convolution sum in (7.7) is obtained simply by translating the origin of the sequence in (7.8) to the n-th position of signal f, multiplying the corresponding terms, and adding. We illustrate this in Figure 7.7.

It is interesting to compute the transfer function of a discrete filter directly from its mask coefficients. We will do the computations for the important case of a finite and symmetric mask of odd order $N = 2M + 1$.

The transfer function is given by the Fourier transform of the mask discretization (h_k), where $k = -M, \ldots, M$. We have

$$\hat{h}(s) = h_0 + \sum_{k \neq 0} h_k e^{-i2\pi ks/N}$$

$$= h_0 + \sum_{k < 0} h_k e^{-i2\pi ks/N} + \sum_{k > 0} h_k e^{-i2\pi ks/N}$$

$$= h_0 + 2 \sum_{k > 0} h_k \cos(2\pi ks/N),$$

where in the last equality we used the mask symmetry and the Euler equation

$$e^{i2\pi ks/N} = \cos(2\pi ks/N) + i\sin(2\pi ks/N).$$

Two-Dimensional Filters

Let's now turn to the two-dimensional case. We start with a convention. Given a digital image in matrix representation $(f(i,j))$, unless we say otherwise, the pixel with coordinates $(0, 0)$ corresponds to the bottom left entry of the matrix:

$$
\begin{array}{cccc}
\vdots & \vdots & \vdots & \\
f(0,2) & f(1,2) & f(2,2) & \cdots \\
f(0,1) & f(1,1) & f(2,1) & \cdots \\
f(0,0) & f(1,0) & f(2,0) & \cdots
\end{array}
$$

This corresponds to the use, in the continuous domain, of the usual Cartesian coordinate system for the plane lattice. The convolution product $f * h$ of the image f with the kernel h is defined at the pixel (n, m) by

$$
(f * h)(n, m) = \sum_{k=-\infty}^{+\infty} \sum_{j=-\infty}^{+\infty} f(k, j)h(n - k, \, m - j), \qquad (7.9)
$$

where we assume that the double series converges. Again, this is certainly the case if one or both of f and h have compact support.

In Figure 7.8 we illustrate geometrically the process of two-dimensional convolution given by Equation (7.9). The sequence $h(-k, -j)$ is obtained from $h(k, j)$ by reflecting about the x-axis and then reflecting about the y-axis (this is equivalent to rotation by 180° about the origin). The sequence $h(n - k, \, m - j)$ is the translate of the sequence $h(-k, -j)$ by the vector (n, m). The pixels common to the images $h(n - k, \, m - j)$ and $f(k, j)$, represented by the darker area in Figure 7.8(c), are multiplied together, and the result is added to yield the value of $f * h$ at (n, m). Note that the filter mask h must have a distinguished origin. We take it as the lower left corner.

Example 7.6. Consider the image $f(k, j)$ of order 2×3 with matrix representation

$$
\begin{array}{|c|c|c|}
\hline
1 & 2 & 3 \\
\hline
2 & 1 & 6 \\
\hline
\end{array}
$$

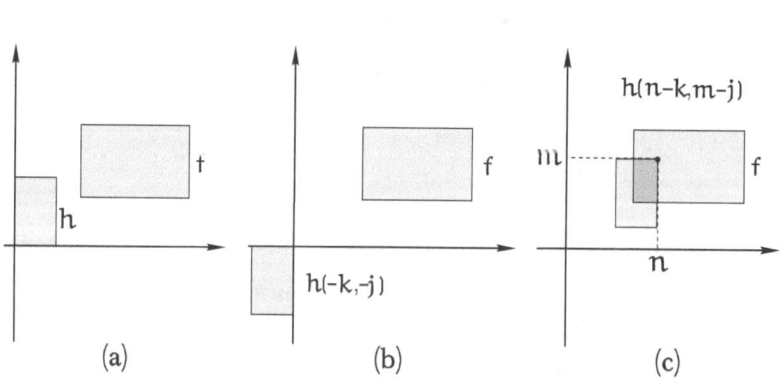

Fig. 7.8. Two-dimensional convolution with FIR filter.

Let h be the filter whose impulse response function is discretized by the matrix

-1	2
2	-1

The convolution product $f * h$ is then given by

-1	0	1	6
0	6	0	9
4	0	11	-6

As in the one-dimensional case, we obtain significant simplifications in the expression for the two-dimensional discrete convolution when the mask of the filter is symmetric and of odd order. If the mask has order $(2k+1) \times (2k+1)$, we can index its matrix as

$(-k,k)$	$\cdots$	$(k,0)$	$\cdots$	(k,k)
$\vdots$		$\vdots$		$\vdots$
$(-k,0)$	$\cdots$	$(0,0)$	$\cdots$	$(k,0)$
$\vdots$		$\vdots$		$\vdots$
$(-k,-k)$	$\cdots$	$(-k,0)$	$\cdots$	$(k,-k)$

The convolution $f * h$ at the pixel (m,n) is obtained by placing the origin $(0,0)$ of the mask on top of the pixel (m,n), multiplying the corresponding elements of mask and image, and adding the results. Figure 7.9 illustrates this operation geometrically for a 3×3 mask.

Computations similar to those performed for the one-dimensional discrete filters allow us to obtain an analytic expression for the transfer function of a finite symmetric filter of odd order $N \times N$, $N = 2M + 1$. In fact, if the mask matrix is (h_{jk}), for $j = -M, \ldots, M$ and $k = -M, \ldots, M$, then the transfer function is given by

$$\hat{h}(u,v) = h_{00} + \sum_{j,k>0} 2h_{jk} \cos[2\pi(ju + kv)/N].$$

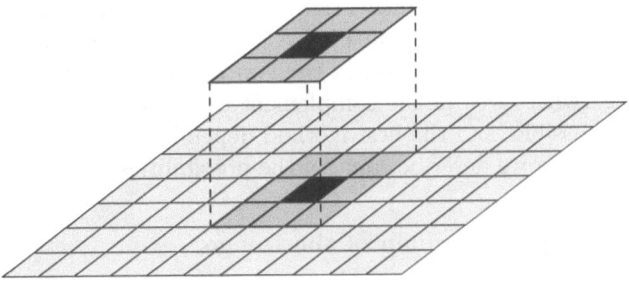

Fig. 7.9. Two-dimensional convolution with symmetric FIR filter of odd order.

Computational Considerations

Before we can implement the convolution product we must take into consideration three problems: nonrealizable colors, computational efficiency, and extension of the image's domain.

Nonrealizable Colors

It may happen that, as we carry out the convolution of an image with the filter mask, the resulting color value at certain pixels is outside the intended color space of the image or of the output device. For instance, some of the values obtained may be negative. Two simple solutions can be adopted. The first is clipping: approximating the color by the nearest point of the desired color space. The second is to apply a global transformation to the output values so they end up within the desired color space. Both approaches, in effect, amount to contracting the color space of the convolved image by applying another filter.

Computational Efficiency

The problem of computational complexity is directly linked to the actual implementation, and we will not discuss it here. We merely mention, as an example, that for a separable filter the implementation of Equation (7.9) can be reduced to the successive application of two one-dimensional convolutions, one along rows and one along columns. This fact, whose proof we leave as an exercise, reduces the number of required operations. Other situations are amenable to other techniques for enhancing computational efficiency.

Extension of the Domain

In applying Equation (7.9) to obtain the convolution of an image with a mask h, we assumed implicitly that the image extends to infinity in all directions. In practice, of course, this is not the case, so we have to deal with the problem of extending the image beyond its original domain in order for (7.9) to make sense. This can be seen clearly in Figure 7.8: including in the sum only the pixels that are in the intersection of the two masks is equivalent to extending the image with zeros in the complement of its domain. Other extension methods are possible, and the best method depends on the problem at hand. In the next section we describe some of the possibilities in common use.

7.3.2 Extending the Domain of the Image

For one to compute, over a domain X, the convolution of a signal with a kernel whose support is B, the signal must be defined at least on the set $X \oplus B$ (this notation is defined in (7.1); it represents the dilation of X by B). Normally we want the filtered image to be defined on the same domain as the original image, so the original image must be extended to the dilation of the domain. There are several ways to perform this extension.

Constant Extension

This method extends the signal by a constant. We illustrate this in Figure 7.10, where we use a continuous one-dimensional signal for ease of understanding. For example, in the discrete one-dimensional case, the signal $f(0)$, $f(1), \cdots, f(n-1)$ can be extended to

$$\cdots f(0) \, f(0) \, \underbrace{f(0) \, f(1) \, \cdots \, f(n-1)}_{\text{original signal}} \, f(0) \, f(0) \, \cdots .$$

In two dimensions, the 3×3 image

$f(0,0)$	$f(1,0)$	$f(2,0)$
$f(0,1)$	$f(1,1)$	$f(2,1)$
$f(0,2)$	$f(1,2)$	$f(2,2)$

might be extended to

$$
\begin{array}{ccc|ccc|ccc}
 & & & & \vdots & & & & \\
f(0,0) & f(0,0) & f(0,0) & f(1,0) & f(2,0) & f(2,0) & \text{f(2,0)} \\
f(0,0) & f(0,0) & f(0,0) & f(1,0) & f(2,0) & f(2,0) & \text{f(2,0)} \\
\hline
f(0,0) & f(0,0) & f(0,0) & f(1,0) & f(2,0) & f(2,0) & \text{f(2,0)} \\
\cdots \, f(0,1) & f(0,1) & f(0,1) & f(1,1) & f(2,1) & f(2,1) & \text{f(2,1)} \, \cdots \\
f(0,2) & f(0,2) & f(0,2) & f(1,2) & f(2,2) & f(2,2) & f(2,2) \\
\hline
f(0,2) & f(0,2) & f(0,2) & f(1,2) & f(2,2) & f(2,2) & f(2,2) \\
f(0,2) & f(0,2) & f(0,2) & f(1,2) & f(2,2) & f(2,2) & f(2,2) \\
 & & & & \vdots & & & &
\end{array}
$$

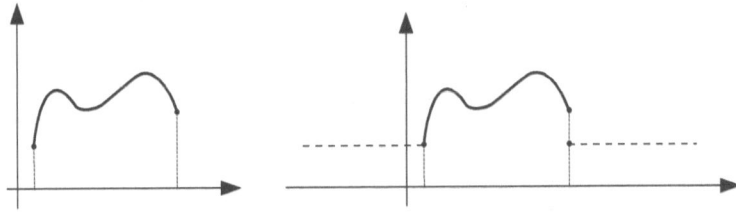

Fig. 7.10. Constant extension of a signal.

which is piecewise constant. It is also very common to extend the image with zeros; this is a particular case of a constant extension.

Periodic Extension

Here we simply repeat the signal, forming a periodic function. Figure 7.11 illustrates this type of extension.

Observe that the signal obtained generally has a discontinuity, unless $f(0) = f(n-1)$. For a discrete one-dimensional signal

$$f(0)\, f(1) \cdots f(n-1),$$

we obtain the extension

$$\cdots f(0)\, f(1) \cdots f(n-1) \underbrace{f(0)\, f(1) \cdots f(n-1)}_{\text{original signal}} f(0)\, f(1) \cdots f(n-1) \cdots.$$

This method can be easily generalized to the two-dimensional case: we repeat the one-dimensional process first for each row of the image and then for each column. We obtain a doubly periodic image, as shown below for a 3×3 image.

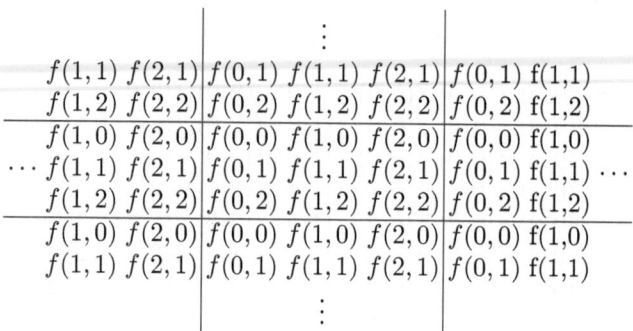

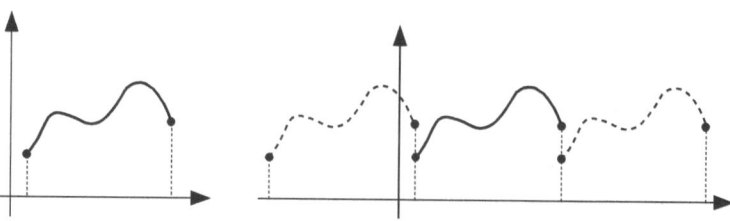

Fig. 7.11. Periodic extension of a signal.

Extension by Reflection

With this method the signal is extended by successive reflections in the edges, as indicated in Figure 7.12. The signal obtained is periodic and has no discontinuities, unlike the methods given above.

In one dimension, the signal $f(0) f(1) \cdots f(n-1)$ is extended to

$$\cdots f(2) f(1) \underbrace{f(0) f(1) \cdots f(n-1)}_{\text{original signal}} f(n-2) f(n-3) \cdots .$$

It is easy to check (see Figure 7.12) that this extension method is equivalent to reflecting the signal once and then performing a periodic extension as described before.

In two dimensions, we may first reflect horizontally in the top and bottom edges, and then reflect the extended image vertically in the left and right edges. This is illustrated in the following extension of a 3×3 matrix:

$$
\begin{array}{ccc|ccc|cc}
 & & & & \vdots & & & \\
f(2,2) & f(1,2) & f(0,2) & f(1,2) & f(2,2) & f(1,2) & f(0,2) \\
f(2,1) & f(1,1) & f(0,1) & f(1,1) & f(2,1) & f(1,1) & f(0,1) \\
\hline
f(1,0) & f(1,0) & f(0,0) & f(1,0) & f(2,0) & f(1,0) & f(0,0) \\
\cdots \; f(2,1) & f(1,1) & f(0,1) & f(1,1) & f(2,1) & f(1,1) & f(0,1) \; \cdots \\
f(2,2) & f(1,2) & f(0,2) & f(1,2) & f(2,2) & f(1,2) & f(0,2) \\
\hline
f(2,1) & f(1,1) & f(0,1) & f(1,1) & f(2,1) & f(1,1) & f(0,1) \\
f(2,0) & f(1,0) & f(0,0) & f(1,0) & f(2,0) & f(1,0) & f(0,0) \\
 & & & & \vdots & & &
\end{array}
$$

Null Extension

For some applications, the order of the matrix defining the mask is much smaller than the spatial resolution of the image. If it is not very important to preserve the values of the filtered image in a small neighborhood of the edge, we may simply decide not to extend the image. In other words, we compute $f * h$ only for pixels (n, m) where Equation (7.9) makes sense. Thus, an $m \times n$

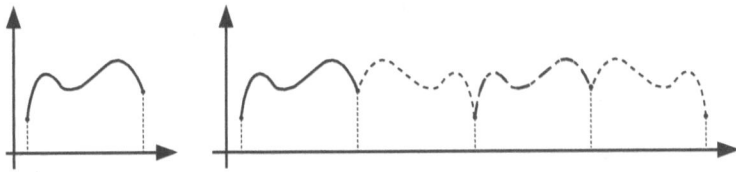

Fig. 7.12. Extension of a signal by reflection.

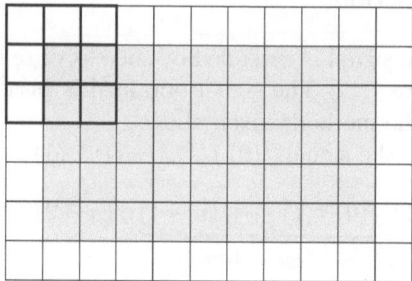

Fig. 7.13. Null extension of an image.

image filtered with an order-k mask, where k is odd, yields an image of order $\left(m - \frac{1}{2}(k-1)\right) \times \left(n - \frac{1}{2}(k-1)\right)$. Figure 7.13 illustrates this situation with a 3×3 mask. In this case the rows and columns of the image whose pixels are shown in white will be undefined in the filtered output (the original pixel's values could be repeated if necessary).

7.4 Examples of Linear Filters

We now introduce the spatially invariant linear filters most commonly used in computer graphics and image processing. In each case we treat the filter first in the continuous domain and then in the discrete domain.

Box Filter

In one dimension, the *box filter* has kernel

$$\text{box}_1(x) = \begin{cases} 1/(2a) \text{ if } & -a \le x \le a, \\ 0 & \text{if } |x| > a. \end{cases}$$

Note that this function differs from the pulse function defined in Chapter 2 by a normalization factor $1/(2a)$.

The definition just given can easily be extended to the two-dimensional case by imposing a separability condition

$$\text{box}_2(x, y) = \text{box}_1(x)\,\text{box}_1(y).$$

The graph of this kernel is shown in Figure 7.14 for the one- and two-dimensional cases.

The box filter is a lowpass filter. This is easy to verify by analyzing the filter in the frequency domain. In fact, the transfer function of the filter box_2 is the two-dimensional sinc function. Indeed, from (7.4) we obtain

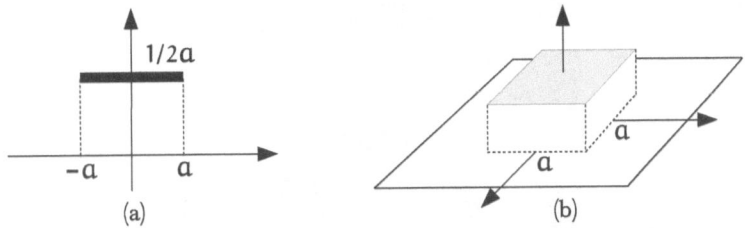

Fig. 7.14. Kernel of the box filter in one and two dimensions.

$$F(\text{box}_2)(u, v) = F(\text{box}_1 \cdot \text{box}_1)(u, v)$$
$$= F(\text{box}_1)(u) \cdot F(\text{box}_1)(v)$$
$$= \frac{\sin x}{x} \frac{\sin y}{y} = \text{sinc}(x, y),$$

whose graph is shown in Figure 7.15, together with its cross section along the x-axis. Note that the cross section $\text{sinc}(x, 0)$ is exactly the graph of the one-dimensional sinc function.

Thus, in the frequency domain, filtering (in the spatial domain) with a box filter corresponds to multiplying the Fourier transform by the sinc function; this clearly dampens high frequencies.

Figure 7.16 shows a contour density image of the transfer function of the box$_2$ filter (white regions correspond to positive values, black areas to negative values, and gray areas to intermediate values). From this picture we conclude that the filter is not isotropic. That is, the dampening of high frequencies depends not only on the frequency values but also on the directions these frequencies occur on the image. Also, the negative values assumed by the transfer function cause a phase shift on some pixels of the filtered image. These problems may give rise to artifacts on the filtered image.

In order to give examples, a discretization of the box filter should be computed. It is very easy to obtain a mask of order n: it is defined by an $n \times n$

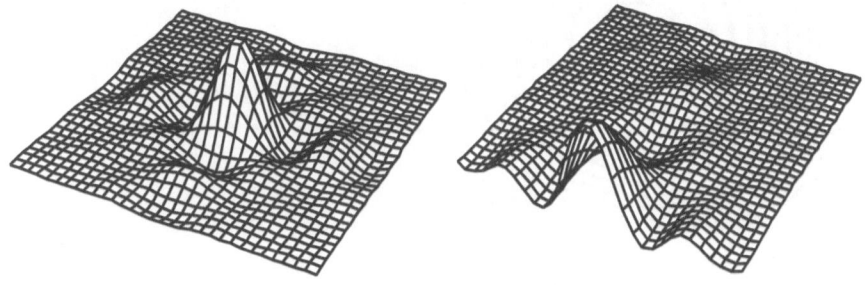

Fig. 7.15. Graph of the function $\text{sinc}(x, y)$ and a cross section thereof.

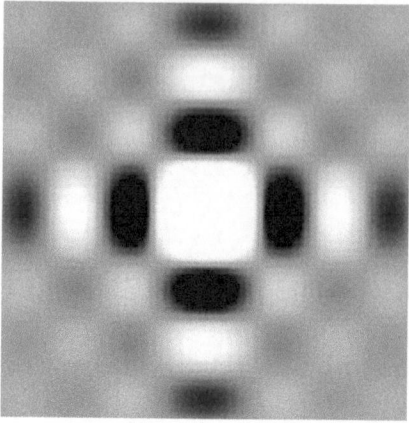

Fig. 7.16. Contour density image of the sinc function.

matrix (h_{ij}) such that $h_{ij} = 1/n^2$. Thus, a discretization of this filter with a 3×3 mask would be

$$\frac{1}{9} \cdot \begin{array}{|c|c|c|} \hline 1 & 1 & 1 \\ \hline 1 & 1 & 1 \\ \hline 1 & 1 & 1 \\ \hline \end{array}.$$

Since the mask is symmetric and of odd order, convolution consists in taking the mean of the pixels in the 8-connected neighborhood of every pixel.

Now consider the test image $f(x, y) = \cos^2(x^2 + y^2)$, shown on the left in Figure 7.17. This image has very high frequencies with a strong directionality.

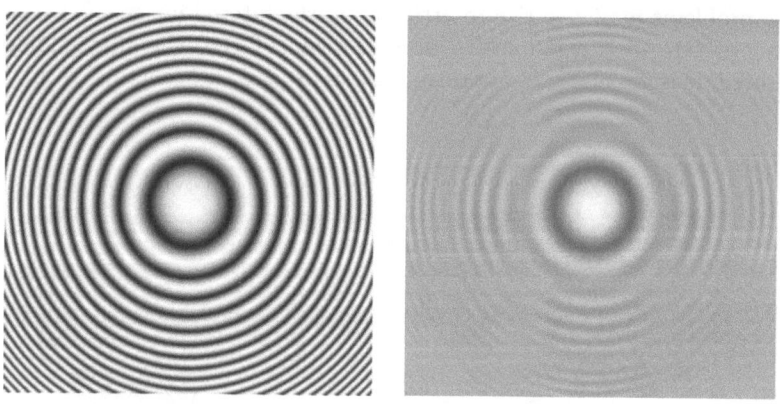

Fig. 7.17. Artifacts of box filtering.

By filtering it using a box filter mask of order 11, we obtain the image on the right. Artifacts due to the filter anisotropy and phase shift are quite perceptible.

When the image does not have strong directionalities in its frequencies, the filtering artifacts are difficult to perceive, especially if we use a small mask. Figure 7.18 shows the effect of a box filter with a 5×5 mask, or about 0.11 cm on a side.

Bartlett Filter

In one dimension, the *Bartlett filter* or *triangular filter* has kernel

$$h_1(x) = \begin{cases} 1 - |x| & \text{if} \quad |x| \leq 1, \\ 0 & \text{if} \quad |x| \geq 1. \end{cases}$$

It is easy to verify that this filter is obtained by the convolution product $h_1(t) = \text{box}(t) * \text{box}(t)$, where $\text{box}(t)$ is the kernel of the box filter defined above with $a = \frac{1}{2}$. In other words, applying the Bartlett filter is equivalent to applying the box filter twice in cascade. Its transfer function is therefore

$$\hat{h}(t) = \text{sinc}^2 t = h_1(x) = \begin{cases} (\sin^2 t)/t^2 & \text{if} \quad t \neq 0, \\ 0 & \text{if} \quad t = 0. \end{cases}$$

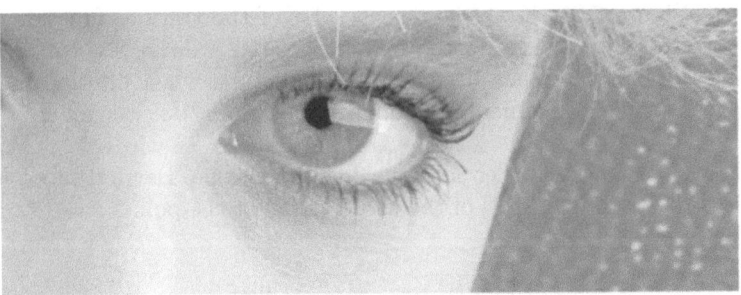

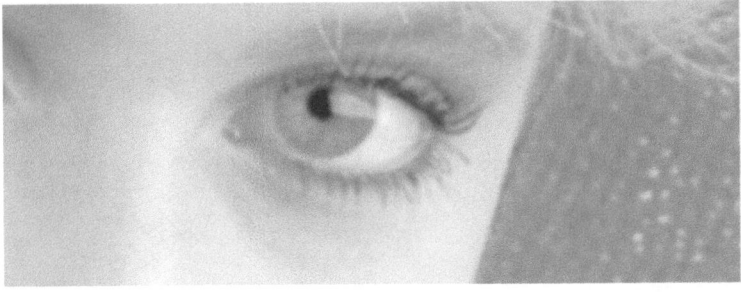

Fig. 7.18. Top: Original image. Bottom: Image after applying a box filter of order 5.

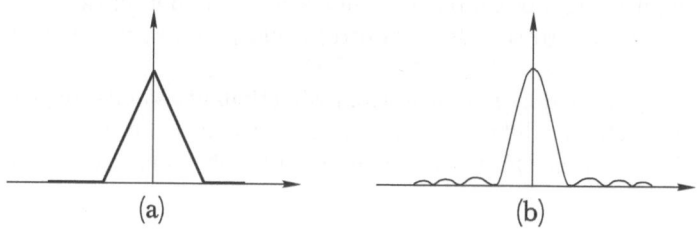

Fig. 7.19. Bartlett filter impulse function (a) and transfer function (b).

Figure 7.19 shows the graphs of the impulse function h and of the transfer function $\hat{h}$ in the one-dimensional case. The Bartlett filter is therefore also a smoothing filter and dampens high frequencies even more than the box filter.

In two dimensions, the Bartlett filter h_2 is defined by separability:

$$h_2(x, y) = h_1(x) \cdot h_1(y).$$

From the separability of the filter kernel h_2, and equation (7.4), we conclude that the transfer function in two dimensions is given by $\mathrm{sinc}^2(x, y)$. The graph of this function is shown in Figure 7.20, together with a cross section along the x-axis.

The graphs in Figures 7.20 and 7.15 have the same domain, $[-12, 12] \times [-12, 12]$. Figure 7.21 shows the density plot of the transfer function (black areas correspond to 0, and lighter areas to positive values). We can conclude that the Bartlett filter has better decay properties when filtering high frequencies and that it assumes only nonnegative values, but it is an anisotropic filter.

To obtain a mask of order 3 for the one-dimensional Bartlett filter, we can discretize the support $[-1, 1]$ of the kernel $h_1(x)$ at the points $x = -\frac{1}{2}$, $x = 0$, and $x = \frac{1}{2}$. We get

$$\tfrac{1}{2} \cdot \boxed{\tfrac{1}{2} \,|\, 1 \,|\, \tfrac{1}{2}} \quad = \quad \tfrac{1}{4} \cdot \boxed{1 \,|\, 2 \,|\, 1}.$$

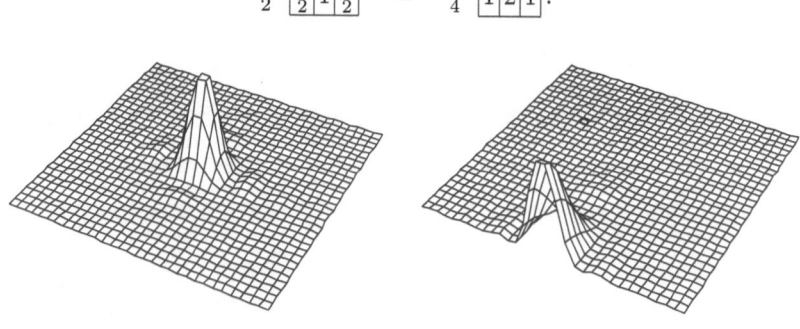

Fig. 7.20. Transfer function of the Bartlett filter in two dimensions, and a cross section thereof.

Fig. 7.21. Density plot of Bartlett's filter transfer function.

To obtain a mask of order 5, we can discretize the support uniformly at the points $-\frac{2}{3}, -\frac{1}{3}, 0, \frac{1}{3}, \frac{2}{3}$. We get

$$\frac{1}{3} \cdot \boxed{\frac{1}{3}\;\frac{2}{3}\;1\;\frac{2}{3}\;\frac{1}{3}} \;=\; \frac{1}{9} \cdot \boxed{1\;2\;3\;2\;1}.$$

In the same way we can get a discretization of the Bartlett kernel of any order. You can check that the order-7 discretization is given by

$$\frac{1}{16} \cdot \boxed{1\;2\;3\;4\;3\;2\;1}.$$

To get a discretization of the two-dimensional Bartlett kernel $h_2(x, y)$, we just have to remember that this kernel is separable, that is, $h_2(i, j) = h_1(i) h_1(j)$ for $i = 1, \ldots, m$ and $j = 1, \ldots, n$. We illustrate this below, deriving the order-5 convolution mask for the Bartlett filter.

1		1	2	3	2	1
2		2	4	6	4	2
3		3	6	9	6	3
2		2	4	6	4	2
1		1	2	3	2	1

$$\boxed{1\;2\;3\;2\;1}$$

We are omitting the normalizing factor $\frac{1}{9}$ for each one-dimensional kernel. Recall that the two-dimensional mask obtained above must be normalized so the entries add up to one, that is, we must divide each entry by 81.

Clearly, in computing these masks we can exploit the symmetry of the convolution kernels.

Figure 7.22 shows an image processed by an order-5 Bartlett filter (about 0.11 cm on a side). Comparing it with Figure 7.18, we see again that the dampening of high frequencies is greater than for the box filter.

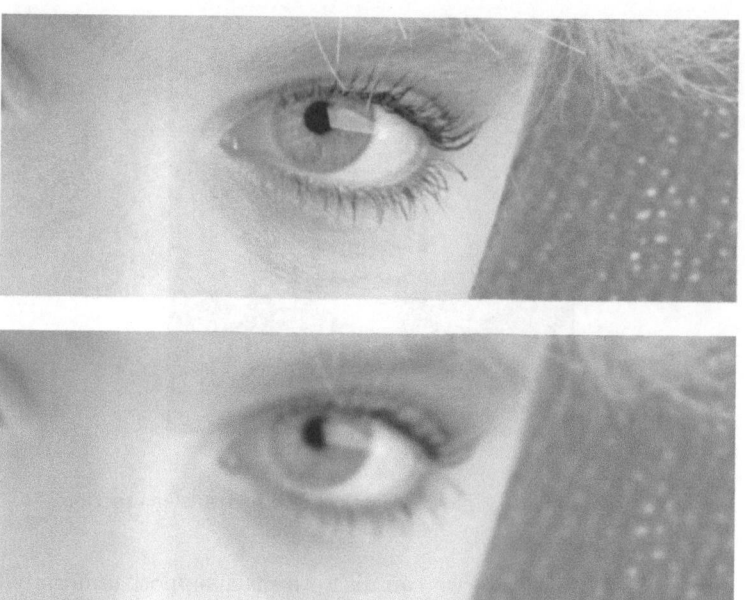

Fig. 7.22. Top: Original image. Bottom: Image after applying a Bartlett filter of order 5.

Piecewise Polynomial Filters

The box filter has a piecewise polynomial kernel of degree 0 and the Bartlett filter has a piecewise polynomial kernel of degree 1. If we continue the process of convolving the box_1 filter successively, we obtain a family of piecewise polynomial filters of increasing degree. In one dimension we would have

$$P_1(t) = \underbrace{box_1(t) * \cdots * box_1(t)}_{n \text{ times}}.$$

The two-dimensional filters are obtained by requiring separability:

$$P_2(x, y) = P_1(x) \cdot P_1(y).$$

If the number of convolution factors n increases arbitrarily, this family of piecewise polynomial filters converges to the gaussian filter. This result follows from the famous central limit theorem.

Since the blurring properties of piecewise polynomial filters improve as the degree increases, we expect that the gaussian filter, as a lowpass filter, has very good properties. This is true, as we will see in the next section.

Gaussian Filter

In one dimension the kernel $G_\sigma(x)$ of the continuous-domain *gaussian filter* is given by the gaussian function

$$G_\sigma(x) = \frac{1}{\sigma\sqrt{2\pi}} e^{-x^2/(2\sigma^2)},$$

where σ is a constant, called the *variance* of the function. In two dimensions the kernel is defined by

$$G_\sigma(x,y) = \frac{1}{2\sigma^2\pi} e^{-(x^2+y^2)/(2\sigma^2)}.$$

This filter is separable: $G_\sigma(x,y) = G_\sigma(x)G_\sigma(y)$. The graph of $G_\sigma(x,y)$ for $\sigma = 2$ is shown in Figure 7.23 together with a cross section along the x-axis. Geometrically, this graph is obtained from the graph of a one-dimensional gaussian curve, by rotation around the vertical axis.

A quick analysis in the continuous domain shows that the gaussian filter is a lowpass filter. We just observe that the Fourier transform of a gaussian distribution is also a gaussian. In other words, the transfer function is gaussian, so that high frequencies in the filtered signal are damped by a factor that grows exponentially with the frequency. The transfer function assumes only nonnegative values, and the rotational symmetry of the gaussian shows that the filter is isotropic.

Compare the graph for the gaussian filter (Figure 7.23) with those for the box and Bartlett filters (Figures 7.15 and 7.20); we observe that the gaussian graph is shown in the square $[-6, 6] \times [-6, 6]$. Figure 7.24 shows the effect of the gaussian filter on the test image used to show the anisotropic effects of the box filter (Figure 7.17).

To discretize the one-dimensional gaussian filter, we could proceed as for the Bartlett filter. However, because of the defining expression of the gaussian kernel, it is not possible to obtain a uniform mask whose entries are rational numbers. This leads to certain problems, mostly of a computational nature. A more elegant and efficient method is to use the central limit theorem to

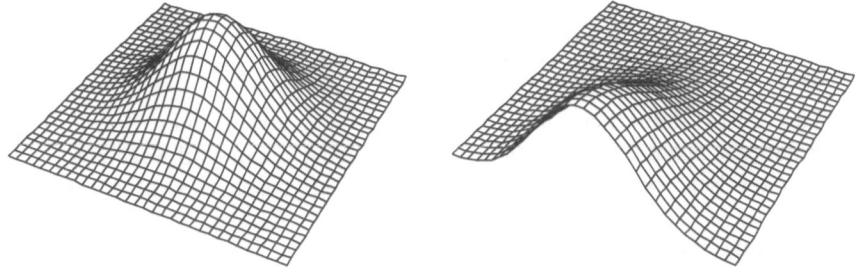

Fig. 7.23. Gaussian distribution function with mean 0 and variance 2.

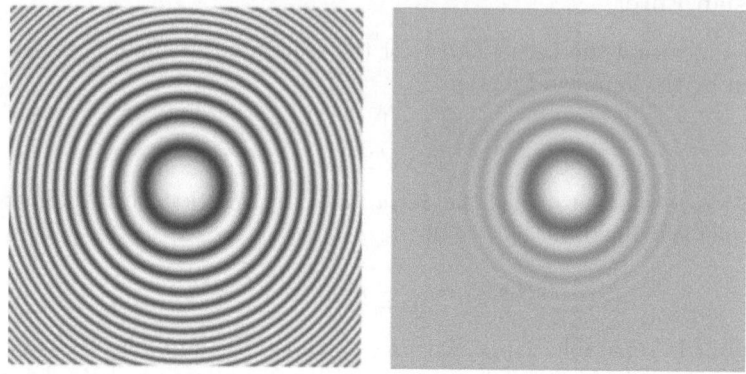

Fig. 7.24. Test image filtered with the gaussian filter.

obtain successive approximations to the gaussian mask. We will do this in the next section.

Binomial Filters

The gaussian function used to define the gaussian filter represents in fact a probability density function with mean zero and variance σ, called the normal probability distribution. A particular case of the central limit theorem, which is well known in statistics, shows that the normal distribution can be approximated by the binomial distribution, which is discrete. We will soon give more details about this.

Consider *an event with two possibilities (success and failure) such that the probability of success is p. The event is repeated independently n times, and we want to measure the probability of s successes occurring.*

The above problem is modeled by the well-known binomial distribution of probability, and the solution is given by the expression

$$b(s) = \binom{n}{s} p^s (1-p)^{(n-s)}, \quad s = 0, 1, \ldots, n. \tag{7.10}$$

The central limit theorem guarantees that, for large values of n, $b(s)$ approximates a normal distribution with the same mean and variance. That is, as n tends to ∞, the function values $b(s)$ approach the value $g(s)$ of a gaussian g.

If we take $p = \frac{1}{2}$ in (7.10), we have

$$b(s) = \frac{n!}{s!(n-s)!} \frac{1}{2^n}, \quad \text{for } s = 0, 1, \ldots, n. \tag{7.11}$$

By varying s, with n fixed, we obtain a mask of order $n + 1$ that is an approximation to a gaussian mask. The table below shows some masks for $n = 1, 2, 3, \ldots, 8$.

n	2^n	mask coefficients
1	2	1 1
2	4	1 2 1
3	8	1 3 3 1
4	16	1 4 6 4 1
5	32	1 5 10 10 5 1
6	64	1 6 15 20 15 6 1
7	128	1 7 21 35 35 21 7 1
8	256	1 8 28 56 70 56 28 8 1

For masks of odd orders $n = 2m + 1$, Equation (7.11) can be rewritten in the form

$$b(s) = \frac{1}{2^{2m+1}} \frac{(2m+1)!}{(m-k)!(m+k)!}, \quad \text{for } k = -R, \ldots, R.$$

As n increases, we obtain discrete convolution masks that approximate the family of piecewise polynomial filters introduced before (box filter, Bartlett filter, and so on). For large values of n we obtain good approximation masks for the gaussian filter.

Using the separability of the gaussian filter, one can easily obtain two-dimensional masks, just as we did for the Bartlett filter in the previous section. We show here two-dimensional masks of order 2, 3, 4, and 5.

$$\frac{1}{4} \cdot \begin{array}{|c|c|} \hline 1 & 1 \\ \hline 1 & 1 \\ \hline \end{array} \qquad \frac{1}{16} \cdot \begin{array}{|c|c|c|} \hline 1 & 2 & 1 \\ \hline 2 & 4 & 2 \\ \hline 1 & 2 & 1 \\ \hline \end{array} \qquad \frac{1}{64} \cdot \begin{array}{|c|c|c|c|} \hline 1 & 3 & 3 & 1 \\ \hline 3 & 9 & 9 & 3 \\ \hline 3 & 9 & 9 & 3 \\ \hline 1 & 3 & 3 & 1 \\ \hline \end{array}$$

$$\frac{1}{256} \cdot \begin{array}{|c|c|c|c|c|} \hline 1 & 4 & 6 & 4 & 1 \\ \hline 4 & 16 & 24 & 16 & 4 \\ \hline 6 & 24 & 36 & 24 & 6 \\ \hline 4 & 16 & 24 & 16 & 4 \\ \hline 1 & 4 & 6 & 4 & 1 \\ \hline \end{array} \cdot$$

The bottom part of Figure 7.25 was obtained from the top part by applying a gaussian filter of order 5, which represents about 0.11 cm of the image side. Note how the loss of high frequencies in this image is more noticeable than in the corresponding box-filtered and Bartlett-filtered images (Figures 7.18 and 7.22, respectively).

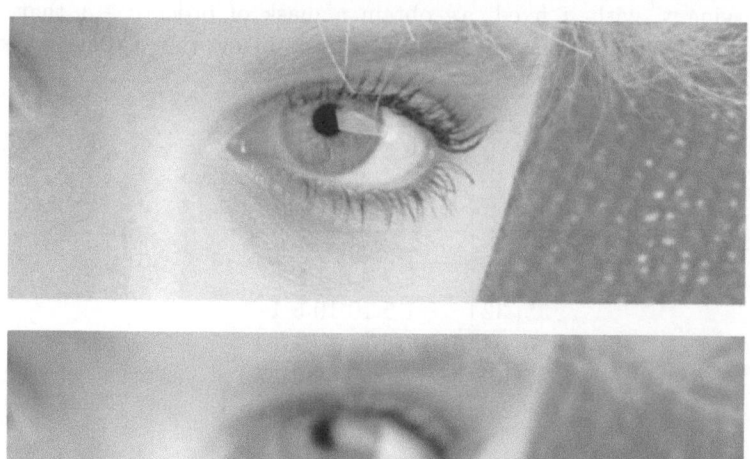

Fig. 7.25. Use of a gaussian filter of order 5.

Laplacian Filter

The *laplacian operator* ∇ is defined on the space of twice-differentiable functions of two variables:

$$\nabla f(x, y) = \frac{\partial^2 f}{\partial x^2} + \frac{\partial^2 f}{\partial y^2}.$$

It defines a linear filter in the continuous domain. In fact, it is a highpass filter. To see this, we first recall the following important property of the Fourier transform:

$$F\left[\frac{\partial f}{\partial x}\right] = 2\pi u \hat{f}(u, v);$$

$$F\left[\frac{\partial f}{\partial y}\right] = 2\pi v \hat{f}(u, v),$$

where $\hat{f}(u, v)$ indicates the Fourier transform of the function $f(x, y)$. It follows that

$$F[\nabla f] = (2\pi)^2 (u^2 + v^2) \hat{f}(u, v),$$

that is, the transfer function of the laplacian filter is

$$H(u, v) = -(2\pi)^2 (u^2 + v^2).$$

The graph of this function is a paraboloid of revolution (Figure 7.26). Thus, low frequencies are damped by the laplacian filter, while high frequencies are

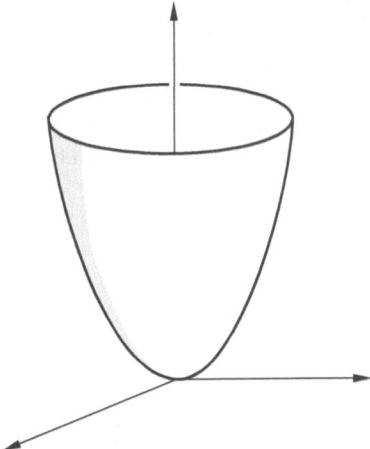

Fig. 7.26. Transfer function of the laplacian filter.

amplified by a factor that grows with the square of the frequency (quadratic modulation).

We now turn to the laplacian filter in the discrete domain. Using Taylor's formula, we have

$$f(x+1) = f(x) + f'(x) + \tfrac{1}{2}f''(x_0), \quad \text{for } x < x_0 < x+1.$$

Thus,

$$f(x+1) \approx f(x) + f'(x) + \tfrac{1}{2}f''(x).$$

Since $f'(x) \approx f(x) - f(x-1)$, it follows from this that

$$f(x+1) \approx 2f(x) - f(x-1) + \tfrac{1}{2}f''(x),$$

so that

$$f''(x) \approx 2f(x+1) - 4f(x) + 2f(x-1).$$

We conclude that, apart from a proportionality factor $\tfrac{1}{2}$, we have

$$\frac{d^2f}{dx^2} \approx \Delta^2 f(x) = f(x+1) - 2f(x) + f(x-1).$$

The discrete laplacian is therefore given by

$$\nabla f(i,j) = \Delta_x^2 f(i,j) + \Delta_y^2 f(i,j)$$
$$= \big(f(i+1,\,j) + f(i-1,\,j) + f(i,\,j+1) + f(i,\,j-1)\big) - 4f(i,j).$$

A 3×3 mask for the laplacian filter is

0	1	0
1	−4	1
0	1	0

.

By analyzing the kernel mask for the laplacian filter obtained above, we see that in regions of the image where we have low variations of intensities (low frequencies), the filtered image is almost black (it is black if the image intensity is constant). It is easy now to devise different masks for highpass filters with behavior similar to the laplacian. An improvement can be obtained by taking the average on an 8-connected neighborhood. The mask is given by

$$\begin{array}{|c|c|c|} \hline 1 & 1 & 1 \\ \hline 1 & -8 & 1 \\ \hline 1 & 1 & 1 \\ \hline \end{array}.$$

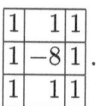

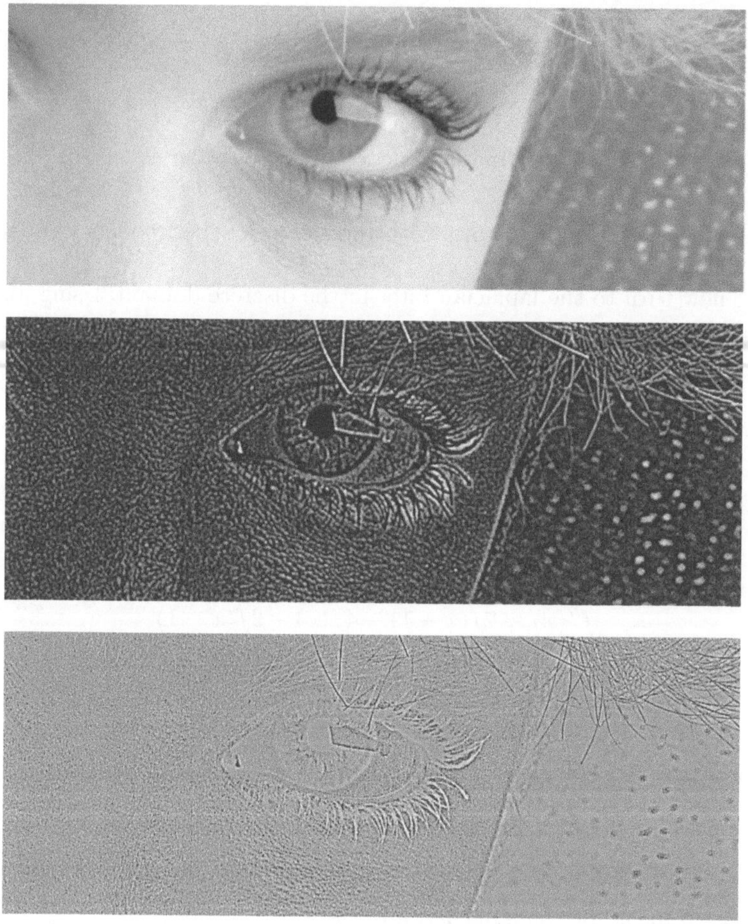

Fig. 7.27. Top: Original image. Middle: filtered with truncation; Bottom: filtered with offset.

When applying the laplacian filter to an image, we may get negative values; in order to visualize the filtered image, we must either clip the negative intensities to 0 or add an offset to the image intensities. This is illustrated in Figure 7.27. The middle figure is the filtered image with negative values clipped. The bottom image shows the same image with an offset of 100 added to the filtered image. The laplacian filter used was the 8-connected neighborhood defined above. Observe that all the low-frequency details of the original (slow variations) are lost in the filtering process.

A Plethora of Highpass Filters

Apart from the proportionality factor $\frac{1}{5}$, the expression just given for $\nabla f(i,j)$ is equal to

$$f(i,j) - \tfrac{1}{5}(f(i+1,j) + f(i-1,j) + f(i,j) + f(i,j+1) + f(i,j-1)),$$

which is exactly the difference between the original image and the output of the image through a (lowpass) box filter with a diamond-shaped mask (the corners of the discretized kernel having zero entries). This means the laplacian filter (apart from a constant factor) can be obtained by lowpass filtering the image and subtracting the blurred image from the original one. This fact can be generalized. If I denotes the identity filter, that is, if $I(f) = f$ for any image f, and B is a blurring (lowpass) filter, then $H = I - B$ is a highpass filter. As an example, taking B as the Bartlett filter, we obtain the highpass filter mask

$$\frac{1}{16} \cdot \begin{array}{|c|c|c|} \hline 1 & 2 & 1 \\ \hline 2 & 4 & 2 \\ \hline 1 & 2 & 1 \\ \hline \end{array} \quad - \quad \begin{array}{|c|c|c|} \hline 0 & 0 & 0 \\ \hline 0 & 1 & 0 \\ \hline 0 & 0 & 0 \\ \hline \end{array} \quad = \quad \frac{1}{16} \cdot \begin{array}{|c|c|c|} \hline 1 & 2 & 1 \\ \hline 2 & -12 & 2 \\ \hline 1 & 2 & 1 \\ \hline \end{array}.$$

7.5 Edge Enhancement Operations

In this section we will use the filters studied earlier in order to obtain several operations that enhance image edges. Edge enhancement is a very important filtering operation. In fact, when manipulating images it is very common that some operations destroy high-frequency information; thus, edge enhancement is used as a way to restore the original image's details.

We will study three operations of edge enhancement: laplacian addition, unsharp masking, and difference of gaussians.

7.5.1 Laplacian Addition

Filtering an image with the laplacian results in an image with information about the high frequencies of the original image. This high-frequency information concentrates around discontinuities of the image function that are perceived as boundaries between the objects present on the image.

Therefore, subtracting the laplacian frequency information decreases the image high contrast, and adding the laplacian frequency information enhances the image details. This is illustrated in Figure 7.28. In (a) we show the original image; in (b) we show the filtered image (with offset); and in (c) we have the image in (a) with the details in (b) added.

Blurring and Diffusion

There is a very interesting physical interpretation of the use of the laplacian filter as an operation for edge enhancement. The operation of blurring an image (decreasing its high frequencies) can be modeled as a diffusion process whose evolution is dictated by the partial differential equation

$$\frac{\partial f}{\partial t} = -k\nabla^2 f, \tag{7.12}$$

where k is a positive constant. The function $f(x, y, t)$ is the blurred image after a time t has elapsed from the beginning of the diffusion process. By expanding f in a Taylor series in time around $t = 0$, we obtain

$$f(x, y, t) = f(x, y, 0) + t\frac{\partial f}{\partial t} + t^2\frac{\partial^2 f}{\partial t^2} + \cdots.$$

By neglecting higher-order terms and using (7.12), we obtain

$$f(x, y, t) = f(x, y) - tk\nabla^2 f,$$

that is,

$$f(x, y) = f(x, y, t) + tk\nabla^2 f.$$

This equation shows that the original, unblurred image $f(x, y) = f(x, y, 0)$ is obtained from the blurred image $f(x, y, t)$ by adding the laplacian of the original image.

Fig. 7.28. (a) Original image. (b) Filtered image. (c) Image with enhanced edges.

Mach Bands and Laplacian

It is interesting to understand geometrically how laplacian addition acts as an edge enhancement filter. For this, consider the step pattern shown in Figure 7.29. The graph on the right shows the intensity values of the pattern along a scanline. At discontinuity points the intensity is the average of the limits on the left and on the right. Suppose that the intensities along each scanline are given by

$$\cdots m \quad m \quad m \quad m \quad \frac{m+n}{2} \quad n \quad n \quad n \quad n \cdots,$$

where m and n are positive constants. Using the one-dimensional laplacian

$$\boxed{-1\,|\,2\,|\,-1}$$

and filtering the scanline, we obtain

$$\cdots 0 \quad 0 \quad 0 \quad m - \frac{m+n}{2} \quad 0 \quad n - \frac{m+n}{2} \quad 0 \quad 0 \quad 0 \cdots.$$

This is better illustrated, in the continuous domain, by Figure 7.30. On the left we have the step intensities before filtering, and on the right we show the filtered steps. By adding the filtered scanline to the original scanline intensities of Figure 7.29, we obtain the scanline intensities shown in the graph of Figure 7.31. The overshoots at the discontinuities are responsible for the perception of enhancement of the edge. The enhanced step pattern is also shown in Figure 7.31. Besides illustrating geometrically edge enhancement by laplacian addition, the preceding example conveys another interesting fact.

The result of the edge enhancement operation is similar to the human vision behavior of the step pattern by the perception of Mach bands discussed in Section 11.2. This remark supports the fact that the human visual system concentrates on edges, ignores uniform regions, and performs an edge enhancement operation similar to the technique of enhancement by laplacian addition.

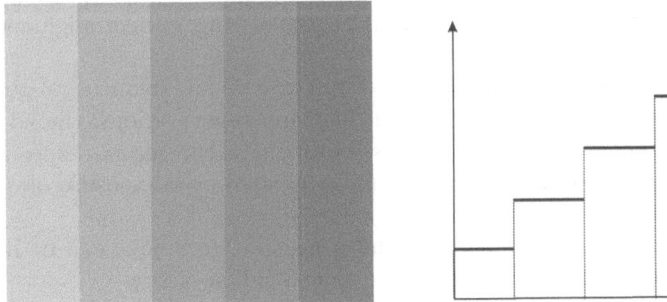

Fig. 7.29. Left: Step pattern. Right: Scanline intensities.

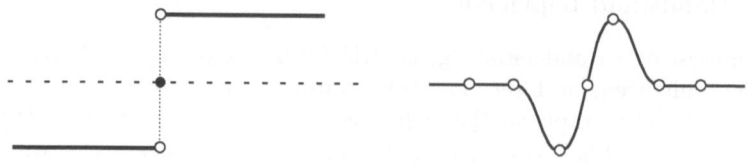

Fig. 7.30. Left: step pattern. Right: filtered step.

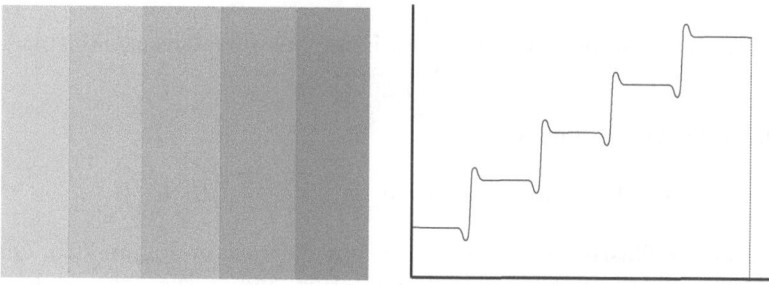

Fig. 7.31. Left: Enhanced ladder. Right: Filtered intensities of scanline.

7.5.2 Unsharp Masking

This section will introduce the most popular edge enhancement operation in use. The unsharp masking technique (USM) is present in most of the image manipulation software on the market. The technique mimics a traditional photographic technique that is very commonly used to enhance images, especially in the field of scientific photography. It is worth describing the analog technique before introducing its digital counterpart.

First we make a blurred contact copy on film. The blurring is obtained by leaving a small gap between the emulsions. After this copy is ready, we produce a new contact print in a two-step exposure. First we expose using the original film, and then we use the blurred negative (with perfect alignment) to expose again.

The rationale behind this process is the following: in low-frequency regions of the image, the blurred film is dark. In high-frequency regions, the edges get blurred and allow some light to pass through in the second exposure. Therefore, the blurred film is used as a mask to allow overexposure on the image details.

Now we describe how the analog unsharp masking technique can be imitated in the digital domain using the operations studied before.

Given an image f, we use a gaussian filter to obtain a blurred image g. We obtain an image h by subtracting the blurred image g from f. We have already seen that the image h could essentially be obtained directly from f

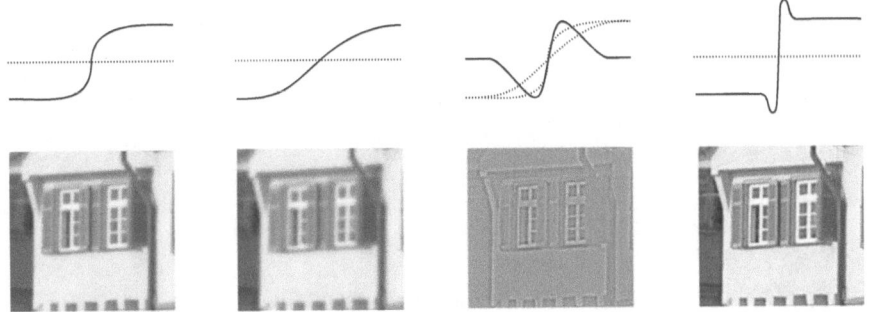

Fig. 7.32. (a) Original image. (b) Blurred image. (c) Difference between first two. (d) Enhanced image.

using a laplacian filter. The edge enhancement on the original image f is obtained by adding the details of the image h to it. Figure 7.32 illustrates the whole process geometrically by using the change of intensities along a scanline under the operation. In (c) we add an intensity offset in order to visualize the negative values.

Even better results in unsharp masking filtering can be obtained by combining the gaussian and laplacian filters: first we blur the image with a gaussian filter, and then we use a laplacian to enhance the edges. This enhanced image is then used to construct the unsharp mask.

Since the above operation

$$\Delta(f * G_\sigma) = f * \Delta G_\sigma,$$

this is equivalent to using a filter with kernel ΔG_σ, whose continuous version is

$$\Delta G_\sigma(x, y) = \frac{x^2 + y^2 - 2\sigma^2}{2\pi\sigma^6} e^{-(x^2+y^2)/(2\sigma^2)}. \tag{7.13}$$

A discrete mask for this filter is given by

0	1	1	0
1	-2	-2	1
1	-2	-2	1
0	1	1	0

7.5.3 Difference of Gaussians

The use of the laplacian filter to improve the unsharp masking technique can be avoided. Very good results are obtained by using the difference between

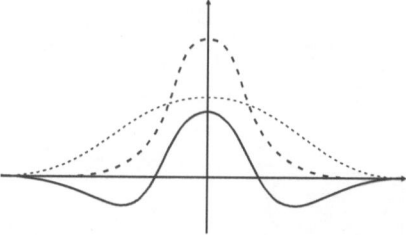

Fig. 7.33. Kernel of the DOG filter.

two gaussian filters with the same mean and distinct variance. This filter is called the *difference of gaussians* (DOG).

Figure 7.33 shows the shape of a DOG filter kernel (the two gaussians are shown in dashed lines) in the one-dimensional case. The two-dimensional kernel is obtained by rotating the one-dimensional kernel around the vertical axis. Notice that the shape is similar to the kernel of the "laplacian-of-gaussian" filter in Equation (7.13).

The DOG filter is widely used in the area of computer vision. Since the eye performs a lowpass filtering operation, it seems reasonable that the DOG filter is a model for the way the human visual system locates boundaries and other high-frequency details on the image.

7.6 Comments and References

The fast Fourier transform is discussed in a great many articles and books; we suggest the reader consult (Lim 1990) and its comprehensive list of references. A fast Fourier transform program, written in C, can be found in (Wolberg 1990).

An elementary approach to general aspects of computer graphics and image processing can be found in (Pavlidis 1982).

The morphological filters introduced in this chapter show only the tip of an iceberg. These filters lie at the foundation of the discipline of mathematical morphology. The classical references for the study of mathematical morphology are (Serra 1982) and (Serra 1988).

A comprehensive discussion on color clipping and color conversion can be found in (Hall 1989), (Cook and Torrance 1981), and (Catmull 1979). See also Chapter 16 of this book.

The space of images defined on a given domain has a natural vector space structure, so it is natural to define algebras on it. This allows the reduction of many problems about images to operations in this algebra. The interested reader can consult (Ritter, Wilson and Davidson 1990).

Image compression using the laplacian pyramid was introduced in (Burt and Adelson 1983). A good survey on the use of pyramid structures can be found in (Adelson et al. 1984). A very accurate and complete discussion of pyramids and their relationship with wavelets can be found in (Meyer 1993). A general overview of the theory of wavelets and its relationship with multiresolution decomposition can be found in (Daubechies 1992).

(Jahne 1993) is a very good and well-illustrated book covering linear filter operations, including detailed discussion of laplacian and gaussian pyramids.

We have tried to illustrate all of the image operations introduced in this chapter. The reader interested in more pictorial examples of the filtering operations should see (Russ 1992), a wonderful book containing hundreds of beautiful images.

The eye image used in Figures 7.18, 7.22, 7.25, 7.27 and 7.35 is a detail from the image "Portrait of a girl in red", by Bob Clemens, from the Kodak PhotoCD, Photo Sampler.

The original image in Figures 7.28 and 7.32 is a detail from "Market Place", by Alfons Rudolph, from the Kodak PhotoCD, Photo Sampler.

References

[Adelson et al. 1984]Adelson, E. H., Anderson, C. H., Bergen, J. R., Burt, P. J., and Ogden, J. M. (1984). Pyramid methods in image processing. *RCA Engineer*, 29(6).

[Burt and Adelson 1983]Burt, P. J. and Adelson, E. H. (1983). The laplacian pyramid as a compact image code. *IEEE Trans. Commun.*, 532–540.

[Catmull 1979]Catmull, E. (1979). A tutorial on compensation tables. *Computer Graphics (SIGGRAPH '79 Proceedings)*, 13(3):1–7.

[Cook and Torrance 1981]Cook, R. L. and Torrance, K. E. (1981). A reflectance model for computer graphics. *Computer Graphics (SIGGRAPH '81 Proceedings)*, 15(3):307–316.

[Daubechies 1992]Daubechies, I. (1992). *Ten Lectures on Wavelets*. Number 61 in CBMS-NSF Series in Applied Mathematics. SIAM Publications, Philadelphia.

[Hall 1989]Hall, R. A. (1989). *Illumination and Color in Computer Generated Imagery*. Springer-Verlag, New York.

[Jahne 1993]Jahne, B. (1993). *Digital Image Processing: Concepts, Algorithms and Scientific Applications*, second ed. Springer-Verlag, New York.

[Lim 1990]Lim, J. S. (1990). *Two-Dimensional Signal and Image Processing*. Prentice-Hall, Englewood Cliffs, NJ.

[Meyer 1993]Meyer, Y. (1989). *Wavelets Algorithms and Applications*. Society for Industrial and Applied Mathematics (SIAM), Philadelphia.

[Pavlidis 1982]Pavlidis, T. (1982). *Algorithms for Graphics and Image Processing*. Computer Science Press, Rockville, MD.

[Ritter, Wilson and Davidson 1990]Ritter, G. X., Wilson, J. N., and Davidson, J. L. (1990). Image algebra: An overview. *Computer Vision, Graphics and Image Processing*, 49:297–331.

[Russ 1992]Russ, J. C. (1989). *The Image Processing Handbook*. CRC Press, Boca Raton, FL.

[Serra 1982]Serra, J. P. (1982). *Image Analysis and Mathematical Morphology*. Academic Press, New York.

[Serra 1988]Serra, J. P. (1988). *Image Analysis and Mathematical Morphology: Theoretical Advances*. Academic Press, London.

[Wolberg 1990]Wolberg, G. (1990). *Digital Image Warping*. IEEE Computer Society Press, Washington, DC.

Sampling and Reconstruction

Analog images must be sampled before being represented on the computer. In order to be visualized they must be displayed on a device that is able to reconstruct color, such as a CRT monitor. The sampling process is called *rasterization*; it is carried out by some sampling device, such as a scanner or TV camera, or by discretizing a continuous mathematical description of a scene, as in the case of the rendering process of image synthesis systems. The display device reconstructs the discrete image, creating an optical-electronic version that is perceived by the eye. Thus, an understanding of sampling and reconstruction is a good foundation for producing good-quality images.

The process of sampling and reconstruction can be translated into operations with images. In this chapter we use the signal theory developed earlier in order to study the problems involved. We restrict our study to uniform point sampling, which is simply called *sampling* in this chapter.

In the first section we make a review of sampling and reconstruction in order to attain some degree of independence from previous chapters.

8.1 Sampling

To keep the notation simple, we will consider one-dimensional signals. Let f be a continuous signal and $\{k\,\Delta t : k \in \mathbb{Z}\}$ a uniform lattice.

8.1.1 Time-Domain Viewpoint

Point sampling of f in the time domain is attained by multiplying f by an infinite Dirac delta impulse train, or comb function, associated to the uniform lattice Δt of the signal domain. That is,

$$f_d = f \cdot \mathrm{comb}_{\Delta t}$$

and

$$\mathrm{comb}_{\Delta t} = \sum_{k=-\infty}^{+\infty} \delta(t - k\,\Delta t), \quad \text{for } k \in \mathbb{Z}.$$

This is illustrated in Figure 8.1.

L. Velho et al., *Image Processing for Computer Graphics and Vision,*
Texts in Computer Science, DOI 10.1007/978-1-84800-193-0_8,
© Springer-Verlag London Limited 2009

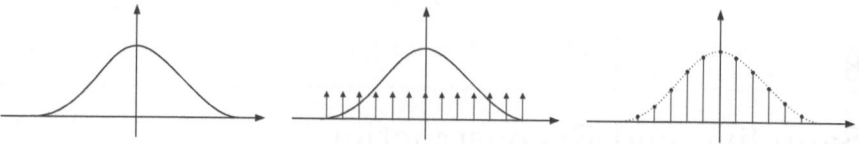

Fig. 8.1. Sampling in the time domain.

8.1.2 Frequency-Domain Viewpoint

Multiplication by the comb signal in the time domain corresponds to convolution of the signal spectrum with the Fourier transform of the comb signal in the frequency domain. The latter is another comb signal associated with the lattice $\text{comb}_{1/\Delta t}$ (with amplitude modulated by $1/\Delta t$). Convolution with this filter corresponds geometrically to translating the signal spectrum along the lattice vertices and summing up the result. This is illustrated in Figure 8.2.

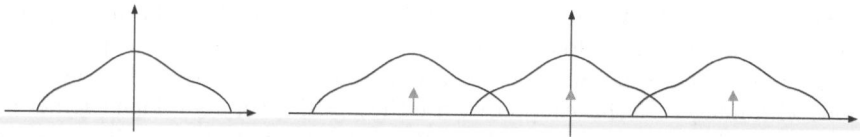

Fig. 8.2. Sampling in the frequency domain.

8.2 Reconstruction

The reconstruction problem is very simple to state: it consists in obtaining the original signal f from the sampled signal f_d. As for sampling, it is instructive to look at this problem in both time and frequency domains.

8.2.1 Frequency Domain Viewpoint

The spectrum of a sampled signal consists of replicas of the spectrum of the original signal, translated along the frequency axis, as illustrated in Figure 8.3. The graph shows that the signal is bandlimited and was sampled within the Nyquist limit of the Shannon-Whittaker theorem.

The sampling process introduces high frequencies not present in the original signal (these frequencies appear as copies of the original spectrum). These high frequencies are introduced by replicating the original spectrum. In Figure 8.3, s_0 separates the frequencies present in the original signal ($s \leq s_0$) from those introduced in the sampling process ($s \geq s_0$). The *cutoff frequency*

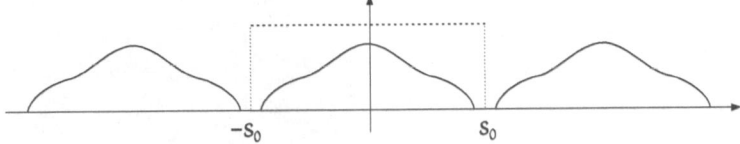

Fig. 8.3. Spectral model of a sampled signal.

s_0 is important because it enables a signal to be exactly reconstructed from the sampled signal by means of a filter that

- eliminates frequencies above the cutoff frequency, and
- leaves unchanged the frequencies below the cutoff (we say it *has unit gain* in the base band $s \le s_0$).

An ideal reconstruction filter is one that satisfies these two conditions. The dashed lines in Figure 8.3 illustrate the transfer function of such a filter.

8.2.2 Time-Domain Viewpoint

In the time domain, reconstructing a signal from its samples amounts to interpolating the samples in order to obtain the continuous signal. This interpolation can be achieved by convolving the time-domain sampled signal with a convenient filter kernel. Indeed, if g is the transfer function of the reconstruction filter in the frequency domain, the reconstructed signal in the time domain is obtained by convolving the sampled signal with $h = F^{-1}(g)$, where F^{-1} is the inverse Fourier transform. The filter h is called a *reconstruction filter* or *reconstruction kernel*.

If the discrete samples are defined by $f_k = f(t_k) = f(k\,\Delta t)$, for $k \in \mathbb{Z}$, the convolution is given by the equation

$$f(t) = f_d * h = \sum_{k=-\infty}^{+\infty} f(t_k)h(t - t_k). \tag{8.1}$$

Geometrically, the reconstruction kernel h is translated to each lattice vertex $t_k = k\,\Delta t$, modulated by the value $f(t_k)$ of the sample at the vertex, and the results are added. Mathematically, the translates $h(t - t_k)$ of the filter kernel constitute the interpolation basis of the reconstruction process. This is illustrated in Figure 8.4.

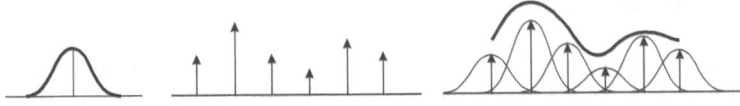

Fig. 8.4. Reconstruction from uniform samples.

Fig. 8.5. Sampling and reconstruction sequence.

Figure 8.5 illustrates the whole process of sampling and reconstruction: part (a) represents the input image; part (b) shows the comb filter; part (c) shows the sampled image; and part (d) shows the reconstructed image from (c) using a box-shaped reconstruction filter.

Here we see an obvious relationship between reconstruction and filtering: *the reconstruction of a signal f can be regarded as the filtering of the discrete signal f_d using a spatially invariant linear filter with kernel h.* In this case the comparison between the various methods of reconstruction reduces to the comparison of interpolation kernels.

Ideal Reconstruction

Shannon's sampling theorem guarantees that it is possible to reconstruct the original signal f from the discrete signal f_d if we take at least one sample for each half-cycle of the signal (Nyquist limit). When this happens, the original

signal can be recovered using the sinc function as a reconstruction kernel. This
is called the *ideal reconstruction filter*. Equation (8.1) in this case becomes

$$f(t) = \sum_{k=-\infty}^{+\infty} 2\Omega \,\Delta t f(k\,\Delta t)\,\mathrm{sinc}\big(2\pi\Omega(t - k\,\Delta t)\big), \qquad (8.2)$$

where Ω is an upper bound for the signal bandwidth; that is, the support of
the Fourier transform $\hat{f}$ is contained in the interval $[-\Omega, \Omega]$. A proof of this
fact can be found on Chapter 2.

8.3 Aliasing

Consider an arbitrary signal f. As just explained, the existence of a cutoff
frequency enables us to separate the base band of the signal from the high
frequencies introduced in the sampling process. When the translated spectrum
overlaps, there is no cutoff frequency: high frequencies in the original signal
are replicated in the sampled signal as low-frequency components. This is
illustrated in Figure 8.6. This phenomenon is called *aliasing*, and the spurious
low-frequency components are called *aliases* of the high-frequency components
they come from.

There are two sources of aliasing:

- the signal is not bandlimited, as in Figure 8.6(a); or
- the signal is bandlimited but was sampled at a lower rate than prescribed
  by the Shannon-Whittaker theorem. See Figure 8.6(b).

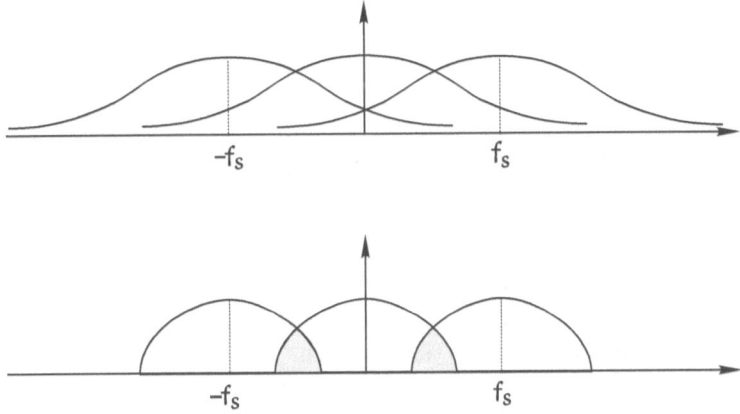

Fig. 8.6. Spectrum of an image sampled with aliasing.

In order to avoid or minimize aliasing, we must therefore work with band-limited signals and sample at a rate equal to or above the Nyquist limit. That is,

$$\frac{1}{\Delta t} > 2\Omega, \qquad \text{or} \qquad \Delta t < \frac{1}{2\Omega},$$

where Ω is an upper bound for the signal bandwidth.

In order to satisfy the above inequality, we should either reduce Δt or decrease Ω. These two possibilities give rise to two distinct methods used to avoid, or at least minimize, aliasing:

- increasing the sampling rate (reducing Δt);
- reducing the high frequencies of the signal (reducing Ω).

The sampling rate can't always be chosen above the Nyquist limit, because it is usually directly related to the resolution of the signal output device. Moreover, a very high sampling rate causes considerable storage and processing problems.

A reduction of the high frequencies can always be performed, using a lowpass filter; but this course of action is only appropriate when the high-frequency information can be discarded without harm to signal perception.

Aliasing Error

When aliasing takes place, the use of an ideal lowpass filter, which allows an ideal reconstruction for signals sampled within the Nyquist limit, no longer works. It is not possible to recover the high frequencies from its low-frequency aliases, because they are combined with the low frequencies present in the original signal. This means that the series in (8.2) does not converge to the signal f. This gives rise to an *aliasing error*, which can be expressed as

$$\varepsilon = \left| f(t) - \sum_{k=-\infty}^{+\infty} 2\Omega \, \Delta t \, f(k \, \Delta t) \, \mathrm{sinc}\big(2\pi\Omega(t - k \, \Delta t)\big) \right|. \qquad (8.3)$$

The reconstructed signal may then be a poor approximation of the original signal.

Perception of Aliasing

It is instructive to understand the way we perceive aliasing. For this we must look at this phenomenon in the time domain. We give an example using the periodic signal $f(t) = \sin(2\pi\omega_0 t)$, with frequency ω_0. The spectral model of f is shown in Figure 8.7(a). The signal f is bandlimited with supp $\hat{f} \subset [-\omega_0, \omega_0]$. Thus the Nyquist limit is $\Delta t < 1/(2\omega_0)$; as already observed, this means we should take at least one sample per half-cycle. See Figure 8.7(b).

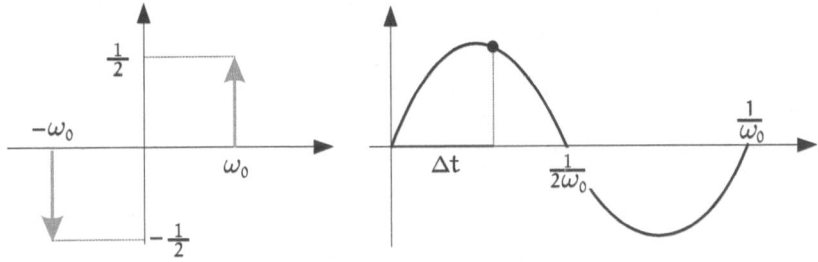

Fig. 8.7. Sampling rate and frequency.

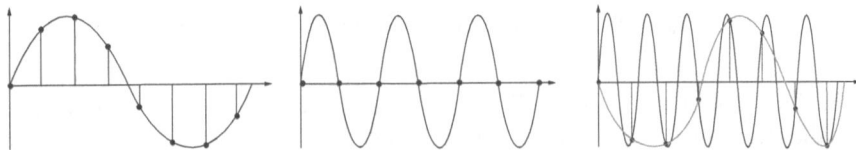

Fig. 8.8. Aliasing in the space domain.

Now we consider what happens when we sample the signal f using different sampling rates. In Figure 8.8(a) the sampling rate is four times the Nyquist limit (four samples per half-cycle), and we clearly see that even a linear interpolation of the samples reconstructs a good approximation of the signal. In Figure 8.8(b) the sampling rate is exactly twice the Nyquist limit (two samples per half-cycle). In this case, a linear interpolation of the samples results in a constant null signal. Finally, in Figure 8.8(c) the sampling rate is less than the Nyquist limit (no samples are taken in some half-cycles). If, in this figure, we interpolate the samples in the most natural way, we obtain a signal (thick curve in the figure) with a much lower frequency than the original.

As we explained, in Figure 8.8(c) high frequencies of the original signal are reconstructed as low frequencies. The reconstructed signal is completely distorted compared with the original signal. You should persuade yourself, by making measurements, that the frequency of the interpolated signal is the sampling frequency minus the frequency of the original signal. This is characteristic of aliasing.

Consider the image $f(x,y) = \cos^2(x^2 + y^2)$ shown in Figure 8.9(a). It has radial bands that get thinner and closer as the "squared radius" $x^2 + y^2$ increases. Thus it has very high frequencies in regions where $x^2 + y^2$ is high. The image in 8.9(b) was undersampled and reconstructed. Notice that "ghosts" of the low frequency information of the image appear in the reconstructed image,

Fig. 8.9. Reconstruction problems due to aliasing.

in regions of high frequencies. This is a drastic reconstruction problem caused by aliasing: high frequencies of the original image are reconstructed as low frequencies.

8.3.1 Aliasing in Computer-Generated Images

In computer graphics we generate an image from a synthetic scene containing objects and light sources. The rendering equation enables us to compute the illumination function f that gives the color for any location in the scene, from the camera point of view. This illumination function is the signal that must be sampled in order to generate the synthetic image of the scene.

You may have heard terms like *point sampling*, *supersampling*, and *area sampling*, all of which are related to the question of avoiding aliasing when sampling a scene in an image rendering system. What do these terms mean in light of the preceding discussion about sampling and aliasing?

Point Sampling

Point sampling simply means what we have been calling so far the sampling of a signal. In our case it means uniform point sampling.

Supersampling

Supersampling means taking a large number of samples $p_1, p_2, \ldots, p_n$ in a pixel p, and computing the final intensity of the pixel as the mean of these intensities:

$$f(p) = \frac{1}{n} \sum_{i=1}^{n} f(p_i).$$

Observe that this is equivalent to using a smoothing filter on the image, and then performing point sampling. As we have seen, this process really does minimize aliasing.

Area Sampling

Area sampling consists in taking the intensity of the illumination function at each pixel as the average of the function intensities over the whole pixel. Thus, the value at a pixel P is taken as

$$f(p) = \frac{1}{\text{Area}(P)} \int_P f(x,y)\, dx\, dy.$$

This is based on the intuitive idea that the "most representative" value of a function over an area is the average.

It can be shown that supersampling converges to area sampling as the number of samples increases to infinity (this is the essential concept behind Monte Carlo methods). See the references in Section 8.9.

The process of taking the average reduces the high frequencies of the signal within the pixel, thus minimizing aliasing.

Analytic Sampling

Another technique, known as *analytic sampling*, is appropriate when the signal has large discontinuity boundaries inside the pixel, which is very common when the signal is the illumination function of a scene. Such discontinuities give rise to high frequencies in the image, which can bring serious aliasing problems. To perform analytic sampling, we partition the pixel P into regions $R_1, R_2, \ldots, R_n$ where the signal is continuous (Figure 8.10). Then we compute an intensity on each set R_i of the partition using area sampling:

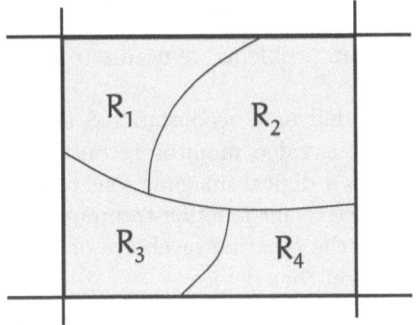

Fig. 8.10. Partition of a pixel into image continuity regions.

$$f_{R_i} = \frac{1}{\text{Area}(R_i)} \int_{R_i} f(x,y) \, dx \, dy.$$

The final intensity $f(P)$ at the pixel is defined as the weighted average of the intensity values for the individual regions:

$$f(P) = \frac{1}{\text{Area}(P)} \sum_{i=1}^{n} \text{Area}(R_i) f_{R_i}$$

Again we observe the same idea: this method is equivalent to applying a smoothing filter and then doing point sampling.

Other Sampling Techniques

Several other methods are used in computer graphics to minimize the effects of aliasing when sampling the illumination function of a scene. Most are variations of the preceding methods, which seek increased efficiency and flexibility in implementation and, in some cases, take advantage of perceptual factors involved in human vision.

A different sampling method for minimizing aliasing artifacts in computer graphics is nonuniform sampling. For this and more information on sampling techniques, see the references in Section 7.9.

8.4 Reconstruction Problems

In the sampling/reconstruction problem we have to deal with three distinct signals: the continuous signal f, the discrete signal f_d, and the reconstructed signal f_r. Ideally we would like $f_r = f$. When this happens we say the reconstruction is *exact*. Exact reconstruction is not always possible. The aim of reconstruction techniques is to minimize the error $|f - f_r|$.

Reconstruction techniques are very important in the manipulation of signals in the computer, for at least three reasons:

- In the solution of certain problems we need a continuous representation of the signal.
- Output devices must deliver a reconstructed signal, to be absorbed by the user's senses: thus a video monitor reconstructs an analog (optical-electronic) image when a digital image is sent to it.
- A good knowledge of the reconstruction techniques used by a given output device is important in the creation or choice of algorithms to process the signal to be displayed on that device.

In the next section we analyze a very simple reconstruction technique in order to have a basic understanding of the problems we may face when reconstructing a signal.

8.4.1 Reconstruction Using a Box Filter

We consider here reconstruction using the filter whose kernel is the pulse function *box filter*

$$p_a(t) = \begin{cases} 1 & \text{if } |t| \le a, \\ 0 & \text{if } |t| > a \end{cases}$$

introduced in Section 2.2.3. The reconstruction series in Equation (8.1) reduces to

$$f_r(t) = f_d * p_{\Delta t/2} = \sum_k f(t_k) p_{\Delta t/2}(t - t_k). \tag{8.4}$$

The reconstructed signal is a sum of pulse functions, modulated by the values of the samples; it is therefore constant on each interval $[t_k - \Delta t/2, \, t_k + \Delta t/2]$ (see Figure 8.11). We say that the original signal is being approximated by a piecewise constant signal. Unless the original signal is piecewise constant, there is no chance of achieving exact reconstruction with this method. The reconstruction process is illustrated in Figure 8.11 for one-dimensional signals. An example of box reconstruction with images was shown in Figure 8.5.

The signal reconstructed with the box filter has discontinuities, which introduce high frequencies in the resulting signal. Indeed, in the time domain the signal is reconstructed by the convolution product appearing in (8.4). In the frequency domain, the spectral model of the reconstructed signal is therefore given by the (usual) product of the spectral model of the discrete signal (f_k) with the transfer function $\hat{p}_{\Delta t/2}(t)$ of the box filter. We know that

$$\hat{p}_{\Delta t/2}(s) = \Delta t \, \text{sinc}(2\pi \, \Delta t \, s).$$

Figure 8.12 shows the spectral model of the sampled signal (dashed curves), superimposed on the graph of the transfer function $\hat{p}_{\Delta t/2}(s)$. As the figure shows, even if the signal is bandlimited and sampling is done according to the Nyquist limit, the spectral model of the reconstructed signal has high frequencies absent from the original signal, because the sinc filter has no cutoff frequency. Such high frequencies manifest themselves perceptually in the reconstructed signal in various ways, depending on the type of signal. When the reconstructed image is displayed on a video monitor, the high frequencies introduced generally impart a jagged appearance to the boundaries of the displayed image.

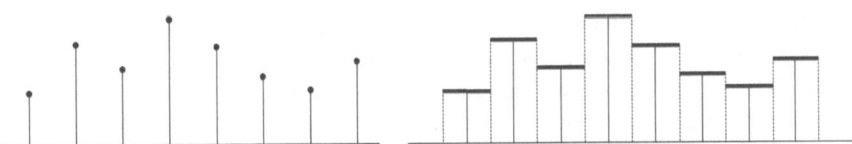

Fig. 8.11. Reconstruction with a box filter.

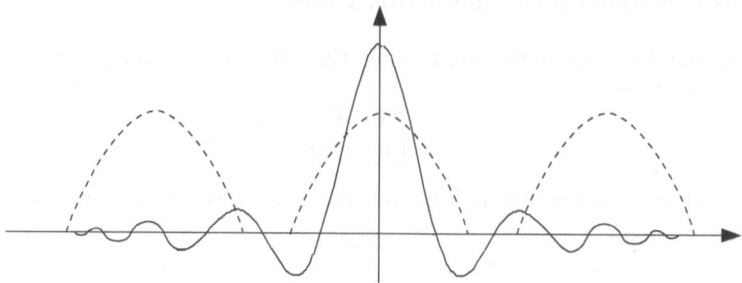

Fig. 8.12. High frequencies introduced by box reconstruction.

It is worth remarking that this jaggedness is usually attributed to aliasing. As just explained, however, the source of the problem is not always aliasing—in a sense, in fact, it is just the opposite. With aliasing we lose the high-frequency information, which gets confused with the low frequencies in the reconstructed signal. The reconstruction process described above, on the other hand, is characterized by the *introduction* of high frequencies into the reconstructed signal.

8.4.2 Analysis of Reconstruction Problems

In a general setting, if a signal f has a linear representation in a basis $\{e_k\}$ of the signal space, we can reconstruct it exactly from the samples $c_k = \langle f, e_k \rangle$ using the equation

$$f(t) = \sum_{k=-\infty}^{+\infty} c_k e_k(t).$$

This equation defines an interpolation among the values of the signal samples c_k. As explained in the beginning of the chapter, when we use uniform point sampling, $c_k = f(k\,\Delta t)$, the latter reconstruction series can be written as

$$f(t) = \sum_{k=-\infty}^{+\infty} f(k\,\Delta t)h(t - k\,\Delta t). \tag{8.5}$$

Thus, the elements of the reconstruction basis $\{e_k\}$ are translates of a fixed reconstruction filter h.

A vector-space interpretation of the reconstruction equation (8.5) is useful in getting a better understanding of the reconstruction process. The translates $h(t - k\,\Delta t)$ of h constitute a set of linear independent functions in the signal space $\mathcal{S}$. In general, however, this set generates only a subspace V of the space $\mathcal{S}$. When $V = \mathcal{S}$, exact reconstruction is always possible. When $V \neq \mathcal{S}$, only signals in V can be reconstructed exactly.

In practice, exact reconstruction may be impossible for any of several reasons:

1. The samples are not an exact linear representation for the signal. This is the case when point sampling is performed disregarding the Nyquist limit.
2. The reconstruction series in (8.5) may have infinitely many nonzero terms, and thus must be truncated when computing the reconstructed signal.
3. The elements of the reconstruction basis may not have compact support. This happens with the sinc function (the ideal reconstruction kernel). Thus, even when the signal is sampled within the hypothesis of the Shannon-Whittaker theorem (Nyquist limit), it may be impossible to compute an exact reconstruction from the sampled signal.

In the first case we have an aliasing error in the signal representation, as already discussed.

In the second case we have a *truncation error*, given by

$$\varepsilon = \left| f(t) - \sum_{k=-N}^{+N} c_k e_k(t) \right| = \left| \sum_{|k|>N} c_k e_k(t) \right|. \tag{8.6}$$

We now analyze the third case, where the reconstruction basis must be clipped to a compact interval in time domain. We will illustrate this with

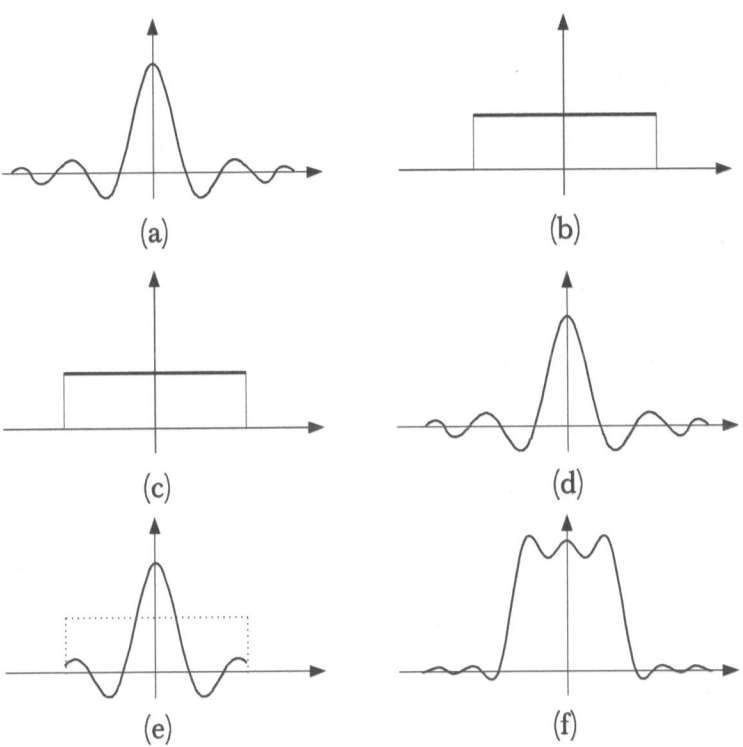

Fig. 8.13. Transfer function of the clipped sinc filter.

Fig. 8.14. Ringing due to sinc filter clipping.

the sinc filter, which gives exact reconstruction according to the Shannon-Whittaker theorem (see Chapter 2). Figure 8.13(a) plots the filter in the time domain, and Figure 8.13(b) shows its transfer function.

Clipping the filter is equivalent to multiplication in the time domain by some box filter. Parts (c) and (d) of Figure 8.13 show the box filter and its transfer function (sinc).

Parts (e) and (f) of Figure 8.13 show, respectively, the clipped reconstruction filter and its transfer function. The transfer function is obtained by convolving the spectra in parts (b) and (d).

Two points should be stressed: the clipping process introduces high frequencies in the clipped filter; these high frequencies give rise to ripple patterns because of the shape of the sinc filter.

If the clipping is performed with very good control, it may cause the accentuation of high frequencies, producing a sharpening of the image details. If not, the high-frequency modulation causes a perceptual artifact called *ringing* in the reconstructed image. This is illustrated in Figure 8.14.

8.5 Some Classical Reconstruction Filters

The box filter used in the reconstruction process of Section 8.4.1 is the simplest of a family of filters called *polynomial*, since their kernels are defined piecewise by polynomial expressions. Other filters in this family are obtained from the box filter by successive convolutions:

box * box = linear;
box * linear = quadratic spline;
box * quadratic spline = cubic spline;

and so on. Because of the correspondence between convolution in the space domain and multiplication in the frequency domain, the transfer function of these filters is a power of that of the box filter, sinc:

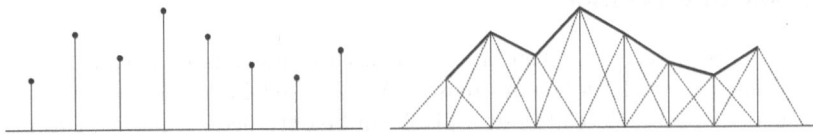

Fig. 8.15. Reconstruction with linear filter.

transfer function	filter
$\text{sinc}(s)$	box
$\text{sinc}^2(s)$	linear
$\text{sinc}^3(s)$	quadratic
$\text{sinc}^4(s)$	cubic
$\vdots$	$\vdots$

A *linear filter* interpolates linearly between samples, so that the reconstructed signal is continuous but has discontinuities in the first derivative. Figure 8.15 shows this method of reconstruction.

Polynomial filters of higher degree have a faster decay rate than those of lower degree. Therefore, the high frequencies introduced in the reconstruction process decrease as the degree of the filter increases.

Signals reconstructed by means of higher-order polynomial filters have a higher order of differentiability, as shown in Figure 8.16. These filters minimize the introduction of high frequencies, but they may eliminate some high frequencies present in the original image. We could say that sharpening is traded for blurring as the degree increases.

High-degree filters have better decay properties and the oscillations of their transfer function are milder, so images processed with these filters are less prone to ringing. We can say that in using higher-degree filters we are trading sharpness for blurring. This is desirable if we cannot control the degree of sharpness to avoid ringing. However, there are other reconstruction filters that allow better control of the trade-off of between sharpness and blurring. See Section 8.9 for references.

Fig. 8.16. Reconstruction with higher-order polynomial filter.

Bilinear Interpolation

The process of reconstructing an image with a linear filter is called *bilinear interpolation*, for the following reason. When used in image reconstruction, the output of a Bartlett filter at a pixel can be obtained by making two linear interpolations, one on the row and one on the column that contain the pixel. For example, if we apply the Bartlett filter with mask

$$\frac{1}{4} \cdot \begin{array}{|c|c|c|} \hline 1 & 2 & 1 \\ \hline 2 & 1 & 2 \\ \hline 1 & 2 & 1 \\ \hline \end{array}$$

to the image

$f(i,j)$	a	$f(i, j+1)$
b	c	d
$f(i+1, j)$	e	$f(i+1, j+1)$

,

where $a, b, c, d, e = 0$, we get the following replacements:

$$a \rightarrow \tfrac{1}{2}(f(i,j) + f(i, j+1));$$
$$b \rightarrow \tfrac{1}{2}(f(i,j) + f(i+1, j));$$
$$c \rightarrow \tfrac{1}{4}(f(i,j) + f(i, j+1) + f(i+1, j) + f(i+1, j+1));$$
$$d \rightarrow \tfrac{1}{2}(f(i, j+1) + f(i+1, j+1));$$
$$e \rightarrow \tfrac{1}{2}(f(i+1, j) + f(i+1, j+1)).$$

Note that polynomial filters seek to perform the reconstruction in a form that approximates the action of the ideal filter given by the sinc function corresponding to the Shannon basis. By the central limit theorem, the process of successive convolutions discussed above converges to the gaussian

$$g(t) = \frac{1}{\sqrt{2\pi\sigma^2}} e^{-x^2/(2\sigma^2)},$$

whose transfer function is also a gaussian: $\hat{g}(s) = e^{-2\sigma^2\pi^2 s^2}$.

In Figure 8.17 we show the kernels of all the reconstruction filters just discussed, together with their transfer functions. We also include the ideal reconstruction filter.

One can extend the filters discussed above to arbitrary dimensions by taking tensor powers of the kernel:

$$p(x_1, x_2, \ldots, x_n) = p(x_1)p(x_2)\cdots p(x_n).$$

We have already used two-dimensional versions of these filters, as examples of lowpass filters in Chapter 6.

The polynomial reconstruction kernels approach the gaussian kernel as the degree increases. The gaussian filter is far from being an ideal reconstruction

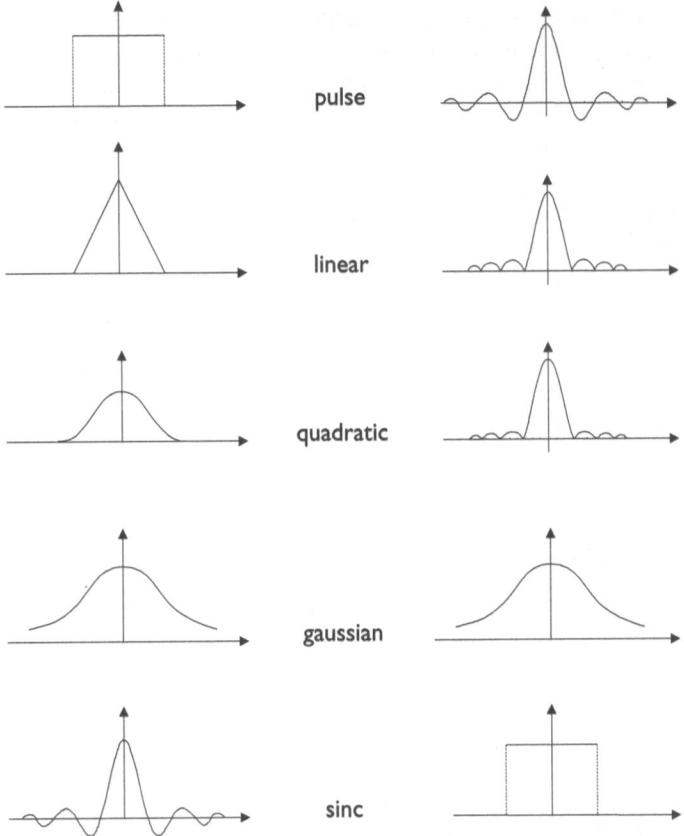

Fig. 8.17. Classical reconstruction filters (left) and their transfer functions (right).

filter, however. In fact, interpolation with Lagrange polynomials, well-known from linear algebra, is a method that approximates exact reconstruction with the Shannon basis as the number of samples tends to infinity.

The literature on reconstruction filters is abundant. See Section 8.9 for references.

8.6 A Study of Reconstruction Problems

The ideal reconstruction filter does not produce good results in practice: the reconstruction series is infinite, and interpolation filter in the space domain (sinc kernel) does not have compact support (it must be clipped to a bounded

interval before being used to interpolate the samples of the image to be reconstructed). Therefore, in trying to achieve a balance between computational
efficiency and the quality of the reconstructed image, we use simpler reconstruction filters, the most common of which are the box filter, the Bartlett
filter, cubic interpolation filters, and approximations to the gaussian filter.

Because the filters used in practice do not satisfy the two defining properties of an ideal reconstruction filter—they don't completely eliminate the
high frequencies introduced by sampling, and don't have unit gain at the frequencies of the original signal—the reconstructed image may present several
reconstruction artifacts.

In this section we will study several cases of reconstruction problems:

- introduction of high frequencies;
- loss of high frequencies;
- base-band modulation;
- hybrid cases, combining the first two above with the third;
- moiré patterns;
- anisotropic effects;
- frequency ripples.

We will illustrate the discussion of some of these problems with the signal
$f(t) = \text{sinc}^2(2\pi \cdot 125t)$, whose graph is shown in Figure 8.18(a). We chose this
signal because it is bandlimited; its Fourier transform is the sawtooth function
whose graph is shown in Figure 8.18(b). These two graphs, like all the others
in this section, have $\omega = 2\pi s$ along the x-axis, where s is the frequency of the
signal.

The highest frequency in the signal f is 250. Thus, if we sample uniformly
at intervals of 10^{-3}, we are within the Nyquist limit imposed by the Shannon–
Whittaker theorem. The discrete signal obtained is shown in Figure 8.18(c),
and its Fourier transform in Figure 8.18(d).

To reconstruct the original signal exactly, we would use an ideal reconstruction filter with cutoff frequency between 250 and 750. Figure 8.19(a)
shows the Fourier transform of the sampled signal together with the graph of
an ideal filter with cutoff frequency 500. Multiplying the two together and taking the inverse transform, we get the reconstructed signal in Figure 8.19(b).
There is no perceptible difference between this signal and the original one
shown in Figure 8.18(a).

We now use the signal f above to study each of the reconstruction problems
just listed.

Introduction of High Frequencies

Spurious high frequencies are present in the reconstructed image when the reconstruction filter's cutoff frequency is not low enough to exclude the frequencies introduced by the sampling process, and these frequencies "leak" into the

(a)

(b)

(c)

(d)

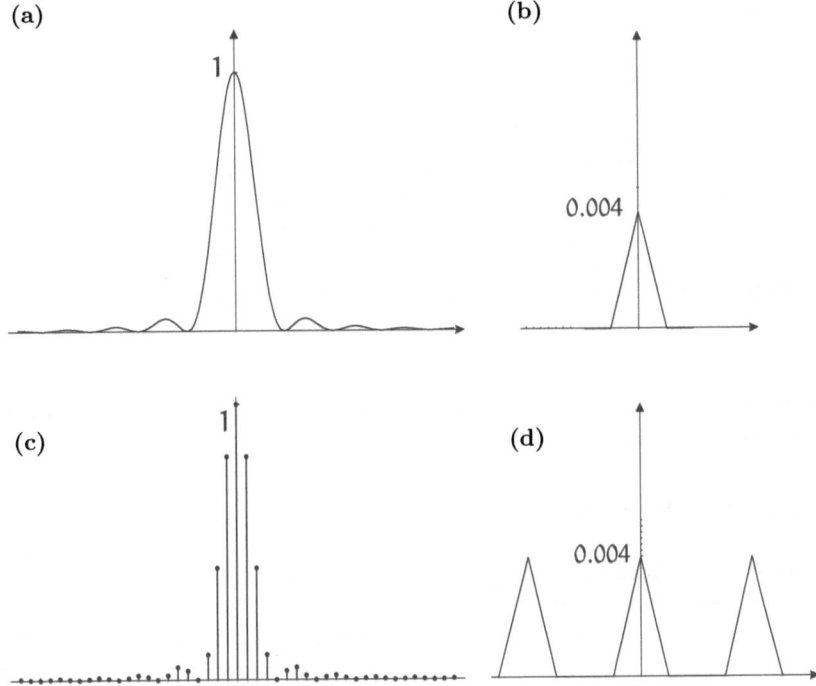

Fig. 8.18. A bandlimited signal (a), its Fourier transform (b), the sampled signal (c), and its Fourier transform (d).

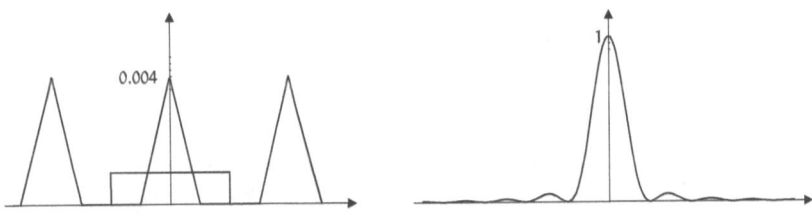

Fig. 8.19. Ideal reconstruction with a sinc filter.

reconstructed signal. To illustrate this problem, we take an ideal reconstruction filter whose support in the frequency domain is the interval $[-850, 850]$: see Figure 8.20(a). Filtering the signal and taking the inverse Fourier transform, we obtain the reconstructed signal, shown in Figure 8.20(b). Observe how the high frequencies introduced in the reconstruction process distort the original signal shown in Figure 8.18(a). In the reconstructed image these

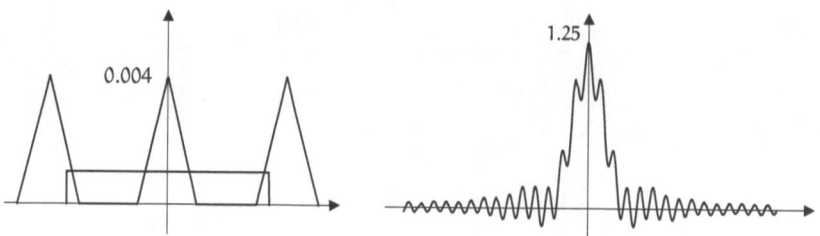

Fig. 8.20. Too high a cutoff frequency leads to the introduction of spurious high frequencies.

high frequencies will be perceived as ringing, a reconstruction artifact already shown in Figure 8.14.

Loss of High Frequencies

The problem opposite the one just discussed occurs when the reconstruction filter has too low a cutoff frequency and causes the loss of high frequencies present in the original signal. To illustrate this, we take a sinc filter whose Fourier transform has support $[-150, 150]$, as in Figure 8.21(a). Filtering the signal and taking the inverse Fourier transform, we obtain the reconstructed signal shown in Figure 8.21(b). Observe the significant loss of high frequencies in comparison with the original signal.

Perceptually, this problem causes blurring (loss of sharpness) in the image. The final result is equivalent to what one would get by reconstructing the image exactly and then applying a lowpass filter.

Base-Band Modulation

Base-band modulation occurs when the reconstruction filter has the right cutoff frequency but does not have unit gain throughout the frequency range of the original signal. The reconstruction filter we will use to illustrate this

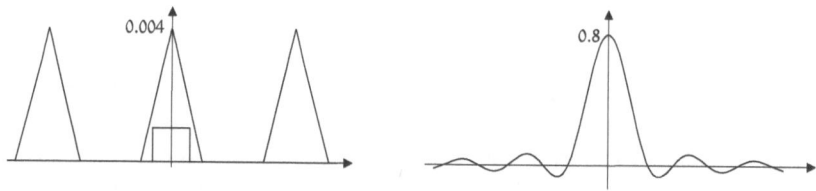

Fig. 8.21. Too low a cutoff frequency leads to a loss of high frequencies.

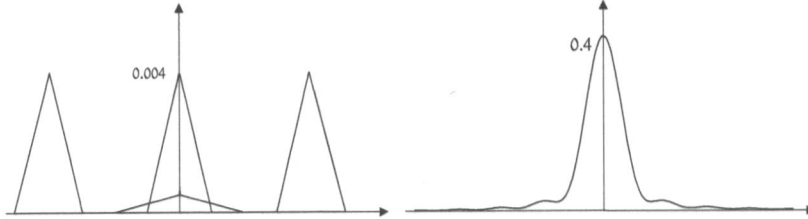

Fig. 8.22. Variable gain leads to frequency modulation.

problem is the sinc^2 filter, whose transfer function is the sawtooth graph shown in Figure 8.18. As shown in Figure 8.22(a), this filter neither introduces new frequencies nor removes frequencies present in the original signal, but it does modulate the original frequencies in the process of reconstruction. Perceptually, this modulation distorts the signal, as we see in the reconstructed signal shown in Figure 8.22(b).

Loss of High Frequencies Plus Modulation

The problem of frequency modulation is commonly found simultaneously with one of two earlier ones: loss of high frequencies or introduction of spurious high frequencies. We look first at the case when the reconstruction filter has a low cutoff and variable gain below the cutoff, as shown in Figure 8.23(a). The reconstructed signal is shown in Figure 8.23(b). One can clearly see the distortion of the signal and the elimination of high frequencies (loss of sharpness) by comparing with the original signal. Perceptually, the reconstructed image will be blurred and the objects on the image will be distorted.

Introduction of High Frequencies Plus Modulation

Here the reconstruction filter has too high a cutoff frequency and also has variable gain over the base band of the original signal. The result is distortion

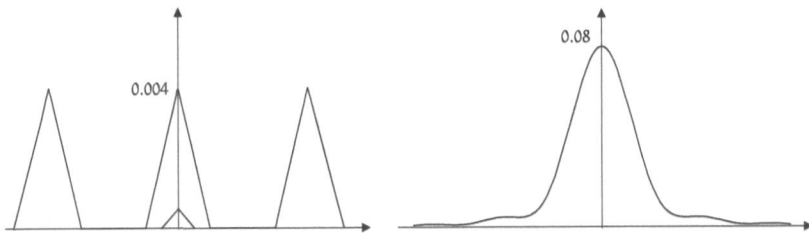

Fig. 8.23. Loss of high frequencies plus modulation.

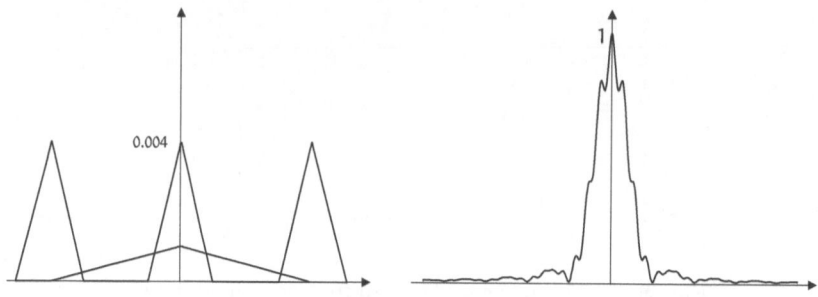

Fig. 8.24. Introduction of high frequencies plus modulation.

of the reconstructed image, together with the introduction of spurious high frequencies, as we see in Figure 8.24. Perceptually, the reconstructed image may present a severe "ringing" effect.

Moiré Patterns

A *moiré pattern* occurs when we use a reconstruction filter with high cutoff frequency and the original image has periodic patterns. It is due to interference between the periodic patterns of the original image and those introduced in the process of reconstruction.

Suppose the image has a periodic pattern with spectrum as in Figure 8.25(a) (a cosine spectrum). The frequency of the image pattern is ξ_0. If we sample the image at a frequency rate f_s slightly greater than $2\xi_0$, we obtain for the sampled signal the spectrum shown in Figure 8.25(b), where gray arrows represent the translated copies of the original spectrum. Note that the frequency ξ_0 of the original signal is very close to the frequency $f_s - \xi_0$ introduced into the sampled signal.

Now consider a reconstruction filter whose cutoff frequency is greater than $f_s - \xi_0$, as shown by the dashed curve in Figure 8.25(b). This filter will introduce the high-frequency component $f_s - \xi_0$ in the reconstructed signal. The signal reconstructed with this filter will have two periodic patterns, corre-

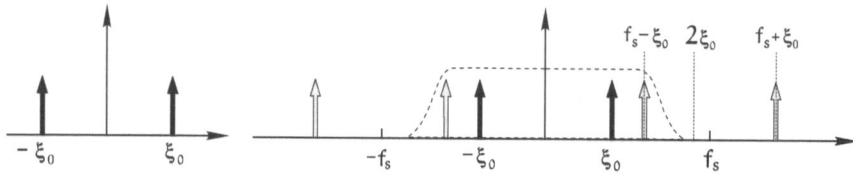

Fig. 8.25. Moiré patterns in the frequency domain.

sponding to the frequencies ξ_0 and $f_s - \xi_0$. Because of the proximity of these frequencies, these patterns get superimposed, creating a periodic pattern that did not exist in the original image (like the beat heard when two very similar tones sound together). This gives rise to a wavy pattern superimposed on the reconstructed image (*moiré*, or *moire*, is a French word applied to fabrics, like watered silk, that display a similar pattern).

Figure 8.9 illustrated this effect. The original image, on the left, has a set of concentric rings of increasing radii and decreasing thicknesses. The image function is $f(x,y) = \cos^2(x^2 + y^2)$. In part (a) the image was sampled at a good frequency rate, and the reconstructed image shows what should be expected from the function f. In part (b) the image was sampled using a very low sampling frequency. Notice that new frequencies are introduced in the reconstruction process, which causes periodic patterns (circles) to appear in the final image. These patterns interfere with the original image, creating small wave patterns.

Anisotropic Effects

When reconstructing images, besides controlling the introduction of high frequencies with the reconstruction filter, we must also control the filter shape—the geometry of the filter support and the filter values. We have seen that ringing artifacts may be caused by the ripples of the sinc filter on high frequency values. The geometry of the filter support may also introduce anisotropic artifacts in the reconstructed image. This is illustrated in Figure 8.26, where the image on the right was obtained by resampling (i.e. sampling and reconstructing) the image on the left. In the reconstruction step we used an elliptically shaped approximation of the gaussian filter.

(a)   **(b)**

Fig. 8.26. Reconstruction with anisotropic effects.

Anisotropic reconstruction artifacts are one of the causes of staircase effects when images are displayed on low-resolution monitors. The reconstruction function of these devices uses "square shaped" filters which produce anisotropic effects on lines that are neither horizontal nor vertical. Anisotropic reconstruction filters are very useful in creating special effects in the reconstructed image.

Frequency Ripples

To understand the problem of frequency ripples, consider the constant signal $f(t) = 1$ shown in Figure 8.27(a). Its spectrum is the Dirac delta function shown in Figure 8.27(b). Part (c) of the figure shows the sampled signal, and part (d) shows its spectrum: a train of Dirac delta functions (the gray arrows in (d) represent the translated copies of the original spectrum). Suppose that in the reconstruction process the impulse at frequency $w_0 = 1/\Delta t$ is not eliminated.

In this case the spectrum of the reconstructed signal f_r is

$$\hat{f}_r(\omega) = \delta(\omega + w_0) + \delta + \delta(\omega - w_0)$$
$$= \delta + \{\delta(\omega - w_0) + \delta(\omega - w_0)\}.$$

The part of the spectrum inside brackets is a cosine spectrum; therefore the reconstructed signal in the time domain is

$$f_r(t) = f(t) + 2\cos(2\pi w_0 t)$$
$$= 1 + 2\cos(2\pi w_0 t).$$

That is, a cosine wave pattern was introduced in the reconstructed signal, causing a ripple effect to appear.

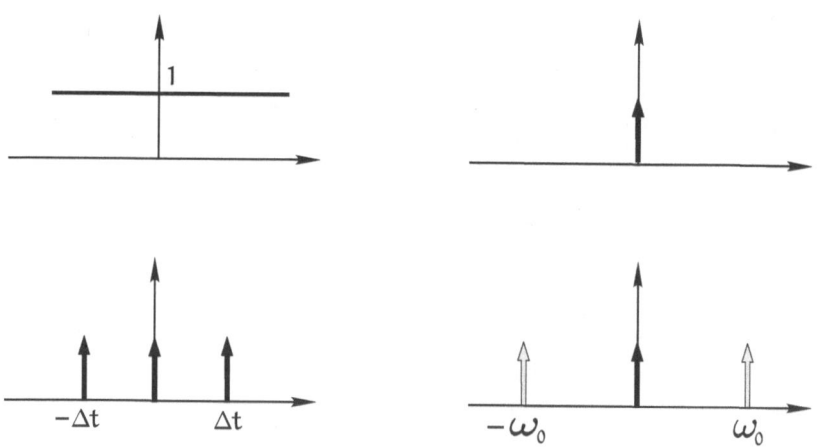

Fig. 8.27. Constant signal and its spectrum.

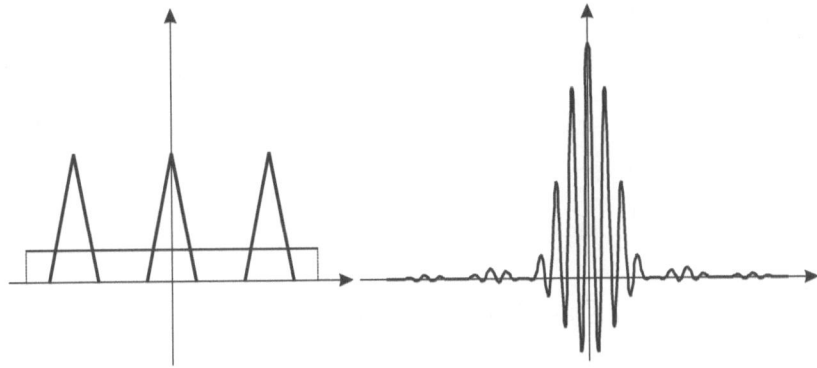

Fig. 8.28. Reconstruction with replicated DC component.

The cause of the ripple effect in the above example is easy: the spectrum of the signal contains only the component of frequency 0 (called the DC component of the signal); The ripples appear because this component is replicated during the sampling process, and its replicas are not filtered during the reconstruction process.

The same ripple effect occurs for arbitrary signals when the DC component of the signal, replicated during the sampling process, is not filtered in the reconstruction. This is illustrated in Figure 8.28, where one replicated DC component is introduced by the reconstruction filter (image on the left). The ripples on the reconstructed image (on the right) are quite perceptible. The correct reconstructed signal for this example is the squared sinc function, shown in Figure 8.18(a).

8.7 Reconstructing After Aliasing

The examples in the previous section show that even a signal sampled according to the Nyquist limit may be improperly reconstructed. In this section we will study a case of reconstructing a signal that has been sampled with aliasing.

Figure 8.29(a) shows a subsampling of the signal $\text{sinc}^2(2\pi\,125t)$ used in the previous section's examples. Figure 8.29(b) shows the Fourier transform of the subsampled signal. The presence of aliasing is quite marked, so problems in reconstructing the original signal are unavoidable.

Indeed, the high frequencies introduced by the sampling process interfere with the low frequencies of the original signal. During the reconstruction, if we try to avoid these high frequencies by eliminating them with the reconstruction filter, we lose high frequencies from the original signal. See Figure 8.30.

On the other hand, if we use a reconstruction filter whose transfer function has bigger support, so as to preserve all of the frequencies from the original

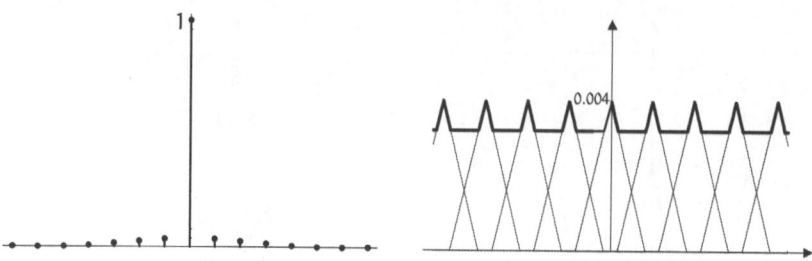

Fig. 8.29. Sampling with aliasing.

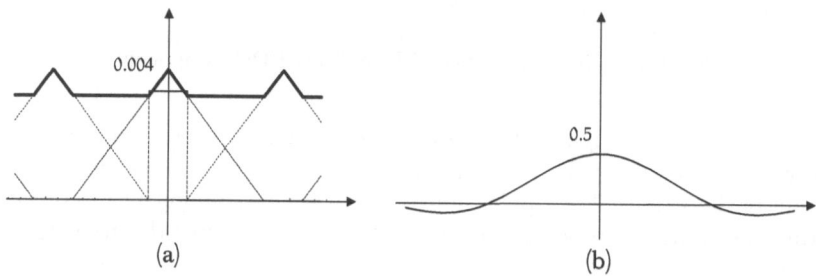

Fig. 8.30. Aliasing and loss of high frequencies.

signal, we do not avoid the high frequencies introduced in the sampling process. These frequencies distort the low frequencies in the reconstructed signal. This is illustrated in Figure 8.31. We can say that the high frequencies that appear in the reconstructed signal in Figure 8.31(b) are caused by aliasing.

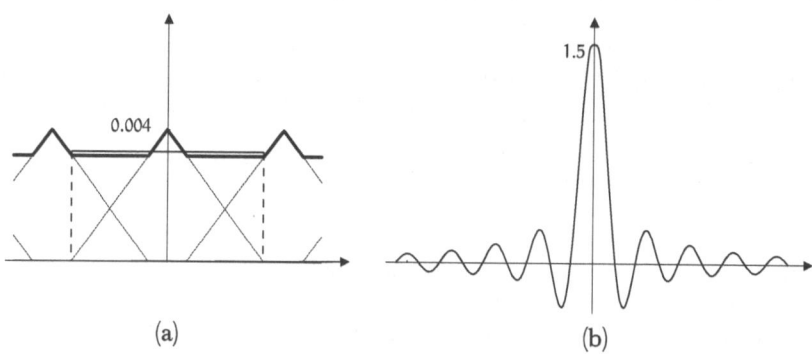

Fig. 8.31. Aliasing and reconstruction with high frequencies.

8.8 A Case Study

In practice, how can one avoid or minimize reconstruction problems? Certainly one obvious answer is immediate: choose a good reconstruction filter. But it is also true that reconstruction problems can be minimized by increasing the sampling rate. Indeed, by increasing the sampling rate, we increase the separation between the translates of the spectrum of the original signal, and this makes it easier for the reconstruction filter to eliminate high frequencies at reconstruction time. (Increasing the sampling rate also helps minimize aliasing. The same solution to remedy two different problems may be the cause of the confusion between aliasing and reconstruction artifacts in the literature.)

It is useful to look at a final example using images instead of one-dimensional signals. Consider the four images shown in Figure 8.32. The image in (a) was sampled and then reconstructed using an appropriate sampling rate. There are no visible reconstruction problems. In (b) the same image was undersampled and then reconstructed using a box filter. Note the distortion in window shades (caused by aliasing) and the introduction of high frequencies (caused by the reconstruction technique), as evidenced by the jagged appearance of details such as the flowers. Image (c) was reconstructed from the undersampled imaged using a Bartlett filter (bilinear interpolation), and

Fig. 8.32. Aliasing and reconstruction.

image (d) was obtained from the same image using a polynomial reconstruction filter of order 3, or bicubic filter.

Note that the jagged edges of the objects in (b) decrease in image (c) and disappear in (d). These artifacts are due to the introduction of high frequencies by the box filter used in the reconstruction; the other two filters do not introduce high frequencies. Instead, they lead to a *loss* of high frequencies in the image, which translates into a loss of sharpness (image blurring).

Note also that the distortion of the window shades does not disappear when we switch to higher-order filters, in (c) and (d), since it is caused by aliasing originating from undersampling. Aliasing artifacts are not eliminated by using good reconstruction filters. They must be eliminated, or minimized, *before* sampling the signal, either by using a lowpass filter or by increasing the sampling rate.

When an image is sampled without aliasing, or in such a way as to minimize aliasing, the quality of the reconstruction depends primarily on the process of reconstruction. When the image is sampled with aliasing, reconstruction problems are unavoidable. The aliased high frequencies introduced at sampling time cannot be filtered out at reconstruction time.

Confusion between aliasing and reconstruction artifacts arises because, in spite of being different, these two phenomena present some subtle similarities. Therefore, it is important to stress the relationship between the artifacts resulting from aliasing and those artifacts originating from poor reconstruction:

- When the image is sampled with aliasing, the reconstructed image will necessarily be distorted (see the window shades).
- When aliasing occurs, high frequencies of the sampled signal are superimposed on the frequencies of the original signal, but these high frequencies are lost in the reconstruction process.
- One of the problems of reconstruction is the introduction of spurious high frequencies in the reconstructed image.
- An increase in the sampling rate minimizes the possibility of aliasing and also minimizes some problems of reconstruction.

In general, when we look at an image, it is hard to tell apart problems due to aliasing from those due to poor reconstruction. In any case this is usually not important. Rather, the point of our study is to allow a better understanding of the relationship between aliasing and reconstruction, so better-quality images can be produced.

It is also worth observing that, for a given reconstruction filter, the quality of the reconstructed image improves when we look at it from far away. Looking at an image from a distance (or, equivalently, with half-shut eyes) amounts to applying a lowpass filter to the image, which palliates the problems arising from spurious high frequencies introduced at reconstruction time (say because the right cutoff frequency could not be determined). Aliasing problems cannot be corrected by dampening high frequencies in the reconstructed image.

8.9 Comments and References

Aliasing problems in signal processing have attracted attention for a long time (Mertz and Grey 1934). Also, early attention was given to the problem in computer graphics (Crow 1977). An exposition of image sampling from the viewpoint of measure theory can be found in (Fiume 1989). In particular, that book contains a proof that supersampling converges to area sampling as the number of samples increases.

It is important to stress that aliasing is a phenomenon closely associated with point sampling. When we use a linear representation arising from a complete orthonormal basis of the signal space, aliasing does not occur.

Nevertheless, for a long time the computer graphics literature did not clearly differentiate between aliasing and reconstruction artifacts in image reconstruction. To our knowledge, the first work to address the difference was (Mitchell and Netravali 1988). In this paper the authors make a clear distinction between sampling artifacts (which they call pre-aliasing) from artifacts introduced by the reconstruction process (which they call post-aliasing). The paper extensively discusses the problem of reconstruction and introduces a parameterized cubic reconstruction filter whose parameters may be tuned in order to minimize some commonly found reconstruction problems.

Another early work that addressed the problem of image sampling and reconstruction was Heckbert's master's thesis (Heckbert 1989), which discusses sampling and reconstruction techniques in the context of image warping. This work and the previously mentioned one did not immediately receive widespread dissemination in the area of computer graphics.

On the other hand, the recent book (Glassner 1995) devotes substantial space to the discussion of reconstruction filters: Chapters 8, 9, and 10 cover the subject of sampling and reconstruction. They contain a rather complete and detailed survey of reconstruction filters and include nonuniform sampling. This reference should be consulted by anyone seriously interested in further pursuing the subject of image reconstruction.

The problem of image sampling and reconstruction is a particular case of the larger problem of sampling and reconstructing an arbitrary graphical object. This more general problem is covered in (Gomes et al. 1996), and, with more details, in (SIGGRAPH 1996).

We mentioned in this chapter that Lagrange interpolation approximates ideal interpolation with the sinc filter. For more information about this the reader should consult (Jain 1989).

Understanding moiré patterns is crucial in the electronic publishing industry. A detailed study of moiré patterns both on the frequency and time domains can be found in (Amidor 1991) and (Amidor et al. 1994).

The sampling and reconstruction of three-dimensional signals is very important because it is directly connected with volumetric visualization and time-varying images. The theory developed in this chapter extends naturally to this case. For time-varying images—in particular, video—we need detailed knowledge of a video signal spectrum (e.g., the NTSC spectrum). See (Dubois 1985) for a well-written treatment.

The image with the ringing effect that appears in Figure 8.14 is taken from (Mitchell and Netravali 1988).

The image that appears in Figure 8.32 is a detail from "Shuttered windows", by Don Cochran, from the Kodak PhotoCD, Photo Sampler.

The authors wish to acknowledge the help of Siome Goldenstein, a student at the Instituto de Matemática Pura e Aplicada (IMPA), in computing several of the examples in this chapter.

References

[Amidor 1991]Amidor, J. P. (1991). *The moiré phenomenon in color separation.* In *Raster Imaging and Digital Typography II, Proceedings of the 2nd Intl. Conf. Raster Imaging and Digital Typography,* Vol. 6, 96–119.

[Amidor et al. 1994]Amidor, J. P., Hersch, R., and Ostromoukhov, V. (1994). *Spectral analysis and minimization of moiré patterns in color separation.* J. Electronic Imaging, 3(3):295–317.

[Crow 1977]Crow, F. (1977). The aliasing problem in computer generated shaded images. *Comm. of the ACM,* 20(11):799–805.

[Dubois 1985]Dubois, E. (1985). The sampling and reconstruction of timevarying imagery with application in video systems. *Proceedings of the IEEE,* 73(4).

[Fiume 1989]Fiume, E. L. (1989). *The Mathematical Structure of Raster Graphics.* Academic Press, New York.

[Glassner 1995]Glassner, A. (1995). *Principles of Digital Image Synthesis,* vol. 2. Morgan Kaufmann Publishers, Inc. San Francisco.

[Gomes et al. 1996]Gomes, J., Costa, B., Darsa, L., and Velho, L. (1996). Graphical objects. *The Visual Computer* 12(6):269.

[Heckbet 1989]Heckbert, P. S. (1989). *Fundamentals of Texture Mapping and Image Warping.* Master's thesis, Dept. of Electrical Engineering and Computer Science, University of California, Berkeley.

[Jain 1989]Jain, A. K. (1989). *Fundamentals of Digital Image Processing.* Prentice-Hall, Englewood Cliffs, NJ.

[Mertz and Grey 1934]Mertz, P. and Grey, F. (1934). A theory of scanning and its relation to the characteristics of the transmitted signal in telephotography and television. *Bell System Tech. J.* 13:464–515.

[Mitchell and Netravali 1988]Mitchel, D. P. and Netravali, A. N. (1988). Reconstruction filters in computer graphics. *Computer Graphics* 22(4):221–228.

[SIGGRAPH 1996]SIGGRAPH (1996). Warping and morphing of graphical objects. *'96 Course Notes ACM/SIGGRAPH.* Also available on CD-ROM.

[Wolberg 1990]Wolberg, G. (1990). *Digital Image Warping.* IEEE Computer Society Press, Los Alamitos, CA.

9

Multiscale Analysis and Wavelets

In this chapter we will introduce the wavelet transform with the purpose of obtaining better representation of images using atomic decompositions in the space-frequency domain.

Most of the derivations will be done for one-dimensional functions that will be extended to images, i.e., two-dimensional functions, afterwards.

9.1 The Wavelet Transform

To analyze an image jointly in the spatial and frequency domains, we must define a transform which is independent of scale. This transform should not use a fixed scale, but the scale should vary.

The scale is defined by the width of the modulation function. Therefore we must use a modulation function which does not have a fixed width. Moreover the function must have good space localization. To achieve this we start from a function $\psi(t)$ as a candidate of a modulation function, and we obtain a family of functions from ψ by varying the scale: We fix $p \geq 0$ and for all $s \in \mathbb{R}$, $s \neq 0$, we define

$$\psi_s(u) = |s|^{-p}\psi(\frac{u}{s}) = \frac{1}{|s|^p}\psi(\frac{u}{s}). \tag{9.1}$$

If ψ has width T (given as the standard deviation) then the width of ψ_s is sT. The modulation of the function ψ by the factor $1/|s|^2$, increases its amplitude when the scale s decreases and vice-versa. In terms of frequencies, we can state: For small scales s, ψ_s has high frequencies, and as s increases the frequency of ψ_s decreases. This fact is illustrated in Figure 9.1.

We need to localize each function ψ_s in space. For this we define for each $t \in \mathbb{R}$ the function

$$\psi_{s,t}(u) = \psi_s(u - t) = |s|^{-p}\psi(\frac{u-t}{s}) = \frac{1}{|s|^p}\psi(\frac{u-t}{s}). \tag{9.2}$$

L. Velho et al., *Image Processing for Computer Graphics and Vision*,
Texts in Computer Science, DOI 10.1007/978-1-84800-193-0_9,
© Springer-Verlag London Limited 2009

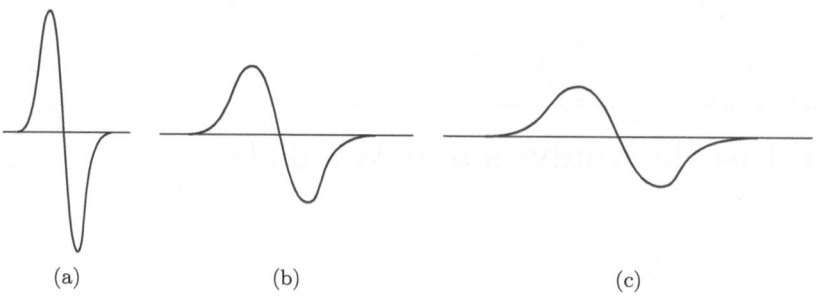

Fig. 9.1. Scales of a function: (a) $s < 1$; (b) $s = 1$; (c) $s > 1$.

Note that if $\psi \in \mathcal{L}^2(\mathbb{R})$, then $\psi_{s,t} \in \mathcal{L}^2(\mathbb{R})$, and

$$||\psi_{s,t}||^2 = |s|^{1-2p}||\psi||^2.$$

By taking $p = 1/2$, we have $||\psi_{s,t}|| = ||\psi||$.

Now we can define a transform on $\mathcal{L}^2(\mathbb{R})$ using functions from the family $\psi_{s,t}$ as modulating functions. More precisely, we have

$$\tilde{f}(s,t) = \int_{-\infty}^{+\infty} f(u)\psi_{s,t}(u)du = \langle \psi_{s,t}, f \rangle. \tag{9.3}$$

This transform is known by the name of the *wavelet transform*.

We can pose the following questions concerning the wavelet transform:

Question 9.1. Is the wavelet transform $\tilde{f}(s,t)$ invertible?

Question 9.2. What is the image of the wavelet transform $\tilde{f}(s,t)$?

9.1.1 Inverse of the Wavelet Transform

By definition we have

$$\tilde{f}(s,t) = \langle \psi_{s,t}, f \rangle = \langle \hat{\psi}_{s,t}, \hat{f} \rangle.$$

Moreover,

$$\hat{\psi}_{s,t}(\omega) = |s|^{1-p}e^{-2\pi i \omega t}\hat{\psi}(s\omega). \tag{9.4}$$

From this it follows that

$$\tilde{f}(s,t) = |s|^{1-p}\int_{infty}^{+\infty} e^{2\pi i \omega t}\hat{\psi}(s,\omega)\hat{f}(\omega)d\omega \tag{9.5}$$

$$= |s|^{1-p}F^{-1}\left(\hat{\psi}(s\omega)\hat{f}(\omega)\right), \tag{9.6}$$

Where F indicated the Fourier transform.

Applying the Fourier transform to both sides of the equation we obtain

$$\int_{-\infty}^{+\infty} e^{-2\pi i\omega t}\tilde{f}(s,t)dt = |s|^{1-p}\hat{\psi}(s\omega)\hat{f}(\omega). \tag{9.7}$$

From the knowledge of $\hat{f}$ we can obtain f using the inverse transform. But we can not simply divide the above equation by $\hat{\psi}$, because it might have zero values. Multiplying both sides of (9.7) by $\hat{\psi}(s\omega)$, and making some computations we obtain the result below:

Theorem 9.3. *If ψ satisfies the condition*

$$C = \int_{-\infty}^{+\infty} \frac{|\hat{\psi}(u)|^2}{|u|} < \infty, \tag{9.8}$$

then

$$f(u) = \frac{1}{C}\int\int_{\mathbb{R}^2} |s|^{2p-3}\tilde{f}_{s,t}(u)\psi_{s,t}(u)dsdt. \tag{9.9}$$

This theorem answers the first question posed at the end of the previous section: The wavelet transform is invertible and equation (9.9) reconstructs f from its wavelet transform.

We can read equation (9.9) of the inverse wavelet transform in two distinct ways:

1. The function f can be recovered from its wavelet transform;
2. The function f can be decomposed as a superposition of the space-frequency atoms $\psi_{s,t}(u)$.

We have seen that the second interpretation is of great importance because, it will lead us to obtain good representations by atomic decompositions of the function f.

9.1.2 Image of the Wavelet Transform

In this section we will discuss the second question we asked before about the image of the wavelet transform.

The wavelet transform takes a function $f \in \mathcal{L}^2(\mathbb{R})$ into a function $\tilde{f}(s,t)$ of two variables. A natural question consist in computing the image of the transform.

The interested reader should consult (Kaiser 1994), page 69. Besides characterizing the image space, this reference brings a proof that the wavelet transform defines an isometry over its image. We will not go into details of the computation here.

9.1.3 Filtering and the Wavelet Transform

Equation (9.3) that defines the wavelet transform can be written as a convolution product

$$\tilde{f}(s,t) = f * \psi_s(u),$$

where $\psi_s(u)$ is defined in (9.1). Thus the wavelet transform is a linear space-invariant filter. In this section we will discuss some properties of the wavelet filter.

The condition (9.8) that appears in the hypothesis of the Theorem 9.3 is called *admissibility condition*. A function ψ that satisfies this condition is called a *wavelet*.

From the admissibility condition it follows that

$$\lim_{u \to 0} \hat{\psi}(u) = 0.$$

If $\hat{\psi}(u)$ is continuous, then $\hat{\psi}(0) = 0$, that is,

$$\int_{-\infty}^{+\infty} \psi(u)\,du = 0.$$

Geometrically, this condition states that the graph of the function ψ must oscillate so as to cancel positive and negative areas in order to have integral zero. Therefore the graph of ψ has the form of a wave. In fact since ψ should have good space localization properties it has a form of a "small wave" (see Figure 9.2). That is why ψ is named by *wavelet*.

Another important conclusion can be drawn from the above computations. Since $\hat{\psi}(u) \in \mathcal{L}^2(\mathbb{R})$, then

$$\lim_{u \to 0} \hat{\psi}(u) = 0.$$

Along with the fact that $\hat{\psi}(0) = 0$, we conclude that the graph of the Fourier transform $\hat{\psi}$ is as depicted in Figure 9.3(a).

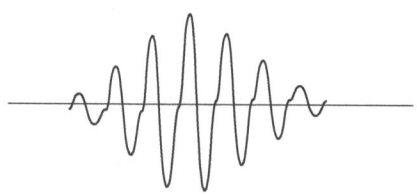

Fig. 9.2. Graph of a wavelet.

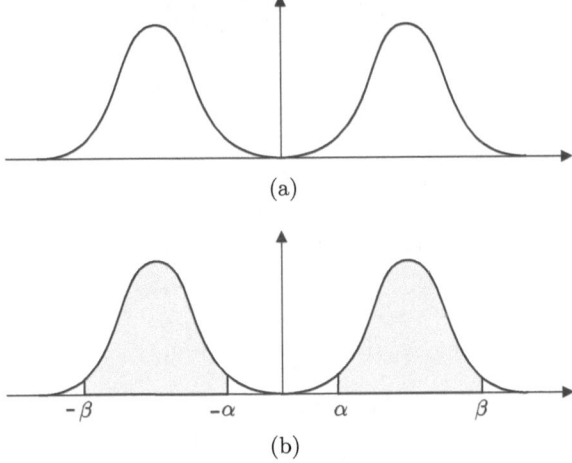

Fig. 9.3. Fourier transform of a wavelet.

If $\hat{\psi}$ has a fast decay when $u \to 0$ and $u \to \infty$, then $\hat{\psi}(u)$ is small outside of a small frequency band $\alpha \leq |u| \leq \beta$ (see Figure 9.3(b)). It follows from equation (9.4) that $\hat{\psi}_{s,t} \approx 0$ outside of the frequency band

$$\frac{\alpha}{|s|} \leq |u| \leq \frac{\beta}{|s|}.$$

Moreover, from equation (9.6) the wavelet transform $\tilde{f}$ does not contain information about f outside of this spectrum interval. In sum, the computations above show that "*the wavelet transform is a linear, space invariant band-pass filter*".

The next two examples are taken from (Kaiser 1994).

Example 1 (Blur Derivative) *Consider a function ϕ of class C^{∞}, satisfying the conditions*

$$\phi \geq 0;$$
$$\int_{\mathbb{R}} \phi(u)du = 1;$$
$$\int_{\mathbb{R}} u\phi(u)du = 0;$$
$$\int_{\mathbb{R}} u^2\phi(u)du = 1.$$

That is, ϕ is a probability distribution with average 0 and variance (width) 1. Suppose that

$$\lim_{u \to +\infty} \frac{\partial^{n-1}\phi}{\partial u^{n-1}}(u) = 0.$$

Defining

$$\psi^n(u) = (-1)^n \frac{\partial^n \phi}{\partial u^n}(u),$$

we have,

$$\int_{\mathbb{R}} \psi^n(u)du = 0.$$

That is, ψ^n satisfies the admissibility condition (9.8). Therefore we can define a wavelet transform

$$\tilde{f}(s,t) = \int_{\mathbb{R}} \psi^n_{s,t}(u)f(u)du, \qquad (9.10)$$

where

$$\psi^n_{s,t}(u) = \frac{1}{s}\psi^n\left(\frac{u-t}{s}\right).$$

(We are taking $p = 1$ in equation (9.2) that defines $\psi_{s,t}(u)$). In an analogous way we define,

$$\phi_{s,t}(u) = \frac{1}{s}\phi\left(\frac{u-t}{s}\right).$$

From the definition of ψ^n we have that

$$\psi^{-n}_{s,t}(u) = (-1)^n s^{-n} \frac{\partial^n \phi_{s,t}}{\partial u^n}(u) = s^{-n} \frac{\partial^n \phi_{s,t}}{\partial t^n}(u). \qquad (9.11)$$

From equations (9.10) and (9.11) it follows that

$$\tilde{f}(s,t) = s^{-n} \frac{\partial^n}{\partial t^n} \int_{\mathbb{R}} \phi_{s,t}(u)f(u)du. \qquad (9.12)$$

The above integral is a convolution product of the function f with the function $\phi_{s,t}$, therefore it represents a low-pass filtering linear space-invariant filtering operation of the function f, which is dependent of the scale s. We will denote this integral by $\overline{f}(s,t)$. Therefore we have

$$\tilde{f}(s,t) = s^{-n} \frac{\partial^n \overline{f}(s,t)}{\partial t^n}, \qquad (9.13)$$

that is, the wavelet transform of f is the n-th space derivative of the average of the function f on scale s. This derivative is known in the literature by the name of blur derivative.

We know that the n-nth derivative of f measures the details of f in the scale of its definition. Therefore, equation (9.13) shows that the wavelet transform $\tilde{f}(s,t)$ gives the detail of order n of the function f, in the scale s. Keeping this wavelet interpretation in mind is useful, even when the wavelet does not come from a probability distribution.

Example 9.4 (The Sombrero Wavelet). We will use a particular case of the previous example to define a wavelet transform. Consider the Gaussian distribution

$$\phi(u) = \frac{1}{\sqrt{2\pi}} e^{-u^2/2},$$

with average 0 and variance 1. The graph of this function is depicted in the image on the left of Figure 9.4. Using the notation of the previous example, we have

$$\psi^1(u) = -\phi'(u) = \frac{1}{\sqrt{2\pi}} u e^{-u^2/2},$$

and

$$\psi^2(u) = \phi''(u) = \frac{1}{\sqrt{2\pi}} (u^2 - 1) e^{-u^2/2}.$$

The function $-\psi^2$ is known as the "sombrero" function, because of the shape of its graph, shown in the right of Figure 9.4.

From the previous example it follows that we can use the sombrero function to define a wavelet transform. We will use this wavelet to illustrate the flexibility of the wavelet transform in analyzing frequencies of a signal. For this, consider the signal whose graph is shown in Figure 9.5.

This signal has high frequencies localized in the neighborhood of $t = 50$, and $t = 150$. From time $t = 280$, the signal has a chirp behavior: a continuum of increasing frequencies. In this region the signal is defined by the function

$$f(t) = \cos(t^3).$$

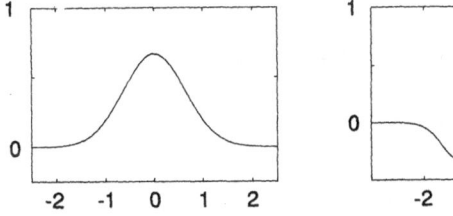

Fig. 9.4. Graph of the sombrero wavelet.

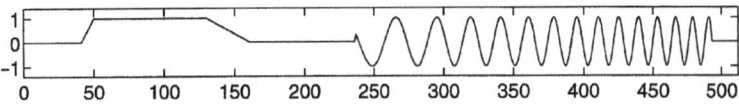

Fig. 9.5. Signal f to be analyzed(Kaiser 1994).

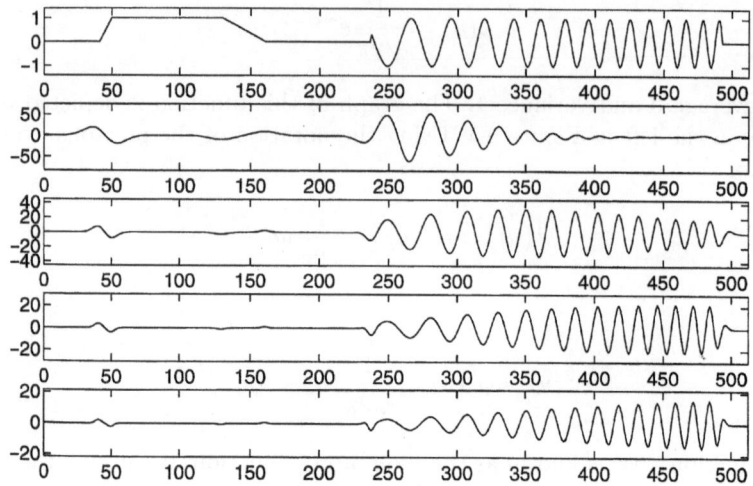

Fig. 9.6. The wavelet transform(Kaiser 1994).

Figure 9.6 shows the graph of the signal and the graph of the wavelet transform for 5 distinct values of the scale s (the scale decreases from top to bottom).

Note that the frequencies associated to the sudden change of the signal at time $t = 50$ and time $t = 150$ are detected by the wavelet transform. Moreover, as the scale s decreases the high frequencies of the chirp signal $\cos(t^3)$ are also detected.

9.2 The Discrete Wavelet Transform

The wavelet transform is defined on the space-scale domain. A natural question is:

Question 9.5. How to discretize the space-scale domain in such a way to obtain a discrete wavelet transform?

We know that the scaling operation acts in a multiplicative way, that is, composing two consecutive scalings is attained by multiplying each of the scale factors. Therefore the discretization of the scaling factor is simple: We fix an initial scale $s_0 > 1$, and we consider the discrete scales

$$s_m = s_0^m, \quad m \in \mathbb{Z}.$$

Positive values of m produce scales larger than 1, and negative values of m produce scales less than 1.

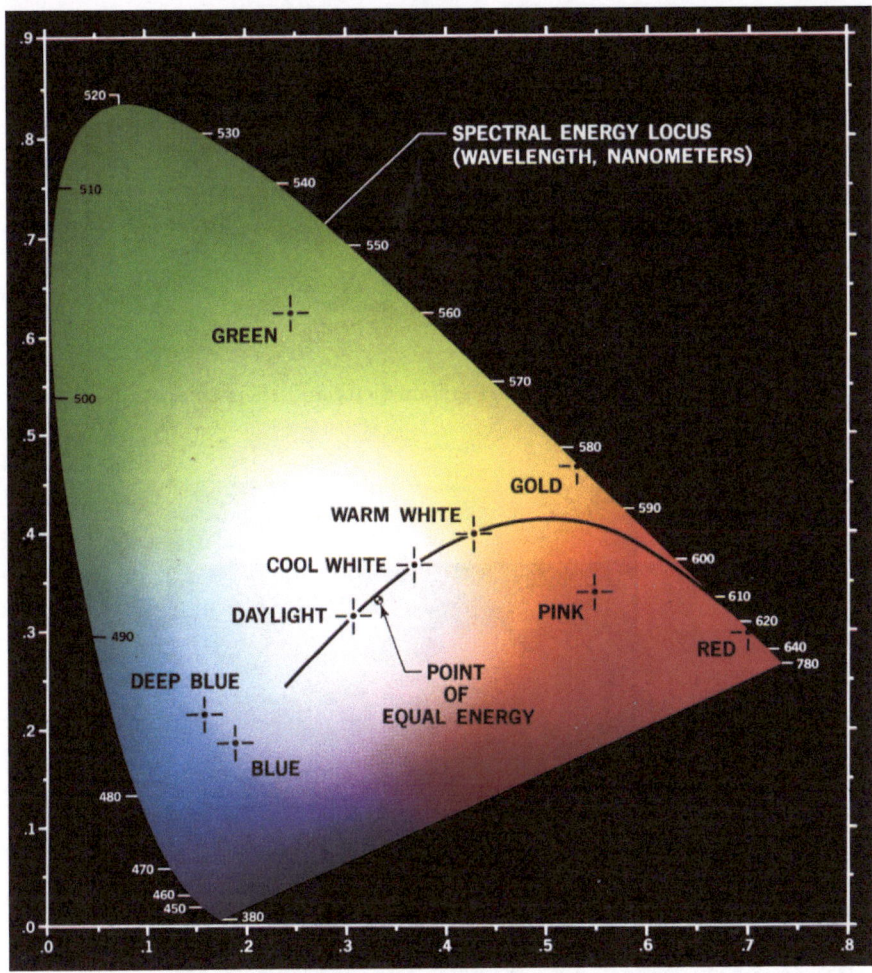

Plate 1. Chromaticity diagram of the CIE-XYZ system (*see page 114*).

Plate 2. Colors in the additive RGB system (left) and their complements (right) (*see page 116*).

Plate 3. Digital color image quantized at 24 bits (*see page 302*).

Plate 4. Uniform quantization at eight bits (left) and four bits (right) (*see page 302*).

Plate 5. Populosity algorithm: result of quantization at eight bits (left) and four bits (right) (*see page 303*).

Plate 6. Median cut algorithm: result of quantization at eight bits (left) and four bits (right) (*see page 307*).

Plate 7. Quantization from 24 to 8 bits, without dithering (left) and with Floyd–Steinberg dithering (right) (*see page 332*).

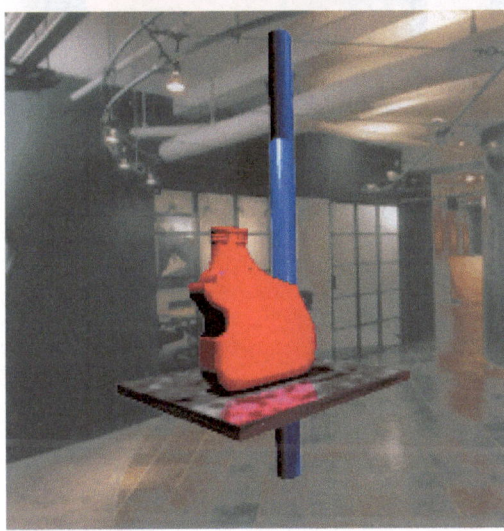

Plate 8. Superimposing a synthetic image on a photograph (*see page 366*).

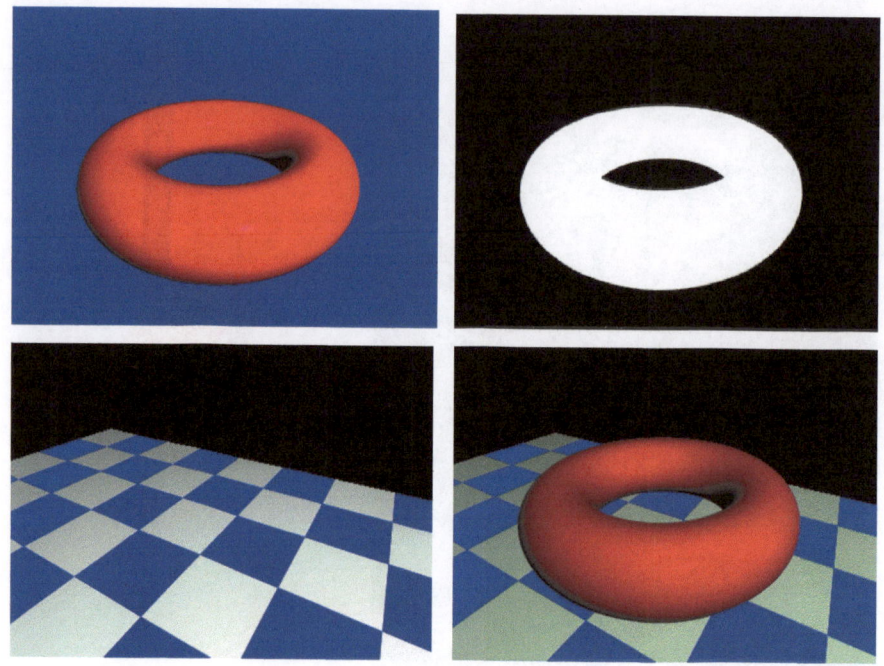

Plate 9. Blue screen and alpha channel (*see page 374*).

Plate 10. Morphing animation sequence (*see page 411*).

Plate 11. Reproduction of a color image (*see page 430*).

Plate 12. CMYK components of the image of Figure 16.10 (*see page 431*).

Plate 13. Enlargement of an offset-printed image. (*see page 432*).

How to discretize the space? Initially we should observe that we must obtain a lattice in the space-scale domain in such a way that when we sample the wavelet transform $\tilde{f}(s,t)$ on this lattice, we are able to reconstruct the function f from the space-scale atoms $\tilde{f}_{m,n}$, with minimum redundancy. As the wavelet width changes with the scale, we must correlate the space with the scale discretization: As the scale increases the width of the wavelet also increases, therefore we can take samples further apart in the space domain. On the other hand, when the width of the wavelet decreases with a reduction of the scale, we must increase the frequency sampling.

To obtain the correct correlation between the scale and space discretization we observe that an important property of the wavelet transform is: *The wavelet transform is invariant by change of scales.* This statement means that if we make a change of scale in the function f and simultaneously change the scale of the underlying space by the same scaling factor, the wavelet transform does not change. More precisely, if we take

$$f_{s_0}(t) = s_0^{-1/2} f\left(\frac{t}{s_0}\right),$$

then

$$\tilde{f}_{s_0}(s_0 s, s_0 t) = \tilde{f}(s,t).$$

Invariance by changing of scale constitutes an essential property of the wavelet transform. It is important that this property be preserved when we discretize the wavelet, so as to be also valid for the discrete wavelet transform. In order to achieve this goal, when we pass from the scale $s_m = s_0^m$ to the scale $s_{m+1} = s_0^{m+1}$, we must also increment the space by the scaling factor s_0. In this way, we can choose a space t_0 and take the length of the sampling space intervals as $\Delta t = s_0^m t_0$. Therefore, for each scale s_0^m the space discretization lattice is

$$t_{m,n} = n s_0^m t_0, \quad n \in \mathbb{Z}.$$

Finally, the discretization lattice in the space-scale domain is defined by

$$\Delta_{s_0,t_0} = \{(s_0^m, n s_0^m t_0) \; ; \; m, n \in \mathbb{Z}\}.$$

Example 9.6 (Dyadic Lattice). We will give a very important example of a wavelet discretization using $s_0 = 2$ (dyadic lattice). We have

$$\Delta_{2,t_0} = \{(2^m, n 2^m t_0) \; ; \; m, n \in \mathbb{Z}\}.$$

The vertices of this lattice are shown in Figure 9.7(a). This lattice is called *hyperbolic lattice* because it is a uniform lattice in hyperbolic geometry (only the points are part of the lattice).

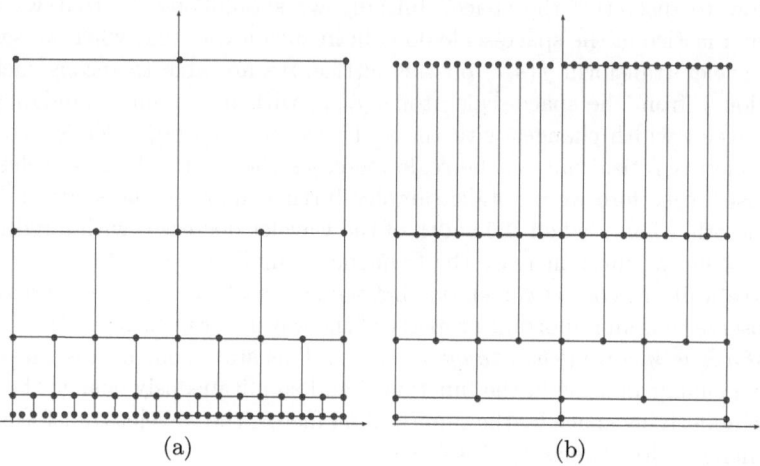

Fig. 9.7. (a) Space-scale lattice.(b) Space-frequency lattice.

To obtain a space-frequency lattice, we must observe that the frequency is the inverse of the scale. In this manner, for a given initial frequency ω_0 the lattice will be given by

$$\Delta_{2\omega_0, t_0} = \{(2^{-m}\omega_0, n2^{-m}t_0) \; ; \; m, n \in \mathbb{Z}\}.$$

The vertices of this lattice are shown in Figure 9.7(b).

9.2.1 Function Representation

From the point of view of atomic decomposition the space-frequency atoms define a tiling of the space-frequency domain in rectangles as shown in Figure 9.8.

The discretization of the wavelet transform $\tilde{f}(s,t) = \langle f, \psi_{s,t}(u) \rangle$ in the space-scale lattice is given by

$$\tilde{f}_{m,n} = \langle f, \psi_{m,n}(u) \rangle,$$

where

$$\psi_{m,n}(u) \quad = \psi_{s_0^m, nt_0 s_0^m}(u) \tag{9.14}$$

$$= s_0^{-m/2} \psi\left(\frac{u - nt_0 s_0^m}{s_0^m}\right) \tag{9.15}$$

$$= s_0^{-m/2} \psi(s_0^{-m}u - nt_0). \tag{9.16}$$

In this context we can pose again the two questions which motivated the process of defining a discrete wavelet transform:

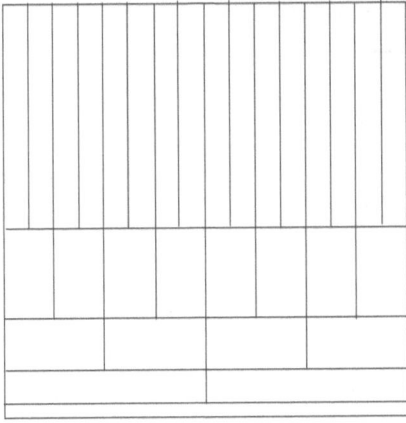

Fig. 9.8. Space-frequency decomposition using wavelets.

Question 9.7. Is the sequence $\langle f, \psi_{m,n} \rangle, m, n \in \mathbb{Z}$ an exact representation of the function f?

Question 9.8. Is it possible to reconstruct f from the family of wavelet space-frequency atoms $\psi_{m,n}$?

A positive answer to these two questions would give us atomic decompositions of the function f using a family $\psi_{m,n}$. of discrete wavelets.

There are several directions we could take to answer the two questions above. Based on the representation theory two natural questions in this direction are:

Question 9.9. Is it possible to define a lattice such that the corresponding family $\{\psi_{m,n}\}$ constitutes an orthonormal basis of $\mathcal{L}^2(\mathbb{R})$?

Question 9.10. Is it possible to define lattices for which the family $\{\psi_{m,n}\}$ is a frame?

If we have orthonormal basis of wavelets or a frame the answer to the two questions posed above are positive.

Chapter 3 of (Daubechies 1992) brings a comprehensive discussion of frames of wavelets. The explicit construction of some wavelet frames is given.

Example 9.11 (Haar Basis). Consider the function

$$\psi(x) = \{\, 1 \text{ if } x \in [0, 1/2) - 1\text{if } x \in [1/2, 1)0\text{if } x < 0 \text{ ou } x > 1.$$

The graph of f is shown in Figure 9.9. This function satisfies the admissibility condition (9.8).

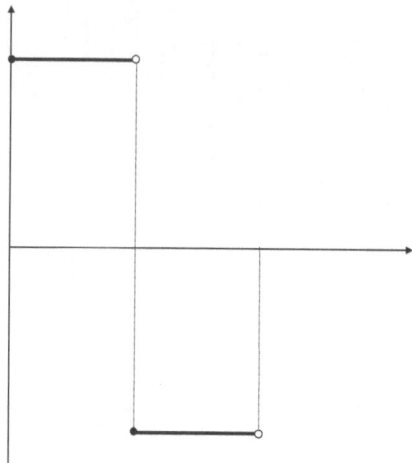

Fig. 9.9. Haar wavelet.

It is possible to show that the set $\psi_{m,n}$, where

$$\psi_{m,n}(u) = 2^{-m/2}\psi(2^{-m}u - n), \quad m, n \in \mathbb{Z},$$

constitutes an orthonormal basis of $\mathcal{L}^2(\mathbb{R})$. Therefore we have an orthonormal basis of wavelets. A direct, and long, proof of this fact is found in (Daubechies 1992), Section 1.3.3. The orthonormality of the set $\psi_{m,n}$ is easy to proof. The fact that the set generates the space $\mathcal{L}^2(\mathbb{R})$ is more complicated. This will follow as a consequence of the theory of multiresolution analysis that we will study next.

9.3 Multiresolution Representation

Our perception of the universe uses different scales: Each category of observations is done in a proper scale. This scale should be adequate to understand the different details we need. In a similar manner, when we need to represent an object, we try to use a scale where the important details can be captured in the representation.

A clear and well known example of the use of scales occurs on maps. Using a small scale we can observe only macroscopic details of the mapped regions. By changing the scale we can observe or represent more details of the object being represented on the map.

Multiresolution representation is a mathematical model adequate to formalize the representation by scale in the physical universe. As we will see, this problem is intrinsically related to the wavelets.

The idea of scale is intrinsically related with the problem of point sampling of a signal. We call *sampling frequency* the number of samples in the unit of time. The length of the sample interval is called the *sampling period*. When we sample a signal using a frequency 2^j, we are fixing a scale to represent the signal: Details (frequencies) of the signal that are outside of the scale magnitude of the samples will be lost in the sampling process. On the other hand, it is clear that all of the details of the signal captured in a certain scale will also be well represented when we sample using a higher scale, 2^k, $k > m$.

These facts are well translated mathematically by the sampling theorem of Shannon-Whittaker that relates the sampling frequency with the frequencies present on the signal.

9.3.1 Scale Spaces

How to create a mathematical model to formalize the problem of scaling representation in the physical universe? The relation between sampling and scaling discussed above shows us the way. For a given integer number j, we create a subspace $V_j \subset \mathcal{L}^2(\mathbb{R})$, constituted by the functions in $\mathcal{L}^2(\mathbb{R})$ whose details are well represented in the scale 2^j. This means that these functions are well represented when sampled using a sampling frequency of 2^j.

The next step consists in creating a representation operator that is able to represent any function $f \in \mathcal{L}^2(\mathbb{R})$ in the scale 2^j. A simple and effective technique consists in using a representation by orthogonal projection. A simple and effective way to compute this representation is to obtain an orthonormal basis of V_j. But at this point we will demand more than that to make things easier: We will suppose that there exists a function $\phi \in \mathcal{L}^2(\mathbb{R})$ such that the family of functions

$$\phi_{j,k}(u) = 2^{-j/2}\phi(2^{-j}u - k), \quad j, k \in \mathbb{Z}, \tag{9.17}$$

is an orthonormal basis of V_j.

Notice that we are using here a process similar to the one we used when we introduced the wavelet transform: We define different scales of ϕ producing the continuous family

$$\phi_s(u) = \frac{1}{|s|^{1/2}}\phi\left(\frac{u}{s}\right).$$

The width of ϕ and ϕ_s are related by

$$\text{width}(\phi) = s \, \text{width}(\phi_s).$$

Thus, as the scale increases or decreases, the width of ϕ_s does the same. Equation (9.17) is obtained by discretizing the parameter s, taking $s = 2^j$, $j \in \mathbb{Z}$. Also, we have demanded that the translated family

$$\phi_{j,k} = \phi_{2^j}(u - k) = 2^{-j/2}\phi(2^{-j}u - k)$$

is an orthonormal basis of V_j. Note that when j decreases, the width of $\phi_{j,k}$ also decreases, and the scale is refined. This means that more features of f are detected in its representation on the space V_j.

The representation of a function $f \in \mathcal{L}^2(\mathbb{R})$ by orthogonal projection in V_j is given by

$$\mathrm{Proj}_{V_j}(f) = \sum_k \langle f, \phi_{j,k} \rangle \phi_{j,k}.$$

We want the representation sequence $(\langle f, \phi_{j,k} \rangle)$ to contain samples of the function f in the scale 2^j. In order to attain this we know that the representation sequence $(\langle f, \phi_{j,k} \rangle)_{j,k \in \mathbb{Z}}$ is constituted by the samples of a filtered version of the signal f. More precisely,

$$\langle f, \phi_{j,k} \rangle = F(k),$$

where F is obtained from f by sampling with a filter of kernel $\phi_{j,k}$: $F = f * \phi_{j,k}$. In order that the elements of the representation sequence are close to the samples of f, the filter kernel $\phi_{j,k}$ must define a low-pass filter. This can be attained by demanding that $\hat{\phi}(0) = 1$, because $\hat{\phi}(\omega)$ approaches 0 when $\omega \to \pm \infty$. The graph of ϕ is depicted in Figure 9.10. With this choice of ϕ, representing a function at scale 2^j amounts to sample averages of f over neighborhoods of width 2^j.

The space V_j is called *space of scale 2^j*, or simply *scale space*.

It is very important that we are able to change from a representation in a certain scale to a representation on another scale. For this we must answer the question: How are the different scale spaces related?

Since the details of the signal which appear on scale 2^j certainly must appear when we represent the signal using a smaller scale 2^{j-1}, we must have

$$V_j \subset V_{j-1}. \tag{9.18}$$

Given a function $f \in \mathcal{L}^2(\mathbb{R})$, a natural requirement is

$$f \in V_j \quad \text{if, and only if,} \quad f(2u) \in V_{j-1}. \tag{9.19}$$

In fact, the scaling of the variable of f by 2 reduces the width of f by the factor of $1/2$ (see Figure 9.11). Therefore the details of f go to a finer scale.

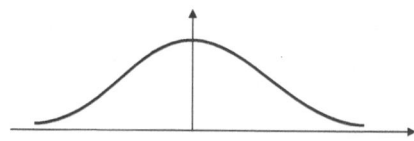

Fig. 9.10. Low-pass filter.

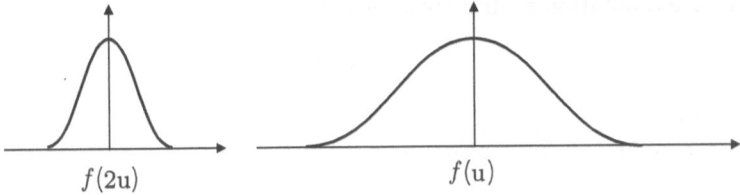

Fig. 9.11. Scaling of f by an scale factor of 2.

Applying successively the condition in (9.19), we obtain

$$f \in V_j \quad \text{if, and only if,} \quad f(2^j u) \in V_0.$$

That is, all spaces are scaled version of the space V_0. In particular, from the fact that $\phi_{j,k}$ in equation (9.17), is an orthonormal basis of V_j, we conclude that

$$\phi_{0,k}(u) = \phi(u - k)$$

is an orthonormal basis of the scale space V_0.

The space $\mathcal{L}^2(\mathbb{R})$, our universe of the space of functions, contains all of the possible scales. This is reflected in the relation

$$\overline{\bigcup_{j \in \mathbb{Z}} V_j} = \mathcal{L}^2(\mathbb{R}).$$

On the other hand, we have

$$\bigcap_{j \in \mathbb{Z}} V_j = \{0\}.$$

In effect, this expression says that the null function is the only function that can be well represented in every scale. In fact it should be observed that any constant function can be represented in any scale, nevertheless the only constant function that belongs to $\mathcal{L}^2(\mathbb{R})$ is the null function.

A Remark About Notation

It is important here to make a remark about the index notation we use for the scale spaces, because there is no uniformity on the literature. We use the notation of decreasing indices

$$\cdots V_1 \subset V_0 \subset V_{-1} \subset V_{-2} \cdots.$$

From the discussion above, this notation is coherent with the variation of the scale when we pass from one scale space to the other: As the indices decrease, the scale is refined, and the scale spaces get bigger.

If we use a notation with increasing indices

$$\cdots V_{-1} \subset V_0 \subset V_1 \cdots,$$

which also appears on the literature, than the base $\phi_{j,k}$ of the scale space V_j should be constituted by the functions

$$\phi_{j,k}(x) = 2^{j/2}\phi(2^j x - k).$$

This is rather confusing because it is not in accordance with the notation used when we discretized wavelets.

9.3.2 Multiresolution Representation

The scale spaces and their properties that we studied above define a multiresolution representation in $\mathcal{L}^2(\mathbb{R})$. We will resume them into a definition to facilitate future references:

Definition 1 (Multiresolution Representation) *We define a multiresolution representation in $\mathcal{L}^2(\mathbb{R})$ as a sequence of closed subspaces V_j, $j \in \mathbb{Z}$, of $\mathcal{L}^2(\mathbb{R})$, satisfying the following properties:*

(M1) $V_j \subset V_{j-1}$;
(M2) $f \in V_j$ if, and only if, $f(2u) \in V_{j-1}$.
(M3) $\bigcap_{j \in \mathbb{Z}} V_j = \{0\}$.
(M4) $\overline{\bigcup_{j \in \mathbb{Z}} V_j} = \mathcal{L}^2(\mathbb{R})$.
(M5) There exists a function $\phi \in V_0$ such that the set $\{\phi(u - k); k \in \mathbb{Z}\}$ is an orthonormal basis of V_0.

The function ϕ is called the scaling function *of the multiresolution representation. Each of the spaces V_j is called* scale spaces, *or, more precisely,* space of scale 2^j.

Example 9.12 (Haar Multiresolution Analysis). Consider the function

$$\phi(t) = \chi_{[0,1]} = \{\, 0 \text{ if } x < 0 \text{ ou } t \geq 1 \quad 1 \text{if } x \in [0,1)$$

It is easy to see that ϕ is a scale function of a multiresolution representation. In this case,

$$V_j = \{f \in \mathcal{L}^2(\mathbb{R}); f|[2^j k, 2^j (k+1)] = \text{constant}, k \in \mathbb{Z}\}.$$

That is, the projection of a function f on the scale space V_j is given by a function which is constant on the intervals $[2^j k, 2^j (k+1)]$. This is the Haar multiresolution representation.

We should notice that conditions (M1), ... (M5), that define a multiresolution representation are not independent. In fact it is possible to prove that condition (M3) follows from (M1), (M2) e (M5). Moreover, condition (M5) can be replaced by the weaker condition that the set $\{\phi(u - k)\}$ is a Riesz basis. Also, the reader might have noticed that we have not imposed that the scale function ϕ satisfies the condition $\hat{\phi}(0) = 1$ (as we know, this condition guarantees that ϕ is a low-pass filter). It can be proved that this low-pass filter condition follows from (M4). For a proof of all of these facts we suggest consulting (Hernandez e Weiss 1996) or (Daubechies 1992). We will return to this problem with more details in next chapter about construction of wavelets.

9.3.3 A Pause to Think

How to interpret geometrically the sequence of nested scale spaces in the definition of a multiresolution representation?

In general, visualizing subspaces of some space of functions is not an easy task. Nevertheless, in this case a very informative visualization of the nested sequence of scale space can be obtained in the frequency domain.

Indeed, the orthogonal projection of a function $f \in \mathcal{L}^2(\mathbb{R})$ in V_j is obtained using a filtering operation of f with the different kernels $\phi_{j,k}$, $k \in \mathbb{Z}$ which define low-pass filters. Indicating the cutting frequency of these filters by α_j (see Figure 9.12), we conclude that each space V_j is constituted by functions whose frequencies are contained in the interval $[-\alpha_j, \alpha_j]$, $\alpha_j > 0$.

When we go from the space V_j to the space V_{j-1} we change from the scale 2^j to a finer scale 2^{j-1}. Therefore the frequency band increases to an interval $[-\alpha_{j-1}, \alpha_{j-1}]$. The graph of the spectrum of $\phi_{j-1,k}$ is the dotted curve in Figure 9.12. The scale space V_{j-1} consists of the set of all the functions whose spectrum is contained in $[-\alpha_{j-1}, \alpha_{j+1}]$.

For each space V_j, with scale 2^j, we have the representation operator $R_j : L^2(\mathbb{R}) \to V_j$, given by the orthogonal projection over V_j

$$R_j(f) = \text{Proj}_{V_j}(f) = \sum_k \langle f, \phi_{j,k} \rangle \phi_{j,k}.$$

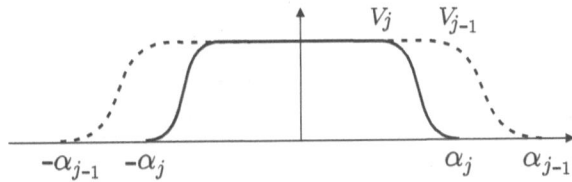

Fig. 9.12. Spectrum of the scaling function.

From condition (M4) of the definition of a multiresolution representation, we have

$$\lim_{j \to \infty} R_j(f) = f, \qquad (9.20)$$

that is, as the scale gets finer we get a better representation of the function f. This is illustrated in Figure 9.13 (from (Daubechies 1992)) we show a function f, and its representation on the spaces of scale V_0 e V_{-1} of the Haar multiresolution representation.

There is a different, and very important way, to interpret equation (9.20) Consider the graph representation of the space V_j on Figure 9.14. We see that the space V_{j-1} is obtained from the space V_j by adding all of the functions from $\mathcal{L}^2(\mathbb{R})$ with frequencies in the band $[\alpha_j, \alpha_{j-1}]$ of the spectrum. We indicate this "detail space" by W_j. It follows immediately that W_j is orthogonal to V_j. Therefore we have

$$V_{j-1} = V_j \oplus W_j.$$

The space W_j contains the details of the signal in the scale V_j. The above equation says that a function represented on a finer scale space V_{j-1} is obtained from the representation on a coarser scale space V_j, by adding details. These details can be obtained using a band-pass filtering, whose passband is exactly the interval $[\alpha_j, \alpha_{j-1}]$. We have seen that the wavelets constitute linear time-invariant band-pass filters. Therefore it seems natural that there might exist some relation between the detail spaces and the wavelets. We will discuss this relation "with details" in next section.

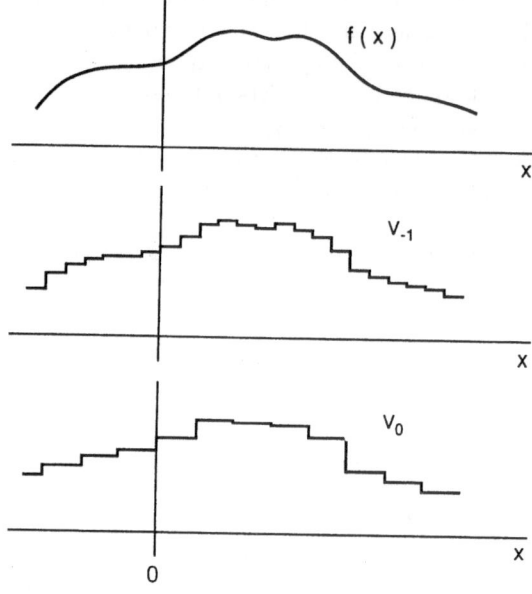

Fig. 9.13. Scale approximations of a function(Daubechies 1992).

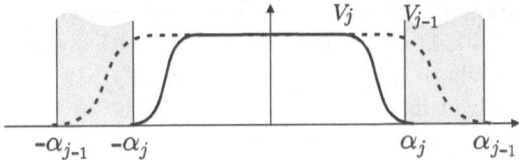

Fig. 9.14. Frequency band between V_j and V_{j-1}.

9.4 Multiresolution Representation and Wavelets

We have proved that given two consecutive scale spaces $V_j \subset V_{j-1}$, the orthogonal complement W_j of V_j in V_{j-1} could be obtained using a band-pass filter defined on $\mathcal{L}^2(\mathbb{R})$. In this section we will show that this complementary space is in fact generated by an orthonormal basis of wavelets.

For every $j \in \mathbb{Z}$, we define W_j as the orthogonal complement of V_j in V_{j-1}. We have

$$V_{j-1} = V_j \oplus W_j.$$

We remind that the best way to visualize the above equality is by observing the characterization of these spaces on the frequency domain (Figure 9.14).

It is immediate to verify that W_j is orthogonal to W_k, if $j \neq k$. Therefore by fixing $J_0 \in \mathbb{Z}$, for every $j < J_0$ we have (see Figure 9.15)

$$V_j = V_{J_0} \oplus \bigoplus_{k=0}^{J_0-j} W_{J_0-k}. \tag{9.21}$$

We should remark that because of the dyadic scales used in the discretization, the frequency bands do not have uniform length, they are represented in the figure using logarithmic scale.

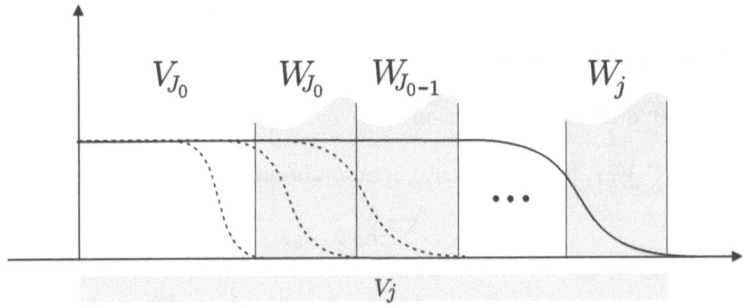

Fig. 9.15. Frequency bands between V_j and V_{J_o-j}.

In sum, equation (9.21) says that the signals whose spectrum is in the frequency band of V_j, is the sum of the signals with frequency band in V_{J_0} with those signals whose frequency band are in W_{J_0}, W_{J_0-1}, ..., W_j. All of the subspaces involved in this sum are orthogonals. If $J_0, k \to \infty$, if follows from conditions (M3) and (M4) that define a multiresolution representation that

$$\mathcal{L}^2(\mathbb{R}) = \bigoplus_{j \in \mathbb{Z}} W_j,$$

that is, we obtain a decomposition of $\mathcal{L}^2(\mathbb{R})$ as a sum of orthogonal subspaces.

We have seen that the projection of a function f in each subspace W_j could be obtained using a band-pass filter. In fact, this filtering process can be computed by projecting f on an orthogonal basis of wavelets. This fact is a consequence of the theorem below:

Theorem 9.13. *For each $j \in \mathbb{Z}$ there exists an orthonormal basis of wavelets $\{\psi_{j,k}, k \in \mathbb{Z}\}$ of the space W_j.*

We will sketch the proof of the theorem because it has a constructive nature which will provide us with a recipe to construct orthonormal basis of wavelets.

Basis of W_0.

Initially we observe that the spaces W_j inherit the scaling properties of the scale spaces V_j. In particular,

$$f(u) \in W_j \quad \text{if, and only if,} \quad f(2^j u) \in W_0. \qquad (9.22)$$

For this reason, it suffices to show that there exists a wavelet $\psi \in W_0$ such that the set $\{\psi(u - k)\}$ is an orthonormal basis of W_0. In fact, in this case, it follows from (9.22) that the set

$$\{\psi_{j,k}(u) = 2^{-j/2}\psi(2^{-j}u - k)\}$$

is an orthonormal basis of W_j.

Low-pass filter and scaling function.

Since $\phi \in V_0 \subset V_{-1}$, and also $\phi_{-1,k}$ is an orthonormal basis of V_{-1}, we have

$$\phi = \sum_k h_k \phi_{-1,k}, \qquad (9.23)$$

where

$$h_k = \langle \phi, \phi_{-1,k} \rangle, \quad \text{and} \quad \sum_{k \in \mathbb{Z}} ||h_k||^2 = 1.$$

Substituting $\phi_{-1,k}(u) = \sqrt{2}\phi(2u - k)$ in (9.23 we obtain

$$\phi(x) = \sqrt{2} \sum_k h_k \phi(2x - k).)$$ (9.24)

Applying the Fourier transform to both sides of this equation, we have

$$\hat{\phi}(\xi) = m_o(\tfrac{\xi}{2})\hat{\phi}(\tfrac{\xi}{2}),$$ (9.25)

where

$$m_0(\xi) = \frac{1}{\sqrt{2}} \sum_k h_k e^{-ik\xi}.$$

Note in equation (9.25) that $\hat{\phi}(\tfrac{\xi}{2})$ there exists a frequency band which has twice the size of the frequency band of $\phi(\xi)$. Therefore, it follows from (9.24) that the function m_0 is a low-pass filter. The function m_0 is called de *low-pass filter of the scaling function* ϕ. It is not difficult to see that m_0 is periodic with period 2π.

Characterization of W_0.

Now we need to characterize the space W_0. Given $f \in W_0$, since $V_{-1} = V_0 \oplus W_0$, we conclude that $f \in V_{-1}$ and f is orthogonal to V_0. Therefore

$$f = \sum_n f_n \phi_{-1,n},$$ (9.26)

where

$$f_n = \langle f, \phi_{-1,n} \rangle.$$

Computations similar to the ones we did to obtain the low-pass filter m_0 of the scaling function, give us the equation

$$\hat{f}(\xi) = m_f(\tfrac{\xi}{2})\hat{\phi}(\tfrac{\xi}{2}),$$ (9.27)

where

$$m_f(\xi) = \frac{1}{\sqrt{2}} \sum_n f_n e^{-in\xi}.$$

After some computations, we can rewrite the equation (9.27) in the form

$$\hat{f}(\xi) = e^{\frac{i\xi}{2}} \overline{m_0 \left(\frac{\xi}{2} + \pi \right)} \nu(\xi)\hat{\phi}(\tfrac{\xi}{2}),$$ (9.28)

where ν is a periodic function of period 2π.

Choosing the Wavelet.

Equation (9.28) characterizes the functions from W_0 using the Fourier transform, up to a periodic function ν. A natural choice is to define a wavelet $\psi \in W_0$ such that

$$\hat{\psi}(\xi) = e^{\frac{-i\xi}{2}} \overline{m_0\left(\frac{\xi}{2} + \pi\right)} \hat{\phi}\left(\frac{\xi}{2}\right). \qquad (9.29)$$

Taking this choice, from equation (9.28), it follows that

$$\hat{f}(\xi) = \left(\sum_k \nu_k e^{-ik\xi}\right) \hat{\psi}(\xi),$$

and applying the inverse Fourier transform, we have

$$f(x) = \sum_k \nu_k \psi(x - k).$$

We need to show that defining ψ by the equation (9.29), $\psi_{0,k}$ is indeed an orthonormal basis of W_0. We will not give this proof here.

Details of the above proof can be found on (Daubechies 1992) or (Hernandez e Weiss 1996).

9.4.1 A Pause... to See the Wavescape

If V_j is the scale space 2^j we have $V_{j-1} = V_j \oplus W_j$. We know that W_j has an orthonormal basis of wavelets $\{\psi_{j,k}, k \in \mathbb{Z}\}$, therefore if R_j is the representation operator on the scale space V_j, we have, for all $f \in \mathcal{L}^2(\mathbb{R})$,

$$R_{j-1}(f) = R_j(f) + \sum_{k \in \mathbb{Z}} \langle f, \psi_{j,k}\rangle \psi_{j,k}. \qquad (9.30)$$

The second term of the sum represents the orthogonal projection of the signal f on the space W_j and it will be denoted by $\mathrm{Proj}_{W_j}(f)$. The terms of this representation sequence are obtained using the discrete wavelet transform.

We know that the wavelet transform is a band-pass filtering operation, ant the scale spaces allow us to represent a function f in different resolutions. When we obtain a representation of f in a certain scale 2^j, we are loosing details of the signal with respect with its representation in the scale 2^{j-1}. The lost details are computed by the orthogonal projection on the space W_j, that is,

$$\mathrm{Proj}_{W_j}(f) = \sum_{k \in \mathbb{Z}} \langle f, \psi_{j,k}\rangle \psi_{j,k}, \qquad (9.31)$$

which is a representation of the signal f in the basis of wavelets of the space W_j.

It is useful to interpret the decomposition $V_{j-1} = V_j \oplus W_j$ in the language of filters. The representation of a signal f in the scale V_j,

$$R_j(f) = \sum_{k \in \mathbb{Z}} \langle f, \phi_{j,k} \rangle \phi_{j,k},$$

is equivalent to filter the signal using the low-pass filter defined by the scaling function ϕ. The representation of the details of f in the space W_j, equation (9.31) is obtained by filtering f with the band-pass filter defined by the wavelet transform associated with ψ.

From the relation $V_{j-1} = V_j \oplus W_j$, we are able to write

$$R_{j-1}(f) = R_j(f) + \text{Proj}_{W_j}(f)$$
$$R_{j-2}(f) = R_{j-1}(f) + \text{Proj}_{W_{j-1}}(f)$$

$$\vdots$$

Note that each line of the equation above represents a low-pass filtering and a band-pass filtering of the signal. Iterating this equation for $R_{j-2}, \ldots, R_{j-J_0}$, summing up both sides and performing the proper cancellations, we obtain

$$R_{j-J_0}(f) = R_j(f) + \text{Proj}_{W_{j-1}}(f) + \cdots \text{Proj}_{W_{j-J_0}}(f). \tag{9.32}$$

The projection $R_j(f)$ represents a version of low resolution (blurred version) of the signal, obtained using successive low-pass filtering with the filters $\phi_j, \phi_{j-1}, \ldots, \phi_{J_0-j}$. The terms $\text{Proj}_{W_{j-1}}(f), \ldots, \text{Proj}_{W_{j-J_0}}(f)$ represent the details of the signal lost in each low-pass filtering. These details are obtained by filtering the signal using the wavelets $\psi_j, \psi_{j-1}, \ldots, \psi_{J_0-j}$. Equation (9.32) states that the original signal f can be reconstructed exactly from the low resolution signal, summing up the lost details.

9.4.2 Two Scale Relation

We now revisit some equations we obtained in the computations of this chapter in order to distinguish them for future references.

Consider an scaling function ϕ associated to some multiresolution representation. Then $\phi \in V_0 \subset V_{-1}$ and $\phi_{-1,n}$ is an orthonormal basis of V_{-1}. Therefore

$$\phi = \sum_{k \in \mathbb{Z}} h_k \phi_{-1,k}, \tag{9.33}$$

with $h_k = \langle \phi, \phi_{-1,k} \rangle$. This equation can be written in the form

$$\phi(x) = \sqrt{2} \sum_{k \in \mathbb{Z}} h_k \phi(2x - k). \tag{9.34}$$

Similarly, given a wavelet ψ associated with a multiresolution representation $\psi \in V_0$, since $V_{-1} = V_0 \oplus W_0$, we have that $\psi \in V_{-1}$, and ψ is orthogonal to V_0, therefore

$$\psi = \sum_{k \in \mathbb{Z}} g_k \phi_{-1,k}, \tag{9.35}$$

or,

$$\psi(x) = \sqrt{2} \sum_{k \in \mathbb{Z}} g_k \phi(2x - k). \tag{9.36}$$

Equations (9.33) and (9.35) (or equivalently (9.34) and (9.36)), are called *two-scale relations,* or *scaling relations* of the scaling function and the wavelet respectively. In several important cases, the sum that defines the two-scale relations is finite:

$$\phi(x) = \sqrt{2} \sum_{k=0}^{N} g_k \phi(2x - k).$$

It is not difficult to see that when this is the case, the support of the scaling function ϕ is contained in the interval $[0, N]$.

Also, note that if ϕ is a solution of the equation defined by the two-scale relation, then $\lambda \phi$, $\lambda \in \mathbb{R}$, is also a solution. In this way, to have uniqueness of the solution we must impose some kind of normalization (e.g. $\psi(0) = 1$).

A priori, it is possible to construct a multiresolution representation and the associated wavelet starting from an adequate choice of the function ϕ. This choice can be done using the two scale relation (9.34). In a similar manner, the two scale equation (9.36) can be used to obtain the associated wavelet.

9.5 The Fast Wavelet Transform

The Fast Wavelet Transform (FWT) algorithm, is the basic tool for computation with wavelets. The forward transform converts a signal representation from the time (spatial) domain to its representation in the wavelet basis. Conversely, the inverse transform reconstructs the signal from its wavelet representation back to the time (spatial) domain. These two operations need to be performed for analysis and synthesis of every signal that is processed in wavelet applications. For this reason, it is crucial that the Wavelet Transform can be implemented very efficiently.

We will see that recursion constitutes the fundamental principle behind wavelet calculations. We will start with a revision of the multiresolution analysis to show how it naturally leads to recursion. Based on these concepts, we will derive the elementary recursive structures which form the building blocks of the fast wavelet transform. Finally, we will present the algorithms for the decomposition and reconstruction of discrete one-dimensional signals using compactly supported orthogonal wavelets.

9.5.1 Multiresolution Representation and Recursion

The efficient computation of the wavelet transform exploits the properties of a multiresolution analysis. In the previous chapters, we have seen that a multiresolution analysis is formed by a ladder of nested subspaces

$$\cdots V_1 \subset V_0 \subset V_{-1} \cdots$$

where all V_j are scaled versions of the central subspace V_0.

From the above structure, we can define a collection of "difference" subspaces W_j, as the orthogonal complement of each V_j in V_{j-1}. That is,

$$V_j = V_{j+1} \oplus W_{j+1}$$

As a consequence, we have a wavelet decomposition of $\mathcal{L}^2(\mathbb{R})$ into mutually orthogonal subspaces W_j

$$\mathcal{L}^2(\mathbb{R}) = \bigoplus_{j \in \mathbb{Z}} W_j$$

Therefore, any square integrable function $f \in \mathcal{L}^2(\mathbb{R})$ can be decomposed as the sum of its projection on the wavelet subspaces

$$f = \sum_{j \in \mathbb{Z}} \mathrm{Proj}_{W_j}(f)$$

where $\mathrm{Proj}_{W_j}(f)$ is the projection of f onto W_j.

From $V_j = V_{j+1} \oplus W_{j+1}$, it follows that any function $f_j \in V_j$ can be expressed as

$$f_j = \mathrm{Proj}_{V_{j+1}}(f) + \mathrm{Proj}_{W_{j+1}}(f).$$

This fact gives us the main recursive relation to build a representation of a function using the wavelet decomposition.

If we denote the projections of f onto V_j and W_j respectively by $f_j = \mathrm{Proj}_{V_j}(f)$ and $o_j = \mathrm{Proj}_{W_j}(f)$, we can write

$$f_j = \underbrace{f_{j+1}}_{f_{j+2} + o_{j+2}} + \quad o_{j+1}$$

Applying this relation recursively we arrive at the wavelet representation

$$f_j = f_{j+N} + o_{j+N} + \cdots + o_{j+2} + o_{j+1}$$

where a function f_j in some V_j is decomposed into its projections on the wavelet spaces $W_{j+1} \ldots W_{j+N}$, and a residual given by its projection onto the scale space V_{j+N}. This recursive process can be illustrated by the diagram in Figure 9.16.

We assumed above that the process starts with a function f_j which already belongs to some scale subspace V_j. This is not a restriction because we can

$$\begin{array}{ccccccccc} f_j & \rightarrow & f_{j+1} & \rightarrow & f_{j+2} & \rightarrow & \cdots & \rightarrow & f_{j+N} \\ & \searrow & & \searrow & & \searrow & & \searrow & \\ & & o_{j+1} & & o_{g+1} & & \cdots & & o_{j+N} \end{array}$$

Fig. 9.16. Wavelet decomposition of a function f.

take the initial j arbitrarily small (i.e. a fine scale). In practice, we work with functions that have some natural scale associated with them.

The wavelet decomposition gives an *analysis* of a function in terms of its projections onto the subspaces W_j. Note that, since by construction $W_j \perp W_l$ if $j \neq l$ and $V_j \perp W_j$, this decomposition of a function is unique once the spaces V_j and W_j are selected.

It is also desirable to reconstruct a function from its wavelet representation using a recursive process similar to the decomposition in Figure 9.16. It turns out that, since $W_j \subset V_{j-1}$ and $V_j \subset V_{j-1}$, the original function can be obtained from the projections, and the wavelet reconstruction is essentially the reverse of the decomposition, as illustrated in Figure 9.17.

The reconstruction gives a mechanism for the *synthesis* of functions from the wavelet representation.

To implement the wavelet decomposition and reconstruction we need to compute the projections onto the spaces V_j and W_j. We know that the set of functions $\{\phi_{j,n}; n \in \mathbb{Z}\}$ and $\{\psi_{j,n}; n \in \mathbb{Z}\}$, defined as

$$\phi_{j,n}(x) = 2^{-j/2}\phi(2^{-j}x - n) \tag{9.37}$$

$$\psi_{j,n}(x) = 2^{-j/2}\psi(2^{-j}x - n), \tag{9.38}$$

are respectively orthonormal basis of V_j and W_j. Therefore, the projections operators Proj_{V_j} and Proj_{W_j} are given by inner products with the elements of these bases

$$\text{Proj}_{V_j}(f) = \sum_n \langle f, \phi_{j,n}\rangle \phi_{j,n} = \sum_n \left(\int f(x)\overline{\phi_{j,n}(x)}dx \right) \phi_{j,n} \tag{9.39}$$

$$\begin{array}{ccccccccc} f_{j+N} & \rightarrow & f_{j+N-1} & \rightarrow & \cdots & \rightarrow & f_{j+1} & \rightarrow & f_j \\ & \nearrow & & \nearrow & & \nearrow & & \nearrow & \\ o_{j+N} & & o_{j+N-1} & & \cdots & & o_{j+1} & & \end{array}$$

Fig. 9.17. Wavelet reconstruction process of a function f.

$$\text{Proj}_{W_j}(f) = \sum_n \langle f, \psi_{j,n} \rangle \psi_{j,n} = \sum_n \left(\int f(x)\overline{\psi_{j,n}(x)}dx \right) \psi_{j,n} \quad (9.40)$$

The problem now is how compute the projection operators Proj_{V_j} and Proj_{W_j} efficiently. In fact, we would like to avoid altogether computing the integrals explicitly. To find a solution we take advantage of the fact that the recursive decomposition/reconstruction processes requires only projections between consecutive subspaces of the multiresolution ladder. For that purpose we will rely on the *two-scale* relations.

9.5.2 Two-Scale Relations and Inner Products

We have seen before that the interdependencies between two consecutive subspaces in a multiresolution analysis are formulated by the equations below, called *two-scale* relations

$$\phi(x) = \sum_k h_k \phi_{-1,k}(x) \quad (9.41)$$

$$\psi(x) = \sum_k g_k \phi_{-1,k}(x) \quad (9.42)$$

Using these two relations, we can express the basis functions of the scale and wavelet spaces, V_j and W_j, at level j in terms of the basis functions of the subsequent scale space V_{j-1}, at finer level $j - 1$. This is possible because, since $V_{j-1} = V_j \oplus W_j$, both $V_j \subset V_{j-1}$ and $W_j \subset V_{j-1}$.

Substituting (9.37) into (9.41), we have

$$\begin{aligned}
\phi_{j,k}(x) &= 2^{-j/2}\phi(2^{-j}x - k) \\
&= 2^{-j/2} \sum_n h_n \, 2^{1/2}\phi(2^{-j+1}x - 2k - n) \\
&= \sum_n h_n \, \phi_{j-1,2k+n}(x) \\
&= \sum_n h_{n-2k} \, \phi_{j-1,n}(x) \quad (9.43)
\end{aligned}$$

Similarly, substituting (9.38) into (9.42), we have

$$\begin{aligned}
\psi_{j,k}(x) &= 2^{-j/2}\psi(2^{-j}x - k) \\
&= 2^{-j/2} \sum_n g_n \, 2^{1/2}\phi(2^{-j+1}x - 2k - n) \\
&= \sum_n g_{n-2k} \, \phi_{j-1,n}(x) \quad (9.44)
\end{aligned}$$

Now, we need to find a way to use the sequences $(h_n)_{n \in \mathbb{Z}}$ and $(g_n)_{n \in \mathbb{Z}}$ to help us compute recursively the inner products $\langle f, \phi_{j,k} \rangle$, and $\langle f, \psi_{j,k} \rangle$. This can be easily done by inserting the expressions obtained for $\phi_{j,k}$ and $\psi_{j,k}$ into the the inner products.

$$\langle f, \phi_{j,k} \rangle = \langle f, \sum_n \overline{h_{n-2k}} \phi_{j-1,n} \rangle = \sum_n \overline{h_{n-2k}} \langle f, \phi_{j-1,n} \rangle \quad (9.45)$$

$$\langle f, \psi_{j,k} \rangle = \langle f, \sum_n \overline{g_{n-2k}} \phi_{j-1,n} \rangle = \sum_n \overline{g_{n-2k}} \langle f, \phi_{j-1,n} \rangle \quad (9.46)$$

9.6 Wavelet Decomposition and Reconstruction

Using the two–scale relations, we showed how to relate the coefficients of the representation of a function in one scale 2^{j-1}, with the coefficients of its representation in the next coarse scale 2^j and with coefficients of its representation in the complementary wavelet space. It is remarkable, that from the inner products of the function f with the basis of V_{j-1}, we are able to obtain the inner products of f with the basis of V_j and W_j, without computing explicitly the integrals! This is the crucial result for the development of the recursive wavelet decomposition and reconstruction method described in this section.

9.6.1 Decomposition

The wavelet decomposition process starts with the representation of a function f in the space V_0. There is no loss of generality here because, by changing the units, we can always take $j = 0$ as the label of the initial scale.

We are given the function $f = \text{Proj}_{V_0}(f)$, represented by the coefficients (c_k) of its representation sequence in the scale space V_0. That is

$$\text{Proj}_{V_0}(f) = \sum_k [\langle f, \phi_{0,k} \rangle \phi_{0,k}(x)] = \sum_k c_{0,k} \phi_{0,k} \quad (9.47)$$

In case we only have uniform samples $f(k)$, $k \in \mathbb{Z}$ of the function, the coefficients (c_k) can be computed from the samples by a convolution operation. This fact is well explained in Section 3.7 of Chapter 3 (see Theorem 2).

The goal of the decomposition is to take the initial coefficient sequence $(c_k^0)_{k \in \mathbb{Z}}$, and transform it into the coefficients of the wavelet representation of the function. The process will be done by applying recursively the following decomposition rule

$$\text{Proj}_{V_j}(f) = \text{Proj}_{V_{j+1}}(f) + \text{Proj}_{W_{j+1}}(f). \quad (9.48)$$

In this way, the process begins with $f^0 \in V_0 = V_1 \oplus W_1$, and in the first step, f^0 is decomposed into f^1/o^1, where $f^1 = \text{Proj}_{V_1}(f)$ and $o^1 = \text{Proj}_{W_1}(f)$.

The recursion acts on f^j, decomposing it into $f^{j+1}+o^{j+1}$, for $j = 0, \ldots N$. The components o^j are set apart. In the end we obtain the wavelet representation of f, consisting of the residual scale component f^N and the wavelet components $o^1, \ldots o^N$.

The core of the decomposition process splits the sequence (c_k^j) of scale coefficients associated with f^j, into two sequences (c_k^{j+1}) and (d_k^{j+1}), of scale and wavelet coefficients associated, respectively with f^{j+1} and o^{j+1}.

We can view this process as a basis transformation where we make the following basis change $(\phi_{j,k})_{k\in\mathbb{Z}} \rightarrow (\phi_{j+1,k}, \psi_{j+1,k})_{k\in\mathbb{Z}}$. Note that both sets form a basis of the space V^j. Equations (9.45) and (9.46) give the formulas to make the transformation on the coefficients of the bases:

$$c_k^{j+1} = \sum_n \overline{h_{n-2k}} c_n^j \tag{9.49}$$

$$d_k^{j+1} = \sum_n \overline{g_{n-2k}} c_n^j \tag{9.50}$$

with the notation $\bar{a} = (\overline{a_{-n}})_{n\in\mathbb{Z}}$.

Note that we are computing the coefficients (c_k^{j+1}) and (d_k^{j+1}) by discrete convolutions, respectively, with the sequences (h_n) and (g_n). Note also, that we are retaining only the even coefficients for the next step of recursion (because of the factor $2k$ in the indices). This is a decimation operation.

In summary, if we start with a sequence (c_n^0), containing $n = 2^J$ coefficients, it will be decomposed into the sequences $(d_{n/2}^1), (d_{n/4}^2), \ldots (d_{n/2^J}^J)$, and $(c_{n/2^J}^J)$. Note that the decomposition process outputs a wavelet representation with the *same number* of coefficients of the input representation.

Another important comment is that, up to now, we implicitly assumed doubly infinite coefficient sequences. In practice, we work with finite representations, and therefore it is necessary to deal with boundary conditions. This issue will be discussed in more detail later.

9.6.2 Reconstruction

The reconstruction process generates the coefficients of the scale representation from the coefficients of the wavelet representation. We would like to have an exact reconstruction, such that the output of the reconstruction is equal to the input of the decomposition. This is possible because we have just made an orthogonal basis transformation.

In order to bootstrap the recursive relations for the reconstruction process, we recall that one step of the decomposition takes a function representation f^{j-1} and splits into the components f^j and o^j.

$$\begin{aligned} f^{j-1}(x) &= f^j(x) + o^j(x) \\ &= \sum_k c_k^j \phi_{j,k}(x) + \sum_k d_k^j \psi_{j,k}(x) \end{aligned} \tag{9.51}$$

We need to recover the coefficients (c_n^{j-1}) from (c^j) and (d^j)

$$c_n^{j-1} = \langle f^{j-1}, \phi_{j-1,n} \rangle \tag{9.52}$$

Substituting (9.51) into (9.52), we obtain

$$c_n^{j-1} = \langle \sum_k c_k^j \phi_{j,k} + \sum_k d_k^j \psi_{j,k}, \quad \phi_{j-1,n} \rangle \tag{9.53}$$

$$= \sum_k c_k^j \langle \phi_{j,k}, \phi_{j-1,n} \rangle + \sum_k d_k^j \langle \psi_{j,k}, \phi_{j-1,n} \rangle \tag{9.54}$$

Because both $\phi_0 \in V_{-1}$ and $\psi_0 \in V_{-1}$, they can be represented as a linear combination of the basis $\{\phi_{-1,n}; n \in \mathbb{Z}\}$. Therefore $\phi_0 = \sum_n \langle \phi_0, \phi_{-1,n} \rangle \phi_{-1,n}$ and $\psi_0 = \sum_n \langle \psi_0, \phi_{-1,n} \rangle \phi_{-1,n}$. Since this representation is unique, using the two scale relations (9.41) and (9.42), we know that

$$h_n = \langle \phi_0, \phi_{-1,n} \rangle \tag{9.55}$$
$$g_n = \langle \psi_0, \phi_{-1,n} \rangle \tag{9.56}$$

The above results provide a reconstruction formula for the coefficients c_n^{j-1} from the coefficient sequences of the decomposition at level j.

$$c_n^{j-1} = \sum_k h_{n-2k} c_k^j + \sum_k g_{n-2k} d_k^j$$

$$= \sum_k \left[h_{n-2k} c_k^j + g_{n-2k} d_k^j \right] \tag{9.57}$$

The reconstruction process builds the final representation (c_n^0), from bottom up. At each step, it combines the sequences (c_n^j) and (d_n^j) to recover the intermediate (c_n^{j-1}), from $j = J, \ldots, 1$.

9.7 The Fast Wavelet Transform Algorithm

The fast wavelet transform (FWT) algorithm is a straightforward implementation of the method described in the previous section. It consists of the recursive application of equations (9.49) and (9.50) for the forward transform, and of equation (9.57) for the inverse transform.

In this section we present the pseudo-code, in C-like notation, of an implementation of the FWT algorithm. The code was structured for clarity and simple comprehension.

9.7.1 Forward Transform

The input of the algorithm is an array v, with 2^{m+1} elements, containing the coefficient sequence to be transformed, and the number of levels m. It uses the global arrays containing the two-scale sequences h and g. There are also global variables associated with these sequences: their number of elements hn and gn; and their offset values ho and go (i.e. the origins h_0 and g_0 of the sequences (h_n) and (g_n)). The main procedure wavelet_fwd_xform executes the iteration of the basic wavelet decomposition.

```
wavelet_fwd_xform(v, m, h, g)
{
    for (j = m; j >= 0; j--)
        wavelet_decomp(v, pow(2,j+1));
}
```

The procedure wavelet_decomp performs the decomposition for just one level, splitting the array v0 of size 2^{j+1}, into two arrays v and w with sizes 2^j. The result is accumulated into the input array v, such that in the end of the decomposition the array v is partitioned into [vN | wN | ... | w2 | w1], with sizes respectively $1, 1, \ldots, 2^m, 2^{m-1}$.

```
wavelet_decomp(v, n)
{
    zero (w, 0, n);
    for (l = 0; l < n/2; l++) {
        i = (2*l + ho) % n;
        for (k = 0; k < hn; k++) {
            w[l] += v[i] * h[k];
            i = (i+1) % n;
        }
        i = (2*l + go) % n;
        m = l + n/2;
        for (k = 0; k < gn; k++) {
            w[m] += v[i] * g[k];
            i = (i+1) % n;
        }
    }
    copy (w, v, n/2);
}
```

The procedure uses a local array w that must have, at least, the same size of v. It calls two auxiliary procedures, zero that fills and array with zeros, and copy that copies one array to another.

9.7.2 Inverse Transform

The inverse transform takes as input an array containing the wavelet representation, in the format produced by wavelet_fwd_xform, and converts it into a scale representation.

The procedure `wavelet_inv_xform` executes the iteration of the basic reconstruction step.

```
wavelet_inv_xform(v, m)
{
    for (j = 0; j <= m; j++)
        wavelet_reconst(v, pow(2, j+1));
}
```

The procedure `wavelet_reconst` performs the reconstruction combining the components vj and wj of the input array to reconstruct vj-1. It replaces [vj wj...] with [vj-1...]. Note that the number of elements of vj and wj is 1/2 of the number of elements of vj-1, therefore they use the same space in the array.

```
wavelet_reconst(w, n)
{
    zero(v, 0, n);
    for (k = 0; k < n; k++) {
        i = floor((k-ho)/2) % (n/2);
        m = (k - h.o) % 2;
        for (l = m; l < hn; l += 2) {
            v[k] += w[i] * h[l];
            i = (i-1) % (n/2);
        }
        i = floor ((k-go)/2) % (n/2);
        m = (k - go) % 2;
        for (l = m; l < gn; l += 2) {
            v[k] += w[i + n/2] * g[l];
            i = (i-1) % (n/2);
        }
    }
    copy(v, w, n);
}
```

9.7.3 Complexity Analysis of the Algorithm

The computational performance of the algorithm is very important. Let's determine what is the computational complexity of the fast wavelet transform.

The computation of each coefficient is a convolution operation with the two-scale sequences. Assuming that these sequences have n coefficients, then the convolution requires n multiplications and $n - 1$ additions.

In order to make the decomposition of a coefficient sequence at level j, from V_j into V_{j+1} and W_{j+1}, we have to compute 2^j new coefficients: 2^{j+1}

for the two components f^{j+1} and o^{j+1}. Since each coefficient requires $2n - 1$ operations, we have a total of $2^j(2n-1)$ operations for one-level transformation.

The full decomposition process is applied for $j\log_2(m)$ levels. Therefore, we have

$$\mathcal{O} = 2^j(2n-1) + 2^{j+1}(2n-1) + \cdots + 2(2n-1)$$

factoring out $(2n-1)$ and noting that $m = 2^j$, we obtain:

$$\mathcal{O}(m(2n-1)[1 + 2^{-1} + 2^{-2} + \cdots + 2^{-j+1}])$$
$$\mathcal{O}(m(2n-1)\tfrac{1-2^{-j}}{1-2^{-1}})$$
$$\mathcal{O}(mn)$$

The above analysis leads us to the following conclusions:

- The complexity is linear with respect to the size of the input sequence;
- The size of the two-scale sequences have a direct relation with the algorithm complexity.

9.7.4 Boundary Conditions

Since in practice we work with finite sequences, it is necessary take special care with the computation near the beginning and the end of the sequences (boundaries).

In order to compute the coefficients in the boundary regions, we have to perform a discrete convolution with the two-scale sequences, and therefore, we may need coefficients that lie beyond the boundaries of the sequence. Note that, for this reason, the boundary region is determined by size of the two-scale sequences. This situation is illustrated in Figure 9.18.

There are some techniques to deal with boundary conditions:

- Extending the sequence with zeros (see Figure 9.19 (a));
- Periodization by translation of the sequence with $x(N + i) \equiv x(i)$ (figure 9.19 (b));

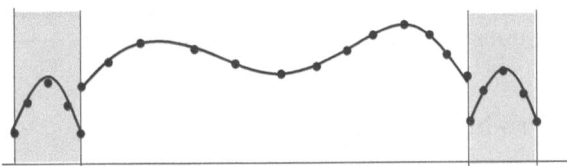

Fig. 9.18. Boundary regions for convolution between finite sequences.

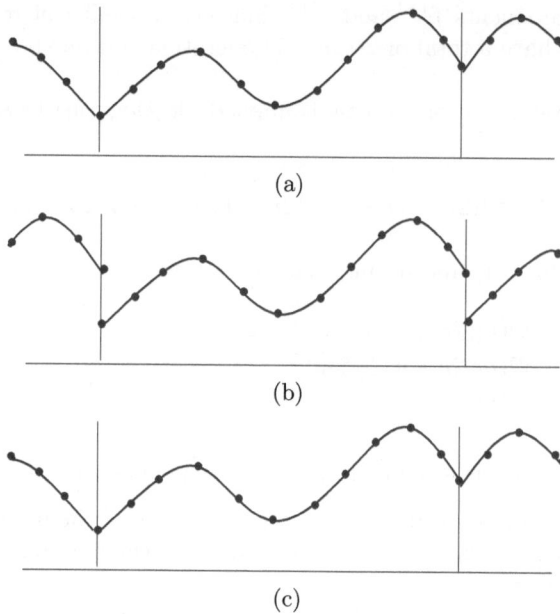

Fig. 9.19. Options for boundary computation (a) Extending with zeros; (b) Periodization; (c) Reflection.

- Periodization by reflection of the sequence with $x(N + i) \equiv x(N - i + 1)$ e $x(-i) \equiv x(i - 1)$ (Figure 9.19 (c));
- Use basis functions adapted to the interval (we are going to discuss this option later).

In the implementation of the fast wavelet transform algorithm presented in Section 9.7, we deal with the boundary problem by a simple periodization of the sequence. This is accomplished using the coefficient with indices i % m.

9.8 Images and 2D-Wavelets

The one-dimensional wavelet transform described in the previous sections can be extended to higher dimensions in several ways. Here we are going to describe only the extension using tensor products.

9.8.1 Tensor Product Extension

A natural way to extend a one-dimensional transformation to two dimensions is using a tensor product structure. This has two main advantages: it is simple and computationally efficient.

The tensor product extension is as follows:

$$V_0 = v_o^h \otimes v_o^v$$

such that

$$F \in V_j \leftrightarrow F(2^j x_1, 2^2 x_2) \in V_0$$

The scale function ϕ_j is then defined as:

$$\phi_j(x_1, x_2) = 2^j \phi(2^j x_1 - k_1)\phi(2^j x_2 - k_2) \tag{9.58}$$

The multiresolution relation now needs to be analyzed more carefully, since it is the result of a tensor product:

$$
\begin{aligned}
V_{j+1} &= v_{j+1}^h \otimes v_{j+1}^v \\
&= (v_j^h \oplus w_j^h) \otimes (v_j^v \oplus w_j^v) \\
&= (v_j^h \otimes v_j^v) \oplus [(v_j^h \otimes w_j^v) \oplus (w_j^h \otimes v_j^v) \oplus (w_j^h \otimes w_j^v)]
\end{aligned}
\tag{9.59}
$$

Note that, as a consequence of the tensor product structure, we now have three types of wavelet functions: two mixed components and a pure component. It is possible to interpret these components as the horizontal wavelets (which detect edges in the horizontal direction), vertical wavelets (which detect edges in the vertical direction) and diagonal wavelets (which detect edges in the main diagonal direction).

$$
\begin{aligned}
\psi^h(x_1, x_2) &= \psi(x_1)\phi(x_2) \\
\psi^v(x_1, x_2) &= \phi(x_1)\psi(x_2) \\
\psi^d(x_1, x_2) &= \psi(x_1)\psi(x_2)
\end{aligned}
\tag{9.60}
$$

9.8.2 The 2D Algorithm

The algorithm for the two-dimensional wavelet transform expoits the separability of the tensor product structure. In this way, both the decompostion algorithm and the reconstruction algoritm can be implemented in two dimensions by the sequential aplication of the corresponding one-dimensional algorithms, to the lines and columns of a 2D matrix representing the image.

The direct 2D wavelet transform is:

```
wavelet_2D_fwd_xform(a, h, g)
{
    for (u=0; u<m; u++)
        wavelet_fwd_xform(a[u,0], h, g);
    for (v=0; v<n; v++)
        wavelet_fwd_xform(a[0,v], h, g);

}
```

The inverse 2D wavelet transform is:

```
wavelet_2D_inv_xform(a, h, g)
{
    for (v=0; v<n; v++)
        wavelet_inv_xform(a[0,v], h, g);
    for (u=0; u<m; u++)
        wavelet_inv_xform(a[u,0], h, g);
}
```

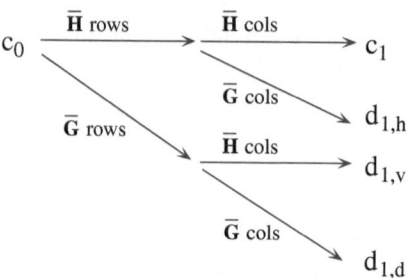

Fig. 9.20. Wavelet transform over a 2D domain: Decomposition scheme.

Fig. 9.21. Wavelet transform over a 2D domain: Original image original, intermediate result (horizontal pass) and final result (horizontal and vertical pass).

We can observe in Figures 9.20 e 9.21 how this decomposition process takes place. In Figure 9.20 we have an schematic diagram of the decomposition and in Figure 9.21 we have the intermediate results and the final image.

9.9 Comments and References

The concept of multiresolution representation and its relation to wavelets was developed by S. Mallat (Mallat 1989b). In the literature it carries different names: *multiscale analysis* or *multiscale approximation*. We have opted for multiresolution representation because it fits better to the emphasis we have been given on function representation.

The material covered in this chapter can be found on (Hernandez e Weiss 1996). Nevertheless the notation of the indices in the scale space differs from the one used here.

For an exposition of the topics in this chapter using the language of operators in function spaces the reader should consult (Kaiser 1994). The approach is algebraically very clear and clean, nevertheless a lot of geometric insight is lost.

The fast wavelet transform algorithm was introduced by Stephane Mallat (Mallat 1989a). One of the first references on the computational implementation of the algorithm appeared in (Press, Teukolsky e Vetterling 1996).

The code for the fast wavelet transform algorithm presented in this chapter was based in the pseudo-code from (Jawerth e Sweldens 1994). This algorithm was implemented in (Bourges-Sévenier 1994).

The book (Wickerhauser 1994) describes a complete system for computation with wavelets, including the fast wavelet transform.

There are several possibilities of extending the wavelet transform to functions of several variables, i.e. $\mathcal{L}^2(\mathbb{R}^n)$. The interested reader should consult (Daubechies 1992), page 33, or (Mallat 1998).

The beautiful examples 1 and 9.4 of this chapter were taken from (Kaiser 1994).

References

[Bourges-Sévenier 1994]Bourges-Sévenier, M. (1994). Réalisation d'une bibliothque c de fonctions ondelettes. Technical report, IRISA – INRIA.

[Chui 1992]Chui, C. K. (1992). *An introduction to wavelets*. Academic Press.

[Costa e Darsa 1992]Costa, B. e Darsa, L. (1992). *Visionaire—Commercial Morphing Software*. Impulse, Inc., Minneapolis.

[Daubechies 1992]Daubechies, I. (1992). *Ten Lectures on Wavelets*. SIAM Books, Philadelphia, PA.

[Fiume 1989]Fiume, E. (1989). *The Mathematical Structure of Raster Graphics*. Academic Press, Boston.

[Gonzalez e Wintz 1987]Gonzalez, R. C. e Wintz, P. (1987). *Digital Image Processing (2nd Edition)*. Addison-Wesley, Reading, MA.

[Hernandez e Weiss 1996]Hernandez, E. e Weiss, G. (1996). *A First Course on Wavelets*. CRC Press, Boca Raton.

[Jawerth e Sweldens 1994]Jawerth, B. e Sweldens, W. (1994). An overview of wavelet based multiresolution analyses. *SIAM Rev.*, 36(3):377–412.

[Kaiser 1994]Kaiser, G. (1994). *A Friendly Guide to Wavelets*. Birkhauser, Boston.

[Lim 1990]Lim, J. S. (1990). *Two Dimensional Signal and Image Processing*. Prentice-Hall, New York.

[Mallat 1989a]Mallat, S. (1989a). Multifrequency channel decomposition of images and wavelet models. *IEEE Transaction on ASSP*, 37:2091–2110.

[Mallat 1989b]Mallat, S. (1989b). Multiresolution approximation and wavelets. *Trans. Amer. Math. Soc.*, 315:69–88.

[Mallat 1998]Mallat, S. (1998). *A Wavelet Tour of Signal Processing*. Academic Press.

[Press, Teukolsky e Vetterling 1996]Press, W. H., Teukolsky, S. A., e Vetterling, W. T. (1996). *Numerical Recipes : The Art of Scientific Computing*, chapter 13, pages 591–606. Cambridge Univ Press.

[Weaver 1989]Weaver, J. (1989). *Theory of Discrete and Continuous Fourier Transform*. John Wiley & Sons, New York.

[Wickerhauser 1994]Wickerhauser, M. V. (1994). *Adapted Wavelet Analysis from Theory to Software*. A. K. Peters, Wellesley, MA.

[Zayed 1993]Zayed, A. (1993). *Advances in Shannon's Sampling Theory*. CRC Press, Boca Raton.

10

Probabilistic Image Models

Nature is a complex scenario, and problem solving is an essential component of human nature. These two ingredients lead to curiosity, the seeking for new useful information among myriads of stimuli.

Deterministic models, as presented in previous chapters, are useful for describing those phenomena without inherent uncertainty and for which all relevant data can be gathered without observation errors. This is seldom the case when dealing with real world data. The latter situation requires models able to cope with the complexity of randomness, and stochastic models are among them.

This chapter presents some of the most successful stochastic models for dealing with image data. The conceptual framework is the image model proposed by (Geman and Geman 1984), that splits image formation into two main components: an unobserved truth and the observed data; this framework is discussed in section 10.1. Models for the observed data, are commented in section 10.2, while models for the classes are presented in section 10.6.

Examples presented here were produced with R, an open source software, freely available for a number of platforms (R 2006).

10.1 Image Formation

As seen in the previous chapter, a discrete image defined on a regular Euclidean grid is a special kind of signal, namely

$$f : \mathcal{U} \subset \mathbb{Z}^2 \to K^m, \tag{10.1}$$

where $K \subset \mathbb{R}$ or $K \subset \mathbb{C}$ and m is called "number of bands". We will deal mostly with images defined on a finite grid so, without loss of generality, we will write $\mathcal{U} = [0, \ldots, n_1 - 1] \times [0, \ldots, n_2 - 1] \subset \mathbb{Z}^2$, and n_1, n_2 will be referred to as the number of rows and the number of columns, respectively.

The main problems that arise when dealing with images are related to processing and analysis, as presented in section 1.2. In order to illustrate

L. Velho et al., *Image Processing for Computer Graphics and Vision*,
Texts in Computer Science, DOI 10.1007/978-1-84800-193-0_10,
© Springer-Verlag London Limited 2009

these two kinds of problems, the models and techniques to tackle them, we will consider them as disjoint, though in many situations they are not.

(Geman and Geman 1984) provided a very useful stochastic framework, that we will now put in the form of equation (10.1). Consider two images f, g as in equation (10.1), such that

$$g = \phi(H(f)) \odot n, \tag{10.2}$$

where H are operations that "blurr" the original information, ϕ are point-wise operations, and n is a random signal called generically "noise" that is composed to the blurred and distorted data by means of a binary operator $\odot$. Image processing deals with the problem of retrieving f from g, as will be presented in Section 10.5

Let us see now an example of image analysis, namely, image classification. Assume that, given the support of the image $\mathcal{U}$, nature chooses a class ξ_u for each coordinate $u \in \mathcal{U}$; by classes we mean "natural" simple targets as, for instance, shallow water, deep water, dry sand, wet sand, forest etc. The number of different classes can be either known or not. A sensor will observe the scene, and will produce a value in K^m for each coordinate as a function of the class in that position and of the imaging technique. The manner in which each class is trasformed into a value is idiosyncratic of the sensor, the class and the conditions under which the observation was made, including the class of neighboring sites. Image classification consists of estimating the true classes $(\xi_u)_{u \in \mathcal{U}}$ from the observed data $(f(u))_{u \in \mathcal{U}}$, possibly using additional information.

A closely related image analysis problem is that of segmentation. Instead of estimating the unobserved classes, an image segmentation is a partition of the support $\mathcal{U}$ into disjoint non-empty sets $\mathcal{U}_k$, $1 \le k \le M$, such that if $u_1, u_2 \in \mathcal{U}_k$ but $u_3 \in \mathcal{U}_\ell$, with $k \ne \ell$, then $f(u_1)$ and $f(u_2)$ share a common property but $f(u_3)$ does not. Typical properties are mean value, texture, color etc.

Edge detection is one of the many tasks that can be formulated such that encompasses both problems: image processing and image analysis.

In the following we will provide a brief account of stochastic models for these problems, and will discuss some of the tools that can be derived under such hypotheses.

10.2 Observed Data

One of the most popular models for describing image data is the multivariate Gaussian law. Among the reasons for this popularity one can mention the central limit theorem, which states that if observations are the result of the sum of infinitely many small loosely related contributions, then the result should follow this law. It is important to notice, though, that the theorem says nothing about actual data and, as will be seen later in this chapter, this rationale should be used with caution in practice.

Another reason for assuming the multivariate Gaussian distribution is that it is tractable from both the theoretical and computational viewpoints, and that it leads to well known restoration and analysis techniques.

This distribution is characterized by the following density:

$$p(\boldsymbol{x}) = \frac{1}{(2\pi)^{m/2}|\Sigma|^{1/2}} \exp\left\{-\frac{1}{2}(\boldsymbol{x} - \boldsymbol{\mu})'\Sigma^{-1}(\boldsymbol{x} - \boldsymbol{\mu})\right\}, \qquad (10.3)$$

where $\boldsymbol{x} = (x_1, \ldots, x_m)'$ describes the observation in $\mathbb{R}^m$, $\boldsymbol{\mu} = (\mu_1, \ldots, \mu_m)' \in \mathbb{R}^m$ is the mean vector and Σ is the $m \times m$ positive definite covariance matrix, so Σ^{-1} exists. The mean vector is the point at which the mode of the distribution is located. The covariance matrix is of the form $\Sigma = (\sigma_{ij})$ where

$\sigma_{ij} = \sigma_{ji}$ is the covariance between components i and j,

$\sigma_{ii} = \sigma_i^2 > 0$ is the variance of component i.

The covariance between components i and j can also be expressed in terms of their correlation $-1 < \rho_{ij} < 1$, namely, $\sigma_{ij} = \rho_{ij}\sigma_i\sigma_j$. More details about this distribution can be found in, among other references, the textbooks by (Krzanowski 1988; Muirhead 1982; Tong 1990).

For simplicity reasons, but without loss of generality, consider the bivariate case. The density given in equation (10.3) reduces to

$$p(x_1, x_2) = \frac{1}{2\pi\sigma_1\sigma_2\sqrt{1 - \rho^2}} \exp\left\{-\frac{1}{2(1 - \rho^2)}\left[\left(\frac{x_1 - \mu_1}{\sigma_1}\right)^2 + \left(\frac{x_2 - \mu_2}{\sigma_2}\right)^2 - 2\rho\left(\frac{x_1 - \mu_1}{\sigma_1}\right)\left(\frac{x_2 - \mu_2}{\sigma_2}\right)\right]\right\}, \qquad (10.4)$$

where $\rho = \rho_{12}$. This distribution will be used to analyze the data that compose Figure 10.1.

As can be seen in Figure 10.1 (the dashed lines denote an area to be further studied, not in the original work), the painting is mainly composed of light blue (the sky), green (the cactus), black (eyes, eyelids, hair and contours), light brown (the skin) and dark brown (the hat) regions. Small white spots are also seen.

Figure 10.2 shows the values of each pixel projected onto the plane formed by the red and blue channels. Each point is shown in the same color as seen in Figure 10.1, and it is clear that a single Gaussian distribution will be inadequate to explain this dataset.

A sample of 14190 pixels were taken from the class 'skin', namely the area shown in Figure 10.1. These values are, at first sight, good candidates for a fit with the bivariate Gaussian law.

The sample mean and covariance matrix, computed with the red and blue channels of the training data, are given by

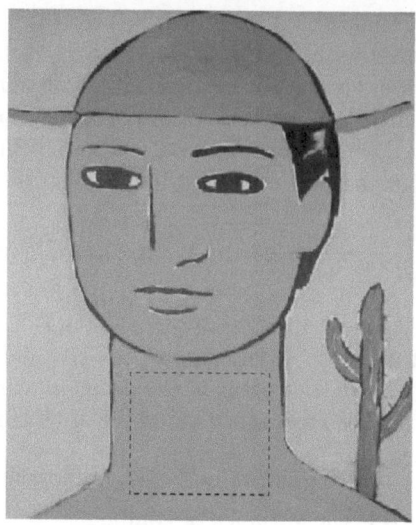

Fig. 10.1. "Moreno Bom" by Enilson Costa, acrylic on canvas, 2007, sampled to 458×372 pixels, with region of interest.

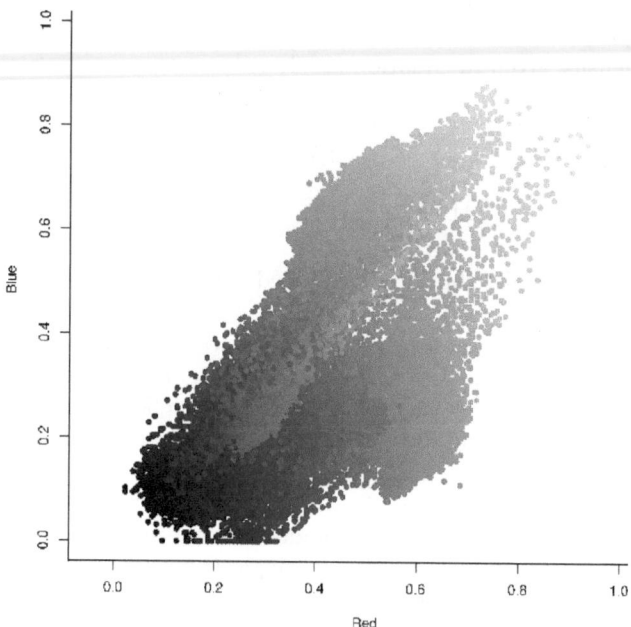

Fig. 10.2. Pixels values projected onto the red-blue plane.

$$\widehat{\boldsymbol{\mu}} = \begin{pmatrix} 0.625 \\ 0.190 \end{pmatrix} \text{ and } \widehat{\Sigma} = 10^{-4} \begin{pmatrix} 2.486 & 2.109 \\ 2.109 & 4.450 \end{pmatrix}.$$

With this, the estimated correlation between the channels is, approximately, $\widehat{\rho} = 0.634$. Using this information, one can plot contour curves of equation (10.3), given these estimates; this is presented in Figure 10.3.

Besides such statistical analysis, which is useful for image processing and analysis, one can use this information in order to synthesize data. The complete estimates from the skin data, i.e., using the red, green and blue channels are

$$\widehat{\boldsymbol{\mu}}_{\text{skin}} = \begin{pmatrix} 0.625 \\ 0.467 \\ 0.190 \end{pmatrix} \text{ and } \widehat{\Sigma}_{\text{skin}} = 10^{-4} \begin{pmatrix} 2.486 & 1.869 & 2.101 \\ 1.870 & 1.823 & 2.162 \\ 2.109 & 2.162 & 4.450 \end{pmatrix}. \tag{10.5}$$

A number of computational platforms can be used to draw samples from the multivariate Gaussian distribution, provided the parameters.

Using this approach, a simulated palette of colors of the painting shown in Figure 10.1 is shown in Figure 10.4. For each color, it was built simulating 100×100 independent outcomes from the Gaussian distribution with the mean

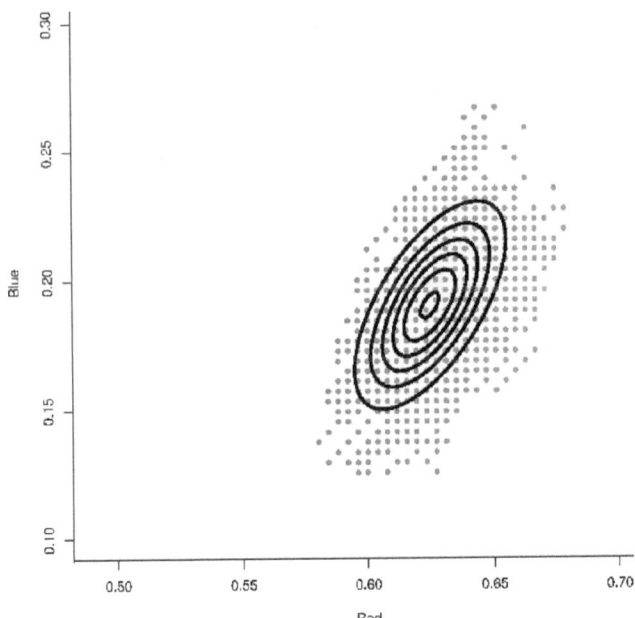

Fig. 10.3. Data from the class 'skin', with contour curves of the estimated bivariate Gaussian density.

Fig. 10.4. Simulated palette of colors: skin, cactus, sky and hat (top to bottom, left to right).

and covariance matrix as estimated from the data. These parameters, besides the ones presented in equation (10.5), are:

$$\widehat{\boldsymbol{\mu}}_{\text{sky}} = \begin{pmatrix} 0.516 \\ 0.677 \\ 0.735 \end{pmatrix}, \widehat{\Sigma}_{\text{sky}} = 10^{-4} \begin{pmatrix} 0.859\ 0.655\ 0.792 \\ 0.655\ 1.035\ 1.187 \\ 0.792\ 1.187\ 1.587 \end{pmatrix}, \tag{10.6}$$

$$\widehat{\boldsymbol{\mu}}_{\text{hat}} = \begin{pmatrix} 0.490 \\ 0.332 \\ 0.182 \end{pmatrix}, \widehat{\Sigma}_{\text{hat}} = 10^{-4} \begin{pmatrix} 7.377\ 4.589\ 1.009 \\ 4.589\ 3.517\ 1.222 \\ 1.009\ 1.222\ 1.589 \end{pmatrix}, \tag{10.7}$$

and

$$\widehat{\boldsymbol{\mu}}_{\text{cactus}} = \begin{pmatrix} 0.281 \\ 0.420 \\ 0.252 \end{pmatrix}, \widehat{\Sigma}_{\text{cactus}} = 10^{-3} \begin{pmatrix} 2.001\ 1.938\ 1.783 \\ 1.938\ 3.001\ 1.247 \\ 1.783\ 1.247\ 2.61 \end{pmatrix}. \tag{10.8}$$

Using simulation introduces variability in the data, a desirable feature when realism is sought. Notice that the cactus and hat colors shown in the synthetic palette in Figure 10.4 vary more than the other two introducing, thus, a more lively effect.

Different applications require different parameters in order to obtain acceptable samples. (Richards and Jia 1999, p. 188) provide the estimated parameters from four different areas: water, fire burn, vegetation and urban. The four bands data they employ come from the Landsat multispectral scanner.

A major weakness of this proposal is the spatial independence among the random variables. This issue can be tackled in two ways: incorporating dependence on the model of the observed data, or providing such structural information in the classes.

Special care must be taken in the choice of the distribution. Though the multivariate Gaussian law is tractable and, under conditions oftern observed in practice, acceptable, this is not the case when leading with imagery obtained with coherent illumination: sonar, ultrasound-B, laser and SAR (*Synthetic Aperture Radar*) data. The departure from the Gaussian law in such cases plays a central role in the development of tools for image processing and analysis. For details on this kind of data, the reader is referred to (Oliver and Quegan 1998) and the references therein.

10.3 Histograms and Estimation

The previous section presented the use of simulation for building synthetic data. As seen, this technique requires the estimation of paramenters that characterize the model for the observations. This section presents definitions leading to procedures that allow the obtainment of such estimates, and other statistically-related techniques for image enhancement.

In the following we will consider $g : S \to K \subset \mathbb{R}$ a single-channel image defined on the regular grid S with values in K, i.e., definition (10.1) with $m = 1$.

One of the most important tools for the analysis of data in general, and of images in particular, is the *histogram*. Consider $\mathcal{I}_K = \{K_0, \ldots, K_{o-1}\}$ a partition of the set K in o elements, i.e.,

1. $K_i \neq \varnothing$ for every $0 \leq i \leq o-1$,
2. $K_i \cap K_j = \varnothing$ for every $i \neq j$, and
3. $\cup_{i=0}^{o} K_i = K$,

then $\mathcal{H}(g, \mathcal{I}_K) = (v_0, \ldots, v_{o-1}) \in \mathbb{N}_0^o$, an o-dimensional vector of natural numbers, is the histogram of the image g with respect to the partition $\mathcal{I}_K$ if

$$v_i = \#\{s \in S : g(s) \in K_i\},$$

in other words, the number of coordinates s where the observed value $g(s)$ belongs to K_i. A very convenient practice consists of using the *histogram of proportions*, given by $h(g, \mathcal{I}_K) = (v_0, \ldots, v_{o-1})/\#S$.

The histogram (of proportions, a denomination which we will omit thereof) is, in some sense, an estimator of the density of the distribution that describes the data and, therefore, it provides important information about the observed values.

In practice, the most frequent partition of K is in o disjoint intervals of equal length. If K is an interval of $\mathbb{Z}$, say $K = [0, k-1]$, then a good starting point is using $K_i = i$. For a discussion on ways of building histograms,

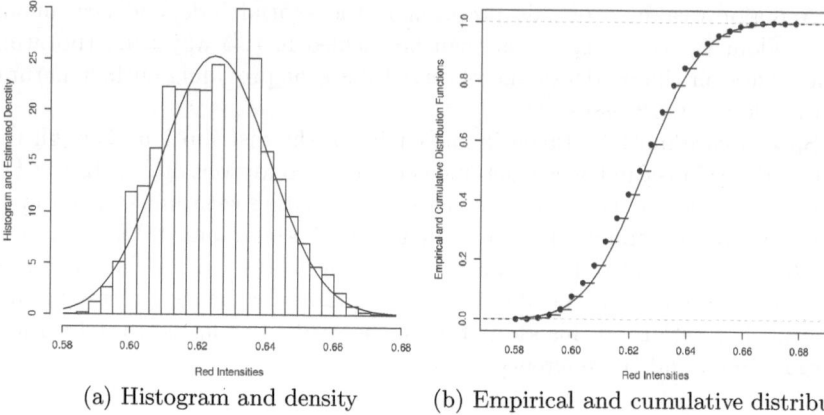

(a) Histogram and density

(b) Empirical and cumulative distribution functions

Fig. 10.5. Histogram, fitted Gaussian density, empirical and estimated cumulative distribution function of the red channel of samples from the skin class.

and other important statistical issues, the reader is referred to (Venables and Ripley 2002).

Figure 10.5(a) presents the histogram of the red channel data of the samples from the skin class, along with the fitted Gaussian density. One can see that the histogram is fairly symmetric, with no evident departure from the assumption of normality.

An important function, closely related to the histogram, is the *empirical function* $\widehat{F} : \mathbb{R} \to [0,1]$, defined as $\widehat{F}(t) = \#\{s \in S : g(s) \le t\}/\#S$. This function computes, in every $t \in \mathbb{R}$, the proportion of coordinates s such that the observed value $g(s)$ is at least t.

The empirical function can be regarded as an estimate of the cumulative distribution function, and the latter contains all the relevant information about the (marginal) process that generated the observed data $\{g(s) : s \in S\}$.

Figure 10.5(b) shows, in steps, the empirical function of the red channel of samples from the skin class. The solid line exhibits the estimated cumulative distribution function of the same data. The fit is visually acceptable.

The empirical function plays a central role in an important class of image transformations: the histogram-based poinwise radiometric operations, being the histogram equalization one of its most used members in practice.

Consider $f, g : S \to \mathbb{R}$ two images, then f is a pointwise radiometric transformation of g if for every $s \in S$ holds that $f(s) = \Upsilon_s(g(s))$, where $\Upsilon_s : \mathbb{R} \to \mathbb{R}$ are real functions. More often than not, $\Upsilon_s = \Upsilon : \mathbb{R} \to \mathbb{R}$, for every $s \in S$, i.e., there is only one radiometric transformation. In this last case one says that f is the result of a location-invariant pointwise radiometric transformation.

There are countless useful radiometric transformations, being the square root and the logarithm two of the most used for positive data. These transformation enhance (expand) low values, at the expense of compacting high values. They are, therefore, useful for visualizing relatively dark images. Another important transformation of this kind is the negative: if $f : S \rightarrow [0,1]$, then g such that $g(s) = 1 - f(s)$ is the digital negative of f.

Figure 10.6(a) shows the blue channel of Figure 10.1, while Figures 10.6(b) and Figures 10.6(c) present the result of applying the square root and the negative, respectively.

In the quest of a pointwise radiometric transformation that enhances every portion of the data, we will now recall an useful result from probability. Consider the continuous random variable X and its cumulative distribution function F. The random variable Y that results from the transformation $Y = F(X)$ has uniform distribution on $(0,1)$.

A perceptually well-contrasted image is close to one with an uniform histogram, so given $f : S \rightarrow \mathbb{R}$, the transformation that produces such image is $g = F(f)$, where F is the cumulative distribution function of the random variable X whose outcomes are the observed values $\{f(s) : s \in S\}$... but one seldom has access to this information.

Instead of relying on the improbable knowledge of F, one can use $\widehat{F}$, which is an estimator. In this fashion, $g = \widehat{F}(f) : S \rightarrow [0,1]$, with $\widehat{F}$ the empirical function of the image f, is the equalized version of f.

Figure 10.7 shows the ingredients and the result of the histogram equalization. Figure 10.7(a) presents the histogram of the blue channel data (Figure 10.6(a)); it is bimodal, corresponding to the two main classes present in this dataset: the sky (high values) and the rest of the picture (low values). Figure 10.7(b) exhibits the empirical function of these dat; the flat region to the middle of the curve corresponds to the intensity levels between the two peaks of the histogram, for which few observations are available. Figure 10.7(c)

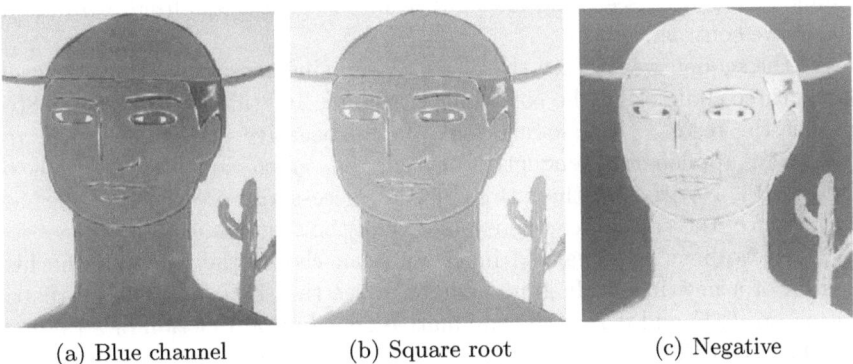

| (a) Blue channel | (b) Square root | (c) Negative |

Fig. 10.6. Original blue channel, square root and digital negative transformations.

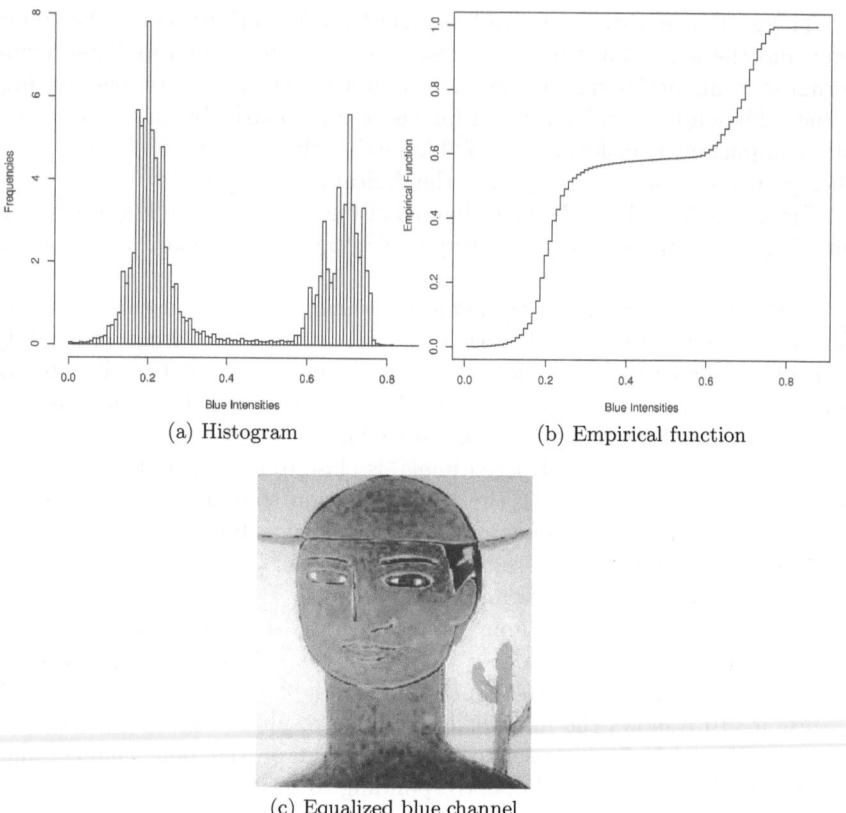

(a) Histogram

(b) Empirical function

(c) Equalized blue channel

Fig. 10.7. Histogram equalization.

presents the result of applying the empirical function to the values of the blue channel. As can be seen, the image with its histogram equalized exhibits an aggressive contrast.

In the sequel, we will see the effect of applying these operations on each of the three channels of the color image. Figures 10.8(a), 10.8(b) and 10.8(c) present the result of the square root, digital negative and equalization, respectively, applied independently on each red, green and blue channels of Figure 10.1. Notice that the latter does not necessarily preserve the hues as, for instance, the color of cactus is bluer than the original.

Once we have an equalized image, we can choose the shape of the histogram of a new image. It is immediate to see that if U has uniform distribution on $(0, 1)$ and if F is the cumulative distribution function of a random variable, then the random variable $Y = F^{-}(U)$ has the distribution characterized by F. F^{-} denotes the pseudoinverse of F, which is given by $F^{-}(t) = \inf\{x \in \mathbb{R} : F(x) \geq t\}$. Notice that if Y is a continuous random variable, then

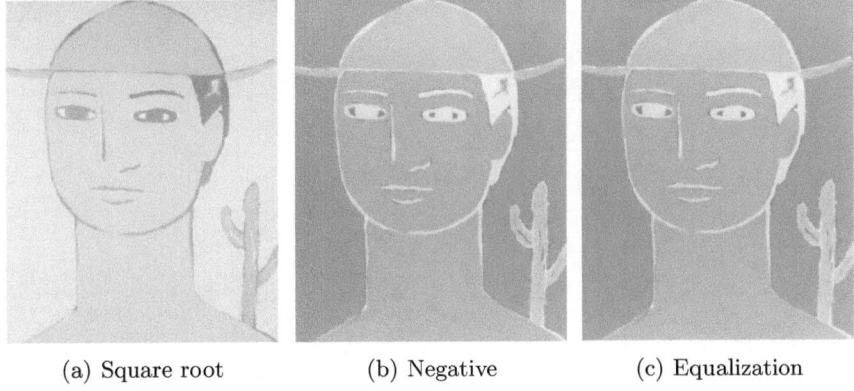

| (a) Square root | (b) Negative | (c) Equalization |

Fig. 10.8. Pointwise radiometric transformation on the three channels.

$F^-(t) = F^{-1}(t)$, but F^- can also be used for discrete random variables. This is known as the 'inversion method' in simulation: a very general procedure for obtaining samples from any distribution, starting from outcomes of the Uniform law.

The Gaussian distribution is usually regarded to as a shape that provides a smooth perception of the data. Since the data in Figure 10.7 are approximately distributed in an uniform fashion, we can apply them Φ^{-1}, with

$$\Phi(x) = \int_{-\infty}^{x} \frac{1}{\sqrt{2\pi}} \exp\{-v^2/2\} \, dv$$

the cumulative distribution function of the standard Gaussian law, in order to obtain $g' = \Phi^{-1}(g)$ the Gaussian version of the original image f. The result of applying this transformation to the three channels of our test image (Figure 10.1) is presented in Figure 10.9, along with the histogram of the data in channel blue.

Histogram specification is a very general and useful technique. It provides means to checking the appearance of images for which the physics of the data acquisition imposes certain distributions, as is the case of, for instance, laser, sonar, ultrasound-B, the already mentioned SAR and night-vision devices.

In the following, we will conduct a simple statistical analysis of the data, with the purpose of building other techniques of image enhancement.

Table 10.1 presents the main statistical descriptors of each band of Figure 10.1. If $x = (x_1, \ldots, x_n)$ denotes a sample of n real values, and the vector $(x_{1:n}, \ldots, x_{n:n})$ is built with the values sorted in non-decreasing order, i.e., $x_{1:n} \leq \cdots \leq x_{n:n}$, then these quantities, assuming n odd for the sake of simplicity, are

- the minimum value: $\min(x) = x_{1:n}$,
- the first quartile: $q_{1/4}(x) = x_{(n+1)/4:n}$,

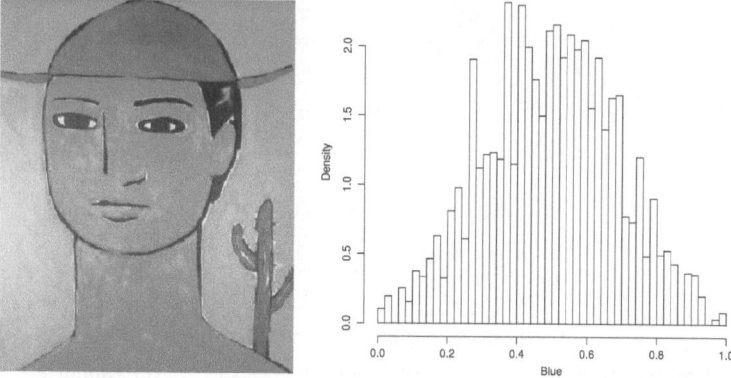

Fig. 10.9. Histogram specification: Gaussian shape.

- the median: $q_{1/2}(\boldsymbol{x}) = x_{(n+1)/2:n}$,
- the mean value: $\overline{\boldsymbol{x}} = n^{-1} \sum_{i=1}^{n} x_i$,
- the third quartile: $q_{3/4}(\boldsymbol{x}) = x_{3(n+1)/4:n}$, and
- the maximum value: $\max(\boldsymbol{x}) = x_{n:n}$.

The sample standard deviation, given by $s_{\boldsymbol{x}} = \sqrt{n^{-1} \sum_{i=1}^{n} (x_i - \overline{\boldsymbol{x}})^2}$, is another important descriptive measure of a dataset. Notice that, as defined, it is the square root of the variance presented in Definition 10.4, page 273.

Notice that these quantities depend only on each data set, i.e., on the values of each band regardless the other bands. These are, then, marginal descriptors, in the sense that are solely related to the marginal properties of the data. They do not describe, for instance, the relationship between channels; this will be discussed later in this chapter.

The *range* of a set of values $\boldsymbol{x}$ is the interval $[\min(\boldsymbol{x}), \max(\boldsymbol{x})]$, and as we can see from Table 10.1, none of the channels under assessment has full (maximum, complete) range, i.e., $[0.0240, 933]$, $[0.051, 0.867]$ and $[0, 0.878]$ are all strictly contained in $[0, 1]$. An equalized image, by definition, has full range but, as previously presented, it does not necessarily preserve the hues.

Another important graphical tool for the assessment of the data are the boxplots. They provide additional, and more quantitative, information to that available in a histogram. It is particularly useful for comparing two or more data sets.

Table 10.1. Main descriptors of each channel of Sertanejo

	min	$q_{1/4}$	$q_{1/2}$	$\overline{\boldsymbol{x}}$	$q_{3/4}$	max	$s_{\boldsymbol{x}}$
Red	0.024	0.455	0.514	0.514	0.620	0.933	0.116
Green	0.051	0.447	0.486	0.508	0.631	0.867	0.132
Blue	0.000	0.196	0.251	0.406	0.682	0.878	0.240

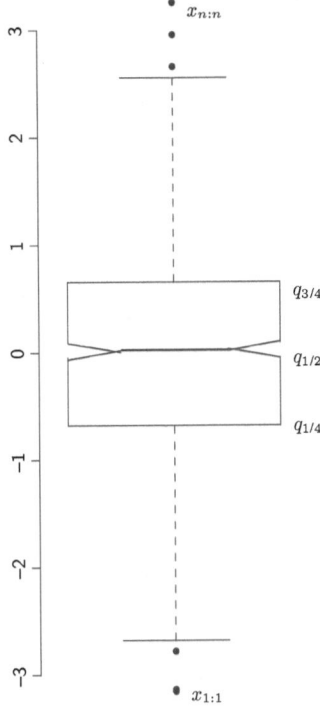

Fig. 10.10. Elements of a boxplot.

Figure 10.10 presents the main elements of a boxplot: the box extends from the lower to the upper quartile, i.e., from $q_{1/4}$ to $q_{3/4}$; the horizontal bars extend from those data that encompass at least $1.5(q_{3/4} - q_{1/4})$, and if there are points beyond these bars they are marked as spots, and they are usually referred to as "outliers". Notice that, in the case of Figure 10.10, there are two outliers: two below (one being the minimum $x_{1:n}$) and one above (one being the maximum $x_{n:n}$). The central bar is the median ($q_{1/2}$). Two notches are drawn besides the median, and if the notches of two plots do not overlap this is strong evidence that the two medians differ significantly; see the `boxplot` reference in the R package (R 2006).

See, for instance, the boxplots presented in Figure 10.11; they exhibit information for the three bands that compose the Sertanejo picture. One readily notices that, though all medians are significantly different, the red and green ones are closer to each other than the blue one; the blue component is lower than the other two and, therefore, may require some additional or stronger enhancement for a well balanced picture.

Another interesting issue of the blue channel is that its mean and median values are, respectively, 0.406 (see Table 10.1) and 0.251. This discrepancy

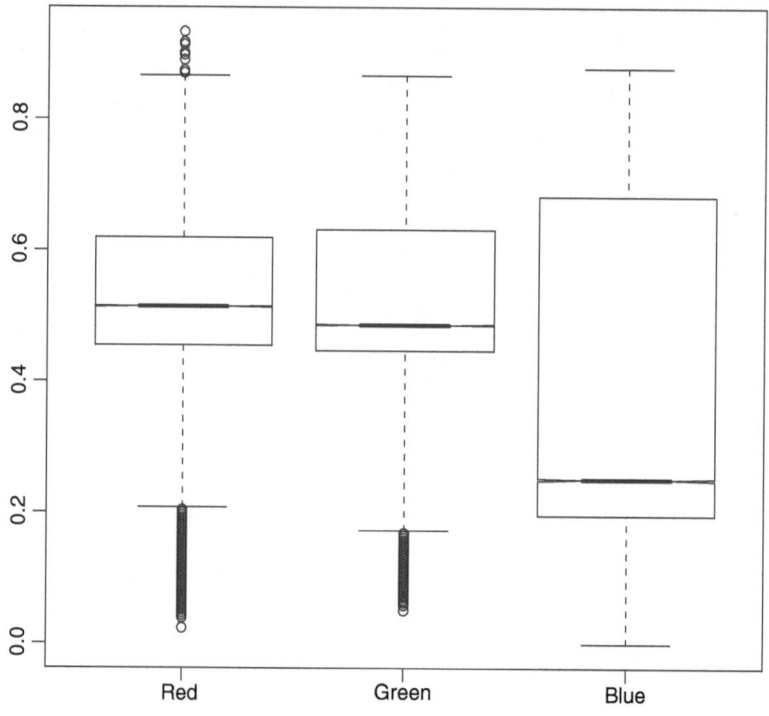

Fig. 10.11. Boxplots of the three bands of Sertanejo.

suggests the presence of highly skewed data and/or more than one population, as already checked in the histogram presented in Figure 10.7(a).

The distribution of observations allows us to conclude that

Red channel: the median is skewed towards lower values, there are outliers both below and above the extreme bars

Green channel: the median is skewed towards lower values, only outliers below the extreme bars are present

Blue channel: does not exhibit outliers and, as the other two channels, the median is skewed towards lower values.

As already presented, graphical and quantitative exploratory analysis of image data leads to relevant information, and provide guidelines for building successful image processing techniques.

A milder pointwise radiometric transformation aiming at improving contrast is the linear stretch. It takes the form of any linear transformation $g(s) = a + bf(s)$, with a, b real numbers. In order to enhance contrast of the data set $\boldsymbol{x}$, with values in $K = [0, 1]$, we will apply the linear transformation given by $x \mapsto (x - x_{1:n})/(x_{n:n} - x_{1:n})$. With this, the transformed data

Fig. 10.12. Result of applying the full range linear stretch.

have full range. Figure 10.12 shows the result of applying this transformation independently to each channel of Figure 10.1.

In the following, we will present the contrast enhancement by decorrelation, a technique based on the use of the covariance matrix of the data. Firstly, consider the plot of all the values of the dataset presented in Figure 10.13 in a 3D cube, where every point is painted with the respective color.

As can be seen in Figure 10.13, the amount of contrast enhancement that can be attained by any linear stretch applied independently to the Red, Green and Blue channels is limited. This is due to the fact that the data are correlated and, thus, the cloud of points is cigar-shaped; no pointwise radiometric linear transformation will fill the cube. We will see that it is possible to compute an adequate rotation, after which pointwise radiometric linear transformations can be applied in order to fill the space and, therefore, provide a very strong image enhancement. The inverse rotation applied to the stretched data will restore, to a certain extent, the original hue of the data.

The first part of the transformation consists of producing a spectral rotation and full stretch. Figures 10.14 and 10.15 show the result of this stage, namely the points in the RGB cube and the resulting image.

In this case, the skin class had its hue more preserved than the others. This is due to a number of factors, related to the transformation, but in this case it is mainly because it is the most numerous class and, then, the covariance matrix of the whole dataset is strongly influenced by these values.

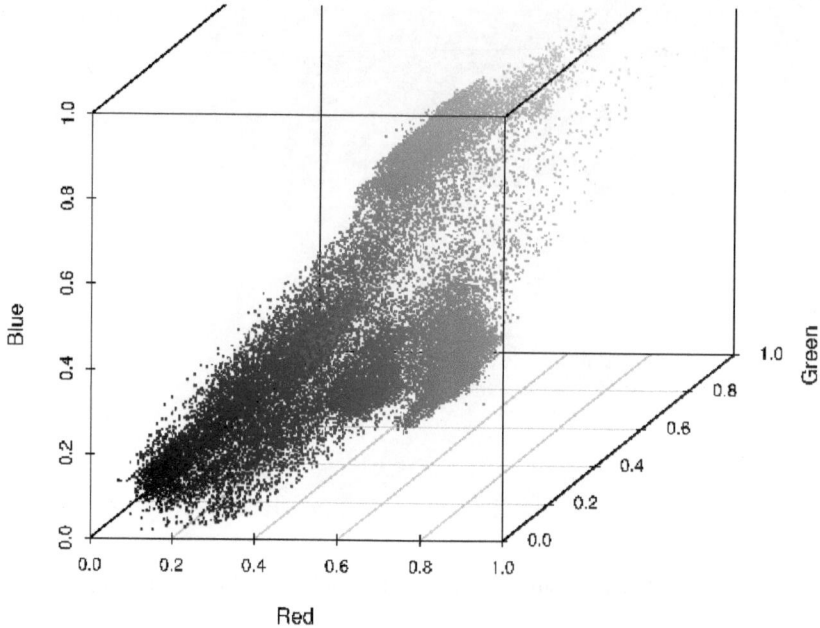

Fig. 10.13. Values of the original image in the RGB cube.

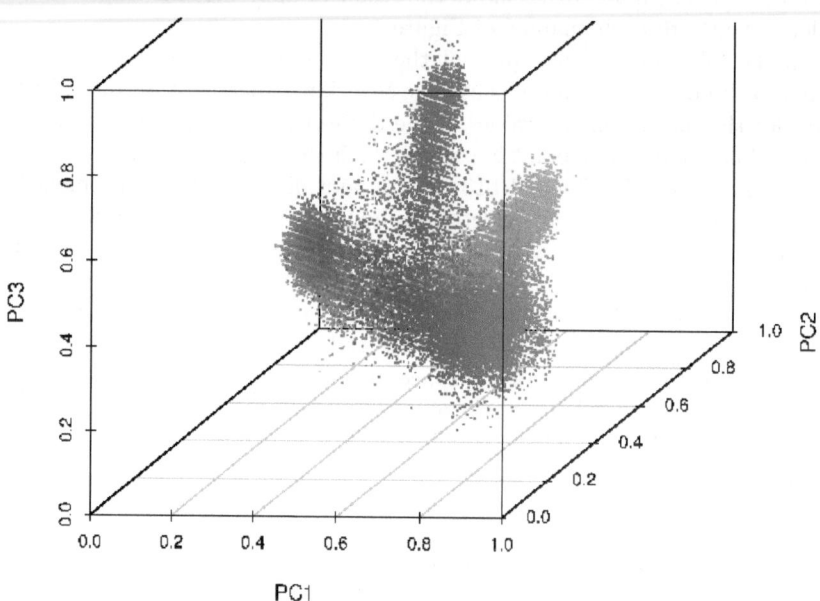

Fig. 10.14. Values of the image after principal components transformation in the RGB cube.

Fig. 10.15. Image after principal components transformation stretch.

The second part of the transformation consists of making the inverse rotation, in order to retrieve, to a certain extent, the original hues. Figures 10.16 and 10.17 show the result of this stage, namely the points in the RGB cube and the resulting image.

In the following, we will provide the technical details of this important transformation, which is closely related to the well know Principal Components Analysis (PCA). We will discuss here only RGB images, but the technique can be applied to data in $\mathbb{R}^m$, with $m \geq 2$.

In order to transform f into g using contrast enhancement by decorrelation, we first need the mean and covariance matrix of f.

Consider $f_1, f_2 : S \to \mathbb{R}$; then $\mathbb{R}^S$ is a vector space and we can define the following operations:

- sum of images: $f_1 + f_2 \in \mathbb{R}^S$, given by $(f_1 + f_2)(s) = f_1(s) + f_2(s)$ for every $s \in S$,
- scalar times an image: $\alpha f_1 \in \mathbb{R}^S$, given by $(\alpha f_1)(s) = \alpha f_1(s)$ for every $\alpha \in \mathbb{R}$ and every $s \in S$,
- the scalar or inner product between two images, given by

$$\langle f_1, f_2 \rangle = \sum_{s \in S} f_1(s) f_2(s),$$

- the L_d norm of an image, given by

$$\|f_1\|_d = \begin{cases} \left(\frac{1}{\#S} \sum_{s \in S} f_1^d(s) \right)^{1/d} & \text{if } d > 0, \\ \max_{s \in S} f_1(s) & \text{if } d = \infty. \end{cases}$$

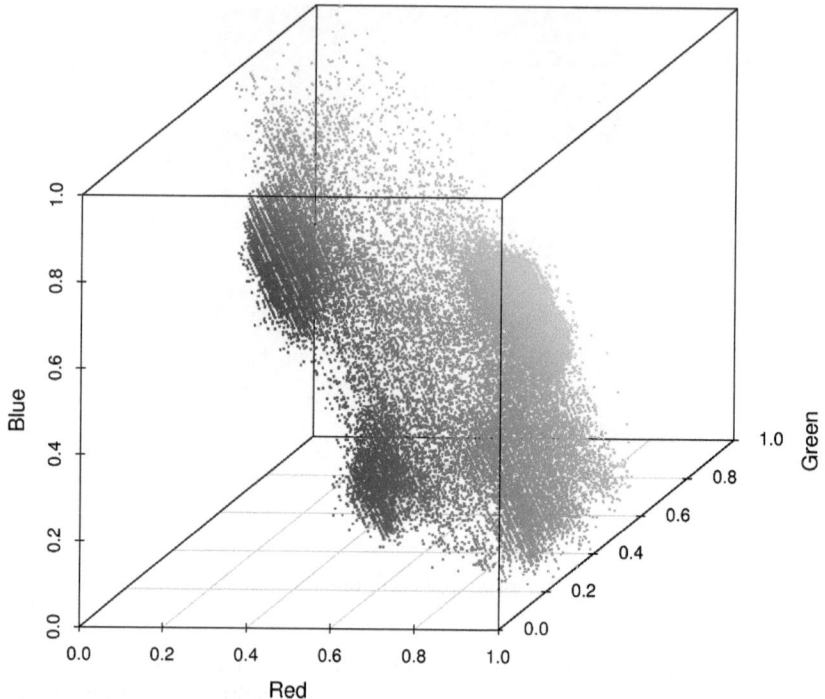

Fig. 10.16. Values of the image after contrast enhancement by decorrelation in the RGB cube.

Fig. 10.17. Image after contrast enhancement by decorrelation.

Usually, L_1 is called "Manhanttan norm" and L_2 "Euclidean norm" and, for the sake of brevity, $\|f_1\|_2$ is simply denoted $\|f_1\|$.

Consider now $\mathbb{I}_S$, the unitary image in $\mathbb{R}^S$, i.e., $\mathbb{I}(s) = 1$ for every $s \in S$.

Definition 10.1 (Mean value). *The mean of f_1 is given by $\langle f_1, \mathbb{I}_s \rangle$:*

$$\overline{f}_1 = \langle f_1, \mathbb{I}_s \rangle = \frac{1}{\#S} \sum_{s \in S} f_1(s).$$

Definition 10.2 (Centered image). *The centered version of f_1 is $\widetilde{f}_1 = f_1 + (-\overline{f}_1)\mathbb{I}_S$.*

Note that $\overline{\widetilde{f}_1} = 0$, i.e., the mean of the centered version of any image is zero.

Definition 10.3 (Covariance). *The covariance between two scalar images f_1 and f_2 is given by the inner product of their centered versions, i.e., $\mathrm{Cov}(f_1, f_2) = \langle \widetilde{f}_1, \widetilde{f}_2 \rangle$.*

Definition 10.4 (Variance). *The variance of a scalar-valued image f_1 is the covariance between f_1 and f_1, i.e., $\mathrm{Var}(f_1) = \mathrm{Cov}(f_1, f_1) = \langle \widetilde{f}_1, \widetilde{f}_1 \rangle = \|\widetilde{f}_1\|$.*

It is immediate that $\mathrm{Cov}(f_i, f_j) = \mathrm{Cov}(f_j, f_i)$ and that $\mathrm{Var}(f_i) \geq 0$ being equal to zero if and only if $f_i(s) = k$, $k \in \mathbb{R}$, for every $s \in S$.

Definition 10.5 (Linear Correlation). *The coefficient of (linear) correlation between f_1 and f_2 is*

$$\widehat{\rho}_{f_1, f_2} = \frac{\mathrm{Cov}(f_1, f_2)}{\sqrt{\mathrm{Var}(f_1)\,\mathrm{Var}(f_2)}} = \frac{\langle \widetilde{f}_1, \widetilde{f}_2 \rangle}{\|\widetilde{f}_1\| \|\widetilde{f}_2\|}.$$

The following property can be easily checked: $-1 \leq \widehat{\rho}_{f_1, f_2} \leq 1$ for every $f_1, f_2 \in \mathbb{R}^S$, and if $\widehat{\rho}_{f_1, f_2} = 0$ we say that f_1 and f_2 are uncorrelated. Notice that $\widehat{\rho}_{f_1, f_2} = 0$ does not imply, in general, that the random variables F_1 and F_2, whose outcomes are f_1 and f_2 respectively, are independent. Independence is more general than zero correlation.

Let us move now back to RGB images as, for instance, $f, g : S \rightarrow \mathbb{R}^3$. Whenever neccessary, each component will be made explicit by $f = (f_1, f_2, f_3)$, where $f_i : S \rightarrow \mathbb{R}$ for $i = 1, 2, 3$. We can now define the covariance matrix of image f as

$$\Sigma_f = \left(\mathrm{Cov}(f_i, f_j) \right)_{1 \leq i \leq j \leq 3}.$$

The diagonal elements of Σ_f are the variance of each component, and the off-diagonal elements are the covariances among them.

Since $\mathrm{Cov}(f_i, f_j) = \mathrm{Cov}(f_j, f_i)$, Σ_f is symmetric and there exists A orthogonal (i.e. for which holds that $A^t A$ is the identity matrix) such that

$A^t \Sigma_f A = \Omega$ is diagonal. The columns of A are called *eigenvectors* of Σ_f, while the elements in the diagonal of Ω are the *eigenvalues* of Σ_f. It is convenient to swap the columns of A such that the diagonal elements of Ω are in non-increasing order.

Defining $g : S \rightarrow \mathbb{R}^3$ as $g = fA$, we say that g is the Karhunen-Loève transform or, more often, the Principal Components transform of f and $g = (g_1, g_2, g_3)$ are the principal components of f. Since g is the result of a linear orthogonal transformation of f, it is a rotation in $\mathbb{R}^3$.

Principal components are uncorrelated by construction, since

$$\mathrm{Cov}(g) = \Sigma_g = \Sigma_{fA} = A^t \Sigma_f A = \Omega = \begin{pmatrix} \omega_1 & 0 & 0 \\ 0 & \omega_2 & 0 \\ 0 & 0 & \omega_3 \end{pmatrix}$$

and, therefore, $\mathrm{Cov}(g_i, g_j) = 0$ when $i \neq j$, so the principal component image is formed by uncorrelated bands.

The diagonal entries of Ω are the variance of each component, i.e., $\mathrm{Var}(g_i) = \omega_i$. The variance can be regarded as a measure of innovation or of information and, therefore,

$$\pi_i = \frac{\omega_i}{\sum_{j=1}^3 \omega_j} \tag{10.9}$$

can be used as a measure of the fraction of information band g_i carries with respect to the whole data set.

As presented, the principal components transformation of image f is obtained using Σ_f, the covariance matrix of f. Arbitrary spectral rotations can be performed on f with respect to any appropriate symmetric matrix Σ, being the only requirement that the dimensions are compatible. In this manner one can use, for instance, the covariance matrices from selected samples of the whole data set as, for instance, the ones computed for skin (equation (10.5)), sky (equation (10.6)), hat (equation (10.7)) and cactus (equation (10.8)). These transformations produce, respectively, Figures 10.18(a), 10.18(b), 10.18(c) and 10.18(d).

One of the most succesful applications of the principal components transformation is related to image compression. For computer graphic applications, it can be used to transform a colored image into gray-tones one. Figure 10.19 presents the three principal components of Figure 10.1 in decreasing order of information content, as measured by equation (10.9).

The proportion of variance each principal component explains in this case is 0.810, 0.185 and 0.005 respectively. The proportion of information the red, green and blue channels carry are, respectively, 0.151, 0.197 and 0.652. Notice how principal components is able to concentrate more information (81%) than the best band with respect to information content (the blue one, 65%).

(a) Using $\hat{\Sigma}_{\text{skin}}$ (b) Using $\hat{\Sigma}_{\text{sky}}$

(c) Using $\hat{\Sigma}_{\text{hat}}$ (d) Using $\hat{\Sigma}_{\text{cactus}}$

Fig. 10.18. Contrast enhancement by decorrelation with respect to selected samples.

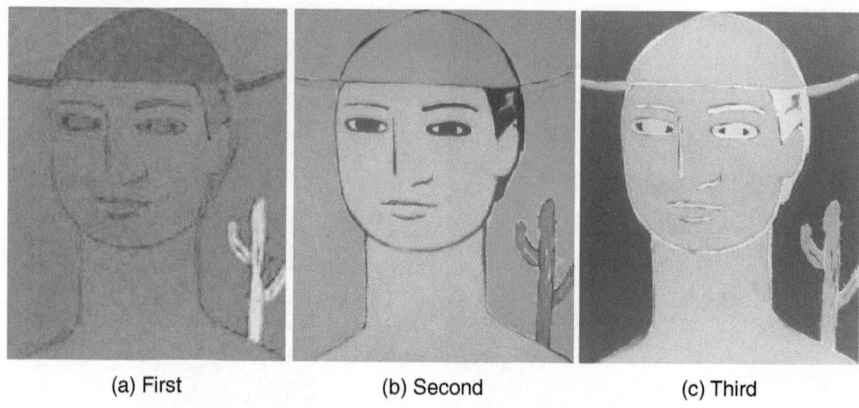

(a) First   (b) Second   (c) Third

Fig. 10.19. The three principal components.

10.4 Correlated Observations

The dependence of observations from site to site has not been considered so far. Spatial correlation is an issue frequently observed in any picture, since close sites tend to have similar values. This spatial dependence is visually perceived as texture, and the modelling and simulation of a particular class of textures is the purpose of this section.

One of the most useful ways to describing, simulating and analyzing structured data, as is the case of images, is by means of Gaussian random fields (discrete random fields will be treated in Section 10.6, for modelling context). These models rely on the specification of two ingredients, namely, marginal laws and correlation structure. More general models require the specification of the complete joint distribution of the process, which is seldom possible in practice, let alone convenient.

Equation (10.3) presents the density of the multivariate Gaussian distribution. Along previous Sections, the meaning of 'multivariate' was strongly tied to the components, i.e., bands of a multispectral image $f : S \to \mathbb{R}^p$. Covariance and correlation were measures of the relationship between observations in different bands i, j but on the same site $s \in S$, in other words, between $f_i(s)$ and $f_j(s)$.

In this section, this multivariate structure will be used to describe the stochastic relationship between the observations in sites $s, t \in S$ of the real-valued image $f : S \to \mathbb{R}$, i.e., between $f(s)$ and $f(t)$.

The mean and covariance functions of the random field F defined on S are, respectively,

$$\mathrm{E}(F(s)) = \mu(s), \text{ and} \qquad (10.10)$$

$$\text{Cov}(F(s), F(t)) = \text{E}(F(s)F(t)) - \mu(s)\mu(t), \qquad (10.11)$$

and we will only consider stationary processes for which holds that

$$\mu(s) = \mu \text{ for every } s \in S, \text{ and}$$
$$\text{Cov}(F(s), F(t)) = C(\|s - t\|),$$

where $C : \mathbb{Z} \to \mathbb{R}$. In such processes, the mean does not change with the position and the covariance function only depends on the distance between the coordinates, so one of the points can be the origin and, then, $C(\|s-t\|) = C(\|s - 0\|) = C(\|s\|)$. Without loss of generality, we can also assume that the random field has unitary variance, so covariance and (auto) correlation functions coincide:

$$\text{Cov}(F(s), F(0)) = \rho(F(s), F(0)) = \varrho(\|s\|).$$

An autocorrelation function is called separable if it admits the following decomposition:

$$\varrho(\|s\|) = \varrho(\|(s_1, s_2)\|) = \varrho_1(s_1)\varrho_2(s_2),$$

where $\varrho_i : \mathbb{Z} \to [0, 1)$, $i = 1, 2$, are suitable one-dimensional correlation functions.

The point now is, given a suitable correlation function ϱ, how can we get samples from the stochastic process F obeying the zero-mean unitary variance Gaussian model with the specified correlation function?

The straightforward approach consists of factoring the correlation function ϱ into the product $\varrho = AA^t$. If F' is a collection of independent identically distributed Gaussian random variables with zero mean and unitary variance, then the convolution between F' and A, $F = A * F'$, has zero mean and correlation matrix

$$\text{E}(FF^t) = \text{E}(AF'(AF')^t) = \text{E}(AF'(F')^t A^t) = A\,\text{E}(F'(F')^t)A^t = A\mathbb{I}A^t = \varrho.$$

The factorization of ϱ can be accomplished by Cholesky decomposition.

Given a symmetric positive definite matriz ϱ, its Cholesky decomposition is an upper triangular matrix A such that $\varrho = A^t A$. It is a special case of a broader class of transformations: matrix factorization, a successful tool for image analysis (Lee and Seung 1999). Consider, for instance, the following correlation matrix ϱ (with only the elements in and above the principal diagonal here shown to unclutter the visualization):

$$\varrho = \begin{pmatrix} 1.000 & 0.992 & 0.621 & 0.465 & 0.979 & 0.991 & 0.971 \\ & 1.000 & 0.604 & 0.446 & 0.991 & 0.995 & 0.984 \\ & & 1.000 & -0.177 & 0.687 & 0.668 & 0.502 \\ & & & 1.000 & 0.364 & 0.417 & 0.457 \\ & & & & 1.000 & 0.994 & 0.960 \\ & & & & & 1.000 & 0.971 \\ & & & & & & 1.000 \end{pmatrix}, \qquad (10.12)$$

and its Cholesky decomposition

$$A = \begin{pmatrix} 1 & 0.992 & 0.621 & 0.465 & 0.979 & 0.991 & 0.971 \\ 0 & 0.129 & -0.086 & -0.111 & 0.156 & 0.096 & 0.161 \\ 0 & 0 & 0.779 & -0.610 & 0.118 & 0.079 & -0.111 \\ 0 & 0 & 0 & 0.632 & -0.002 & 0.024 & -0.069 \\ 0 & 0 & 0 & 0 & 0.054 & -0.015 & -0.043 \\ 0 & 0 & 0 & 0 & 0 & 0.036 & 0.090 \\ 0 & 0 & 0 & 0 & 0 & 0 & 0.067 \end{pmatrix}.$$

Notice the presence of a first-order negative correlation coefficient in equation (10.12).

Applying the aforementioned direct transformation on a set of 128×128 independent identically distributed Gaussian random variables with zero mean and untary variance, one obtains the field shown in Figure 10.20.

As pointed out by (Dietrich and Newsam 1997), this direct procedure is computationally expensive due to the need of factoring the autocorrelation matrix ϱ and then computing each convolution on the independent Gaussian field. The simplest way to accomplish this task is using the Fourier transform.

The spectral density function of the random field F is the Fourier transform of its covariance function, and if F' is the independent zero-mean unitary variance Gaussian field as above, then the spectral density of $A * F'$ is the Fourier transform of $\varrho = AA^t$. Assume in the following that ϱ is a separable

Fig. 10.20. Correlated Gaussian random field by direct transformation.

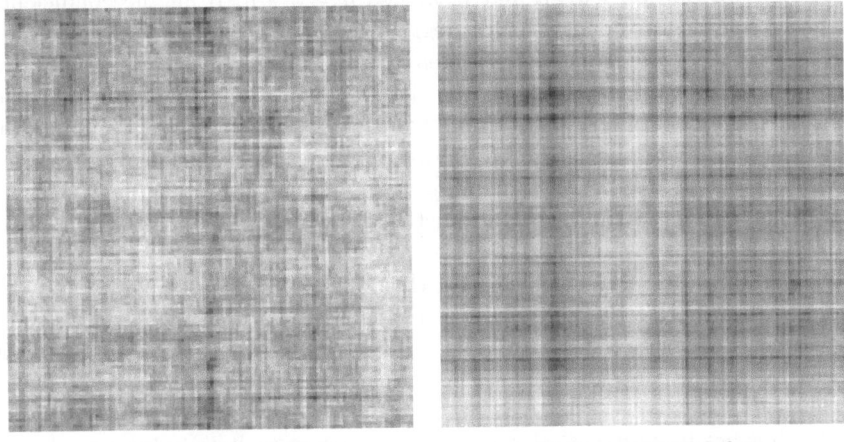

(a) Exponential correlation   (b) Gaussian correlation

Fig. 10.21. Gaussian Random fields.

function for which holds that $\varrho(s,t) = \varrho_1(s)\varrho_1(t)$. Then, an algorithm for generating F with specified correlation function ϱ is the following:

1. Simulate F', as defined above
2. Compute the frequency mask $\Psi(u,v) = \sqrt{\mathcal{F}(u)\mathcal{F}(v)}$
3. Calculate the resulting field $F = \mathcal{F}^{-1}(\Psi \cdot \mathcal{F}(F'))$,

where '$\cdot$' denotes the element-by-element product and $\mathcal{F}$ the Fourier transform operator.

The easiest way to specify valid correlation functions for this method is through functional forms. Usual functional forms include the log-linear or exponential ($\varrho_1(t) = \exp\{-t\}$) and log-quadratic or Gaussian ($\varrho_1(t) = \exp\{-t^2\}$) functions. Figure 10.21 presents two outcomes of such fields on a grid of size 128×128.

(Schlather 1999) provides a detailed discussion of the properties of positive definite functions, which are good candidates for building covariance functions.

Using the inversion method, and the technique based on the frequency mask, (Bustos et al 2001) present a general technique for obtaining correlated fields with a wide variety of marginal distributions.

10.5 Filtering

The discussion that ended the last section, namely the generation of textures by means of convolutions, provides a clue for image restoration.

Back to the model presented in equation (10.2), consider the situation of having an unobserved image f that is transformed into the available image

g by the convolution with a matrix h, i.e., $g = f * h$. If all the coefficients in h are non-negative, his kind of degradation is called "blurring" or "low-pass". Figure 10.22(a) shows the green channel of the painting presented in Figure 10.1, denote it f. Notice the detail of the eye; it will be useful to compare the effect of the forthcoming transformations with the original data.

Figure 10.22(b) shows the effect of applying a convolution mask of size 9×9 with values $1/81$ (i.e., it shows $g = h * f$). The detail of the eye clearly shows the extent of the degradation: the light line below the eyelid is now barely visible, for instance.

Assume also that h is known and that one wants to retrieve f from the observed data g. It would suffice to convolve g with a function h' that ellimi-nates the effect of h. When available, h' is known as the inverse filter and it is defined as the function that satifies $f = \Psi(g) = \Psi(f * h) = f * h * h'$, so $h * h'$ is the identity.

Such exact inversion is seldom possible in practice, mainly because degra-dation is frequently associated to noise. Figure 10.22(c) presents the effect of adding Gaussian white noise to the blurred image ($g = h * f + n$). Though at first glace it might seem that this picture has retrieved part of the detail lost in figure 10.22(b), this is an effect of noise on blurred imagery; notice that in the detail of the eye there is no edge information, in the contrary, the degradation is now stronger. In such cases, one has to rely on estimators.

The general purpose of restoration through filtering is finding an estimate of f, say $\hat{f}$, which is a function of the available data and, eventually, of addi-tional information about the degradation process.

There are many approaches to restoration. In the following we will com-ment a few of those that stem from statistical ideas.

The image shown in figure 10.22(c) suffers from additive Gaussian noise. In this case, it was built in that way; in practice, this is the first hypothesis to be tested when there is no information suggesting other model.

A first idea to combat this kind of noise is taking local means. The ratio-nale behind this technique is that if $X_1, \ldots, X_n$ are independent identically distributed random variables with mean μ and variance σ^2, then the mean $\overline{X} = n^{-1} \sum_{i=1}^{n} X_i$ has the same mean and variance $n^{-1/2}\sigma^2$. Since the vari-ance is a measure of dispersion, the smaller the variance the more concentrated the data and, hopefully, the less noisy the image. Figure 10.23(a) presents the result of applying a Gaussian low-pass filter of size 13×13 to the blurred and noisy image.

A Gaussian low-pass filter of size $K \times K$ (K odd) and scale $s > 0$ is defined by a convolution mask with entries

$$a_{i,j} = Z_{K,s} \exp\left\{-\frac{i^2 + j^2}{2s^2}\right\},$$

where $-(K-1)/2 \leq i, j \leq (K-1)/2$ and $Z_{K,s}$ is the constant that grants $\sum_{i,j} a_{i,j} = 1$. Values decrease as they are further apart from the center of the mask, so the variance reduction is not a factor of K, but the filter introduces

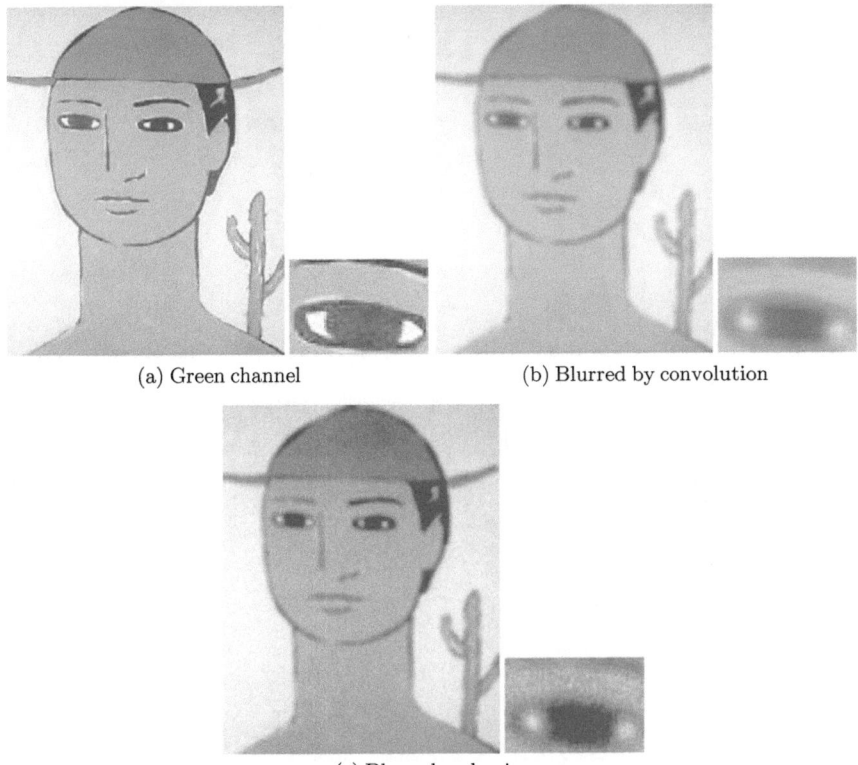

(a) Green channel   (b) Blurred by convolution

(c) Blurred and noisy

Fig. 10.22. Original image, blurred and noisy versions.

less blurring than one with constant values. Figure 10.23(b) shows the effect of applying a mask of side 13 with constant values equal to $1/13^2$; the intense blurring is noticeable, as well as the noise reduction.

Though not a convolutional, i.e., linear filter, the median is an interesting trade-off between noise reduction and edge preservation. Figure 10.23(c) presents the result of applying this filter over a window of size 13×13 to the blurred and noisy data. As can be seen, the noise reduction is less effective than the one obtained by the constant filter, but edges are better preserved.

The filters presented so far employ a fixed rule on the data, i.e., they are invariant. With the exception of the median, they are also linear. From the observation of the data, it is clear that the data do not follow the same model over the whole image: there are edges dividing dark and light areas, for instance, besides different textures. In order to cope with this local informatin, that can hardly be dealt with invariant filters, the literature abounds with the so-called adaptive techniques.

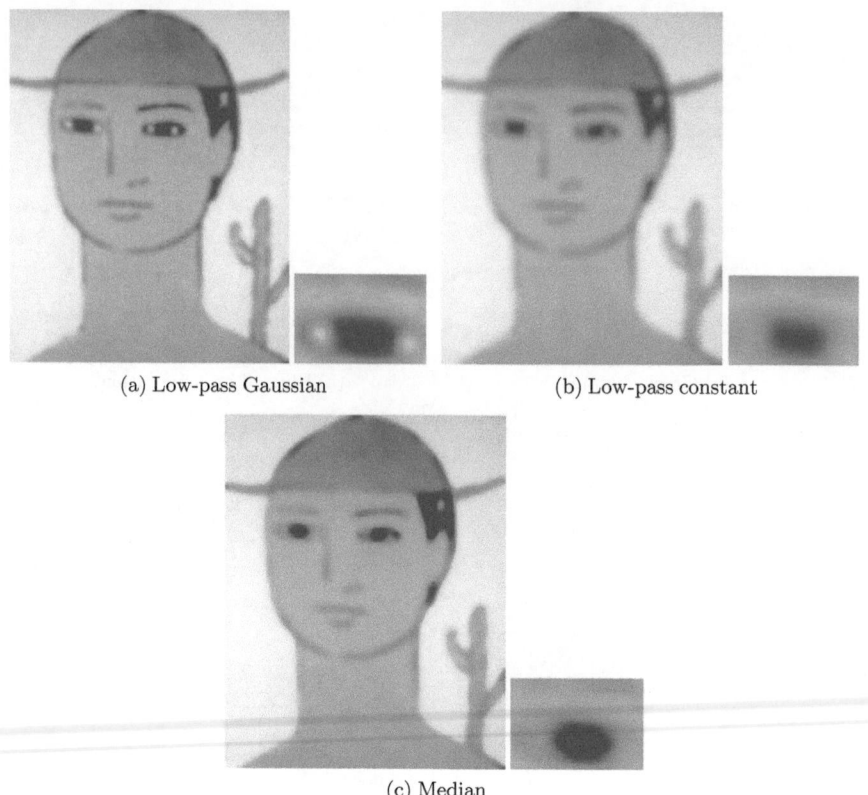

(a) Low-pass Gaussian   (b) Low-pass constant

(c) Median

Fig. 10.23. Filtered images with windows of size 13×13.

Consider, for instance, the Local-σ filter with parameter $\alpha > 0$. It computes a new value by first calculating the standard variation $\widehat{\sigma}$ of the data around the pixel f_s. The new value is the mean of those observations that lie within the interval $[f_s - \alpha\widehat{\sigma}, f_s + \alpha\widehat{\sigma}]$. See figure 10.24(a) for the result of applying such filter on the noisy and blurred data, over windows of size 9×9 with $\alpha = 1.5$.

The Local-σ filter belongs to the class of trimmed mean estimators, i.e., those where the estimate is computed as a weighted mean of the observations $\sum_{i=1}^{M} a_i x_s$, and the coefficients a_i typically depend on the whole data set. This class is known in the literature of quantitative robustness as T estimators (Maronna et al 2006).

An interesting and pioneering approach to adaptive image smoothing is the Nagao-Matsuyama filter (Nagao and Matsuyama 1979). It computes the mean on subwindows within the main window, and retains the value which is

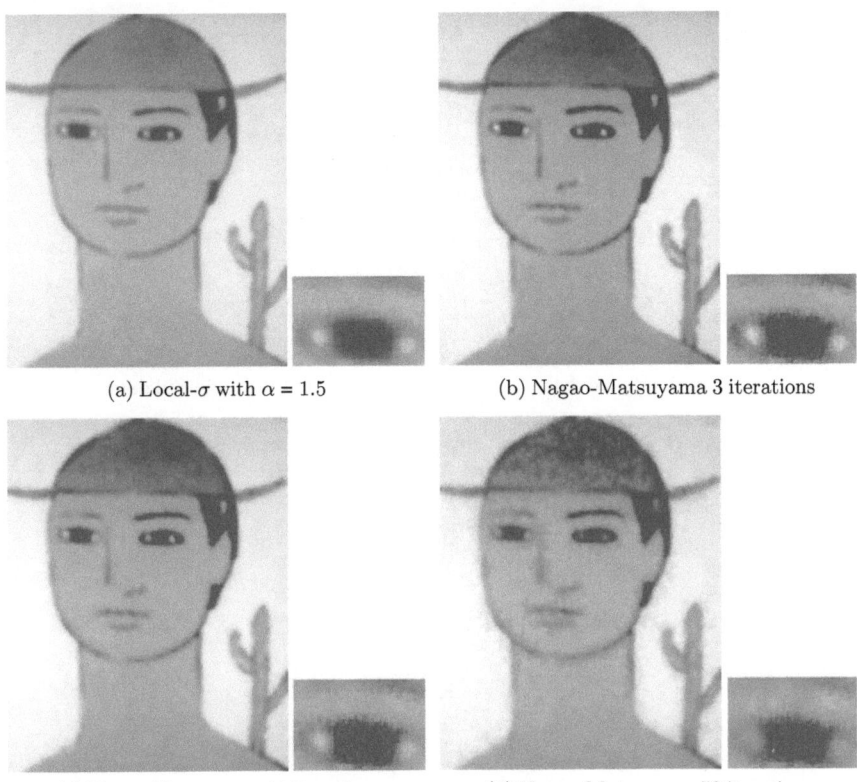

(a) Local-σ with $\alpha = 1.5$ (b) Nagao-Matsuyama 3 iterations

(c) Nagao-Matsuyama 11 iterations   (d) Nagao-Matsuyama 50 iterations

Fig. 10.24. Adaptive filters.

closer to the central one. In this manner, it adds geometrical information to the computation of the coefficient of this T estimate.

Figures 10.24(b), 10.24(c) and 10.24(d) present the result of applying the Nagao-Matsuyama filter to the blurred and noisy data from figure 10.22(c) iteratively 3, 11 and 50 times, respectively. Besides the noise reduction, the edge preservation is clear, as is the (often undesired) the blocking effect. In our implementation, nine windows of size 3 × 3 were used around the central pixel. Other window sizes, subwindow shapes and comparison criteria can be used depending on the acceptable hypothesis that govern the data formation.

Edge detection is an important tool in several image applications. In particular, many non-photorealistic rendering techniques rely on the detection of edges for the generation of cartoon models. Figure 10.25 presents the result of applying the laplacian filter to the images shown in figure 10.22. The laplacian filter is the discrete version of the second derivative of the image in each point.

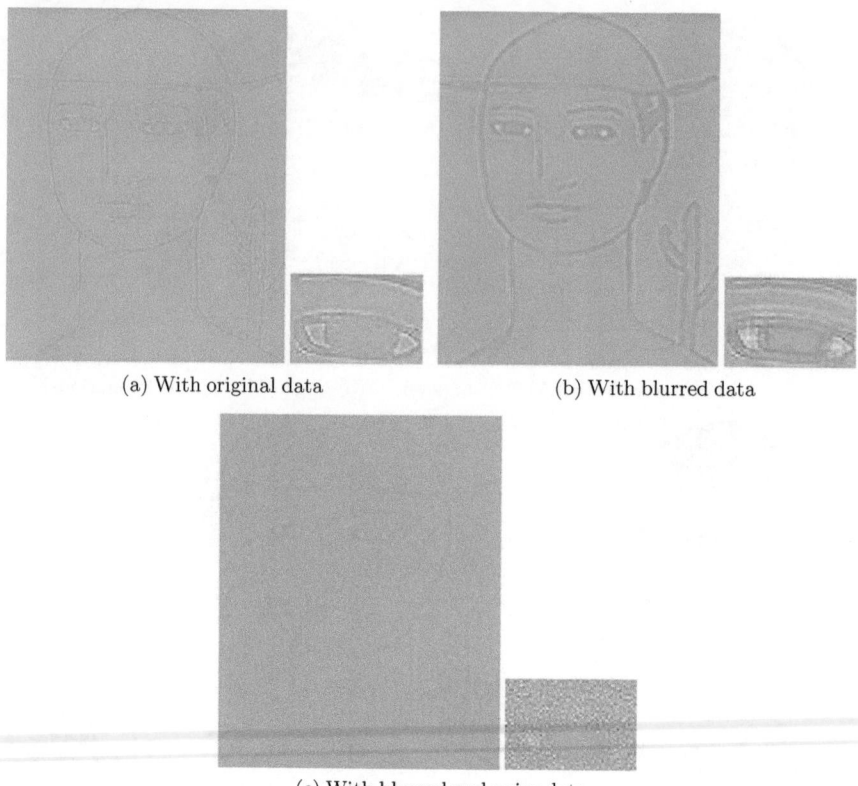

(a) With original data   (b) With blurred data

(c) With blurred and noisy data

Fig. 10.25. Edge detection by convolution with the laplacian filter.

It enhances edges regardless the orientation, and it is defined as a convolution filter with mask

$$\begin{pmatrix} 0 & -1 & 0 \\ -1 & 4 & -1 \\ 0 & -1 & 0 \end{pmatrix}.$$

This simple filter does not only enhance edges, but also noise. As can be seen in figures 10.25(a) and 10.25(b), it tolerates a fair amount of blurring in the input data, but checking figure 10.25(c) one concludes that when applied to noisy data the results are of little use.

An iterative procedure for the detection of edges under the presence of noise is presented by (Tupin et al 1998). The technique starts smoothing the data and detecting edges; these edges are used for further smoothing, but avoiding using data from different sides of an edge. After this selective smoothing has been applied, a new edge detection is performed and so on.

The literature about image filters is vast, stemming from classical approaches based on signal processing (Lim 1989), statistical methods (Kay 1993; Shiavi 1999), mathematical morphology (Serra 1988), wavelets and multiscale decomposition (Mallat 1999; Meyer 1993) among others.

10.6 Classes

Previous sections have dealt with properties of the observed data. In order to have a complete view of Geman and Geman model, c.f. equation (10.1), this section presents a brief discussion about one of the most important models for classes, namely, the Potts model.

Imagine we want to provide a stochastic model for the observation of maps, as the one presented in Figure 10.26(b). Such a model, if both adequate and tractable, would lead to two important outcomes. The first is the development of tools that take into account what is expected as a good result; this leads to Bayesian techniques. The second is the ability to generate believable maps that can be used as input for Monte Carlo experiments with the purpose of quantifying the quality of image processing procedures. (Moschetti et al 2006) provide a detailed discussion on the importance of using a Monte Carlo approach for speckle filters.

Figure 10.26 presents a typical situation where simulation is used for generating believable images. In this case, discussed by (Lucca et al 1998), the purpose is simulating an arbitrary number of images from a SAR sensor over the same area in the Amazon forest. Since this is technically and economically unfeasible, a map of edges is built from real data (Figure 10.26(a). Figure 10.26(a) presents the result of this preliminar analysis; this can be assumed as the ground truth. Figure 10.26(b) presents these areas after labelling them into classes; notice that there are disconnected segments with the same class as, for instance, the two red areas: a big one and a smaller, linear one.

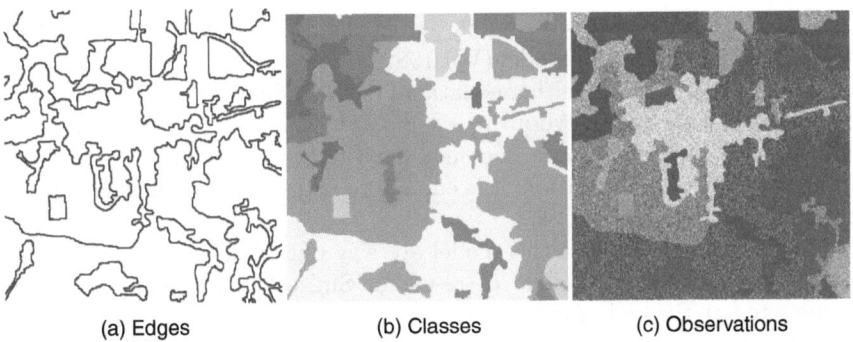

(a) Edges   (b) Classes   (c) Observations

Fig. 10.26. Edges, classes and observations.

Figure 10.26(b) presents the result of sampling data from those distributions that are able to characterize each type of class for a certain kind of sensor; for details about these distributions, the reader is referred to (Frery et al 1997; Frery et al 2007).

This approach grants the adequacy of both classes as observations, but the conclusions drawn from it are only pertinent for areas of the same type. When one wants to consider a fully random situation, classes have to be modeled as a stochastic process.

The Ising model was proposed in the early 20s as a means of explaining the magnetic properties of certain materials. The idea behind the model is describing the interaction of particles with two possible spin states; when most spins point to the same direction, there is magnetization.

The beauty of the Ising model is related to the complex behavior it presents, albeit its simple formulation. Instead of dealing directly with the joint distribution of all the random variables, one may start by proposing conditional laws governing the behavior of the random variable indexed by sites $s \in S \subset \mathbb{Z}^2$ in the following manner:

$$\Pr(X_s = \xi \mid X_{S \setminus \{s\}} = x_{S \setminus \{s\}}) = \Pr(X_s = \xi \mid X_{\partial_s} = x_{\partial_s}) \quad (10.13)$$

$$\propto \exp\{\beta \#\{t \in \partial_s : x_t = \xi\}\}, \quad (10.14)$$

where ξ is one of the possible states, β is a real number, $\partial_s \subset S \setminus \{s\}$ is the neigborhood of site $s \in S$, '$\propto$' denotes equality up to a constant that does not depend on ξ and '$\setminus$' denotes set substraction, i.e., $A \setminus B = A \cap B^c$.

Equations (10.13) and (10.14) imply two important issues of the model, respectively

1. that the probability of observing class ξ at site s given all other observations $x_{S \setminus s}$, depends only on the state of neighboring sites x_{∂_s}, and
2. that the log-probability depends linearly, through the factor β only, on the number of neighboring sites that chose class ξ.

Assuming a finite number of possible states, instead of just two as in the Ising model, one has the Potts model (Wu 1982).

A relevant question is either equation (10.14) induces a unique joint distribution for the classes on every site $s \in S \subset \mathbb{Z}^2$. The answer, when S is finite, is always affirmative, so one does not have to worry about computing the cumbersome joint distribution; it is enough to specify the conditional laws, as presented in equation (10.14).

Technical details about this model are way out of the scope of this book, since it is still an active field of research. Suggested additional references commented at the end of this Chapter.

A good repertoire of techniques includes the judicious use of algorithms for the simulation of the Potts model. Unfortunately, there are no straight ways of sampling from the joint distribution induced by equation (10.14). There are, though, at least four important iterative procedures that lead to obtaining such samples.

There are two main classes of techniques for obtaining samples from the joint distribution induced by equation (10.14): pointwise and cluster-based. The Gibbs sampler belongs to the former, while the Swendsen-Wang and Wolff algorithm belong to the latter; the Metropolis technique belongs to either class, depending on its implementation. In the following, we will briefly see the Gibbs sampler and the Wolff algorithm.

Starting from an arbitrary initial configuration, the Gibbs sampler consists of replacing the observation in each coordinate by the outcome of a random variable obeying the conditional distribuition specified in equation (10.14). This sampling should be performed infinitely many times for the whole set of coordinates in order to obtain a true sample from the correct distribution but, in practice, a finite number of iterations will do the trick.

The Wolff algorithm requires an additional structure. Starting from an arbitrary initial configuration, it first finds the connected clusters of neighboring sites that exhibit the same value. Once these clusters are found, each link is deleted with probability $e^{-\beta}$. Among the newly formed clusters, one is randomly chosen with a probability proportional to their sizes. All the observations in this last cluster are changed into a new value, uniformly sampled among the other allowable values. Similarly to the previous algorithm, this procedure should be repeated infinitely many times in order to obtain a genuine sample from the desired distribution, but a finite number of times will suffice in practice.

Figure 10.27 presents five outcomes of the Ising model on a squared support of size 64. Figures 10.27(a), 10.27(b) and 10.27(c) present outcomes of the model with two classes ($N = 2$), the indepent case ($\beta = 0$) and two positive values of β. Figures 10.27(d) presents the Potts model with three classes ($N = 3$) and $\beta = 0.88$, while Figure 10.27(e) presents the four classes ($N = 4$) model with $\beta = 0.96$.

Notice that the bigger β the tighter the clusters are, but there is also the tendency of having just one class with isolated spots. In fact, the theory of Potts random fields states that, for the case with infinite support, values of $\beta > \log(1 + \sqrt{N})$ yield outcomes of mostly one class.

Though these figures may not look too appealing of believable as prior outcomes for maps, when the Potts model is incorporated into a Bayesian context, the resulting techniques are much better than the ones obtained by merely dealing with marginal information. The reader is referred to the works by (Frery et al 2007) and (Frery et al in press) for a quantitative assessment of the improvement of classification results of using the Potts model for two quite different data sets, namely, polarimetric and optical.

Details about these simulation techniques can be found in (Metropolis et al 1953; Geman and Geman 1984; Swendsen and Wang 1987; Wolff 1989a; Wolff 1989b).

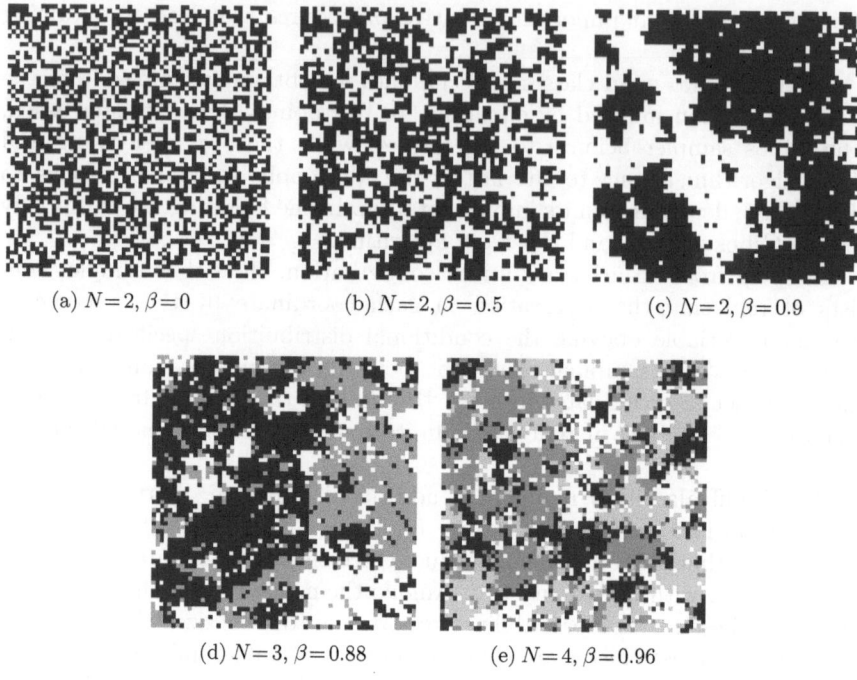

(a) $N=2, \beta=0$ (b) $N=2, \beta=0.5$ (c) $N=2, \beta=0.9$

(d) $N=3, \beta=0.88$ (e) $N=4, \beta=0.96$

Fig. 10.27. Samples from the Ising model.

10.7 Comments and References

The paper by (Geman and Geman 1984) is one of the most influential works in the area. Until the preparation of this book, it counted more than 3600 citations in the ISI Web of knowledge. Though less cited, the work by (Carnevalli et al 1985) is also important and seminal. The formal approach to image formation, analysis and restoration was presented in the former, and extended in (Bustos and Frery 1992). Edge detection combining image processing and analysis is well presented, with excellent results for difficult to deal with nongaussian data, in (Tupin et al 1998).

Regarding the multivariate Gaussian distribution, there are many textbooks that present it. The three we cited here provide different approaches. (Krzanowski 1988) is directed towards applications, with emphasys on classification and discriminant analysis. (Tong 1990) is brief and more theoretical thant the former. (Muirhead 1982) is quite rigorous, and provides a comprehensive account of the theory.

R (R 2006) is one of the most successful initiatives in computational statistics. It can be freely downloaded, for a variety of hardware and software platforms, from `www.r-project.org`. This was the choice because (quoting the authors' site):

> One of R's strengths is the ease with which well-designed publication-quality plots can be produced, including mathematical symbols and formulæ where needed. Great care has been taken over the defaults for the minor design choices in graphics, but the user retains full control.

The book by (Venables and Ripley 2002) is an excellent reference for both the language and its use in the statistical analysis of data.

Among the books on image processing, the following provide either a classical viewpoint or a specific approach. (Jain 1989) is one of the most complete and classical textbooks on this subject, while (Lim 1989) tackles image processing from a signal processing viewpoint; these are essential and complementary books. Two other very important books on image processing are those by (Gonzalez and Woods 1992), with emphasis on computer vision, and by (Richards and Jia 1999), which provides an approach particularly useful for remote sensing. The work by (Barrett and Myers 2004) is a monumental book with strong bias towards image formation and low-level image processing.

The work by (Schlather 1999) is an excellent starting point for the study of random fields. He discusses the general problem of characterizing valid covariance functions as positive definite (complex) functions, provides examples, ways of building such functions and of checking candidates, and also discusses simulation techniques and computational issues. He mostly discusses Gaussian random fields, but other laws are also considered. The works by (Bustos et al 2001) and by (Tough and Ward 1999) also study techniques for the generation of non-Gaussian random fields with specified correlation structure.

Additional references about the use of robust statistics in the design of image filters are (Allende et al 2006; Bustos et al 2002; Frery et al 1998).

The paper by (Wu 1982) is one of the most complete and authoritative works about the Potts model. This, along with the book by (Kinderman and Snell 1980) are excellent starting points for the study of this important model for image processing and analysis. The book by (Winkler 2006) provides an updated account of applications of this model in image analysis. (Pickard 1987) presents, from a probabilistic viewpoint, relevant issues of the Ising model, while (Besag 1986; Besag 1989) propose a statistical framework for image analysis and restoration.

References

[Allende et al 2006]H. Allende, A. C. Frery, J. Galbiati, and L. Pizarro. M-estimators with asymmetric influence functions: the GA0 distribution case. *Journal of Statistical Computation and Simulation*, 76(11): 941–956, 2006.

[Barrett and Myers 2004]H. H. Barrett and K. J. Myers. *Foundations of Image Science*. Pure and Applied Optics. Wiley-Interscience, NJ, 2004.

[Besag 1986]J. Besag. On the statistical analysis of dirty pictures (with discussion). *Journal of the Royal Statistical Society B*, 48(3):259–302, 1986.

[Besag 1989]J. Besag. Towards Bayesian image analysis. *Journal of Applied Statistics*, 16(3):395–407, 1989.

[Bustos et al 2001]O. H. Bustos, A. G. Flesia, and A. C. Frery. Generalized method for sampling spatially correlated heterogeneous speckled imagery. *EURASIP Journal on Applied Signal Processing*, 2001(2):89–99, June 2001.

[Bustos and Frery 1992]O. H. Bustos and A. C. Frery. A contribution to the study of Markovian degraded images: an extension of a theorem by Geman and Geman. *Computational and Applied Mathematics*, 11(1): 17–29, 281–285, Jan., Sept. 1992.

[Bustos et al 2002]O. H. Bustos, M. M. Lucini, and A. C. Frery. M-estimators of roughness and scale for GA0-modelled SAR imagery. *EURASIP Journal on Applied Signal Processing*, 2002(1):105–114, 2002.

[Carnevalli et al 1985]P. Carnevalli, L. Coletti, and S. Patarnello. Image processing by simulated annealing. *IBM Journal of Research and Development*, 29(6):569–579, Nov. 1985.

[Dietrich and Newsam 1997]C. R. Dietrich and G. N. Newsam. Fast and exact simulation of stationary Gaussian processes through circulant embedding of the covariance matrix. *SIAM Journal on Scientific Computing*, 18(4):1088–1107, 1997.

[Frery et al 2007]A. C. Frery, A. H. Correia, and C. C. Freitas. Classifying multifrequency fully polarimetric imagery with multiple sources of statistical evidence and contextual information. *IEEE Transactions on Geoscience and Remote Sensing*, 45(10):3098–3109, 2007.

[Frery et al in press]A. C. Frery, S. Ferrero, and O. H. Bustos. The influence of training errors, context and number of bands in the accuracy of image classification. *International Journal of Remote Sensing*, in press.

[Frery et al 1997]A. C. Frery, H.-J. Müller, C. C. F. Yanasse, and S. J. S. Sant'Anna. A model for extremely heterogeneous clutter. *IEEE Transactions on Geoscience and Remote Sensing*, 35(3):648–659, may 1997.

[Frery et al 1998]A. C. Frery, S. J. S. Sant'Anna, N. D. A. Mascarenhas, and O. H. Bustos. Robust inference techniques for speckle noise reduction in 1-look amplitude SAR images. *Applied Signal Processing*, 4:61–76, 1997.

[Geman and Geman 1984]D. Geman and S. Geman. Stochastic relaxation, Gibbs distributions and the Bayesian restoration of images. *IEEE Transactions on Pattern Analysis and Machine Intelligence*, 6(6): 721–741, 1984.

[Gonzalez and Woods 1992]R. C. Gonzalez and R. E. Woods. *Digital Image Processing*. Addison-Wesley, MA, 1992.

[Jain 1989]A. K. Jain. *Fundamentals of Digital Image Processing*. Prentice-Hall International Editions, Englewood Cliffs, NJ, 1989.

[Kay 1993]S. M. Kay. *Fundamentals of statistical signal processing*. Prentice Hall Signal Processing Series. Prentice Hall, NJ, 1993.

[Kinderman and Snell 1980]R. Kinderman and J. L. Snell. *Markov Random Fields and Their Application*. Contemporary Mathematics. AMS, Providence, Rhode Island, 1980.

[Krzanowski 1988]W. J. Krzanowski. *Principles of Multivariate Analysys: a User's Perspective*. Oxford Statistical Science Series. Claredon Press, Oxford, 1988.

[Lee and Seung 1999]D. D. Lee and H. S. Seung. Learning the parts of objects by non-negative matrix factorization. *Nature*, 401:788–791, 1999.

[Lim 1989]J. S. Lim. *Two-Dimensional Signal and Image Processing*. Prentice Hall Signal Processing Series. Prentice Hall, Englewood Cliffs, 1989.

[Lucca et al 1998]E. V. D. Lucca, C. C. Freitas, A. C. Frery, and S. J. S. Sant'Anna. Comparison of SAR segmentation algorithms. In *Second Latinoamerican Seminar on Radar Remote Sensing: Image Processing Techniques*, pages 123–130, Santos, SP, Brazil, Sept. 1998. European Space Agency (ESA).

[Mallat 1999]S. Mallat. *A wavelet tour of signal processing*. Academic, San Diego, CA, 2 edition, 1999.

[Maronna et al 2006]R. A. Maronna, R. D. Martin, and V. J. Yohai. *Robust Statistics: Theory and Methods*. Wiley series in Probability and Statistics. Wiley, England, 2006.

[Metropolis et al 1953]N. Metropolis, A. W. Rosembluth, M. N. Rosembluth, A. H. Teller, and E. Teller. Equations of state calculations by fast computing machines. *Journal of Chemical Physics*, pages 1087–1092, 1953.

[Meyer 1993]Y. Meyer and R. D. Ryan. *Wavelets: Algorithms & Applications*. SIAM, Philadelphia, 1993.

[Moschetti et al 2006]E. Moschetti, M. G. Palacio, M. Picco, O. H. Bustos, and A. C. Frery. On the use of Lee's protocol for speckle-reducing techniques. *Latin American Applied Research*, 36(2):115–121, 2006.

[Muirhead 1982]R. J. Muirhead. *Aspects of Multivariate Statistical Theory*. Wiley Series in Probability and Mathematical Statistics. Wiley, New York, 1982.

[Nagao and Matsuyama 1979]M. Nagao and T. Matsuyama. Edge preserving smoothing. *Computer Graphics and Image Processing*, 9:394–407, 1979.

[Oliver and Quegan 1998]C. Oliver and S. Quegan. *Understanding Synthetic Aperture Radar Images*. Artech House, Boston, 1998.

[Pickard 1987]D. K. Pickard. Inference for discrete Markov fields: The simplest nontrivial case. *Journal of the American Statistical Association*, 82(1):90–96, 1987.

[R 2006]R Development Core Team. R: A language and environment for statistical computing, 2006. ISBN 3-900051-07-0.

[Richards and Jia 1999]J. A. Richards and X. Jia. *Remote Sensing Digital Image Analysis*. Springer, Berlin, 1999.

[Serra 1988]J. P. F. Serra. *Image Analysis and Mathematical Morphology: Theoretical Advances*, volume 2. Academic Press, London, 1988.

[Shiavi 1999]R. Shiavi. *Introduction to Applied Statistical Signal Analysis*. Academic Press, San Diego, 1999.

[Schlather 1999]M. Schlather. Introduction to positive definite functions and to unconditional simulation of random fields. Technical Report ST-99-10, Department of Mathematics and Statistics, Lancaster University, UK, 1999.

[Swendsen and Wang 1987]R. Swendsen and J. Wang. Nonuniversal critical dynamics in Monte Carlo simulations. *Physical Review Letters*, 58(2):86–88, 1987.

[Tong 1990]Y. L. Tong. *The Multivariate Normal Distribution*. Springer Series in Statistics. Springer-Verlag, New York, 1990.

[Tough and Ward 1999]R. J. A. Tough and K. D. Ward. The correlation properties of gamma and other non-Gaussian processes generated by memoryless nonlinear transformation. *Journal of Physics D: Applied Physics*, 32:3075–3084, 1999.

[Tupin et al 1998]F. Tupin, H. Maitre, J.-F. Mangin, J.-M. Nicholas, and E. Pechersky. Detection of linear features in SAR images: application to road network extraction. *IEEE Transactions on Geoscience and Remote Sensing*, 36(2):434–453, 1998.

[Tupin et al 1998]F. Tupin, H. Maitre, J.-F. Mangin, J.-M. Nicholas, and E. Pechersky. Detection of linear features in SAR images: application to road network extraction. *IEEE Transactions on Geoscience and Remote Sensing*, 36(2):434–453, 1998.

[Venables and Ripley 2002]W. N. Venables and B. D. Ripley. *Modern Applied Statistics with S*. Statistics and Computing. Springer, New York, 4 edition, 2002.

[Winkler 2006]G. Winkler. *Image Analysis, Random Fields and Markov Chain Monte Carlo Methods: A Mathematical Introduction*. Stochastic Modelling and Applied Probability. Springer, 2 edition, 2006.

[Wolff 1989a]U. Wolff. Collective Monte Carlo updating for spin systems. *Physical Review Letters*, 62(4):361–364, 1989.

[Wolff 1989b]U. Wolff. Comparison between cluster Monte Carlo algorithms in the Ising model. *Physics Letters B*, 228(3):379–382, 1989.

[Wu 1982]F. Y. Wu. The Potts model. *Reviews of Modern Physics*, 54(1): 235–268, 1982.

11

Color Quantization

We have mentioned that the color discretization process is known as *quantization*. Quantization allows the conversion of an image having a continuous color gamut into one having a discrete color gamut.

In this chapter we will study in more detail the various problems that arise in color discretization. In order to do this, we must characterize more precisely the notion of quantization. Let $R_k = \{v_1, v_2, \ldots, v_k\}$ be a finite-dimensional subset of $\mathbb{R}^n$. A *quantization* of $\mathbb{R}^n$ is a map $q : \mathbb{R}^n \to R_k$. We are interested in quantization of color spaces. In this case, $\mathbb{R}^n$ is a finite-dimensional color space, and R_k is a finite subset of colors of the space. The map q is called a *quantization map*. The set R_k is called the *codebook* of the quantization map.

Frequently we need to define a quantization map between two finite sets of colors. This occurs, for instance, when we need to define a map $q : C \to C'$ from a color solid C, where colors are represented by M bits, into a color solid C', where colors are represented by N bits, with $N < M$. We consider this quantization problem as a particular case of the general definition above, by including the color space C into $\mathbb{R}^n$, for some n. Such a q is called an *N-bit quantization map*.

The quantization of a digital image consists in discretizing the image's color gamut, which implies the quantization of the color information associated with each pixel in the image. More precisely, if $f : U \to C$ is a discrete-continuous or discrete-discrete image, the result of the quantization of $f(x, y)$ is a discrete-discrete image $f' : U \to C'$ such that $f'(x, y) = q(f(x, y))$, where q is the quantization map. Thus, quantization changes the color resolution of the image.

When the color spaces C and C' have dimension 1, the quantization process is called *one-dimensional quantization*. When the color spaces have dimension n, and the quantization of each color vector $c = \{c_1, c_2, \ldots, c_n\}$ is performed by quantizing each component c_i separately, we have *scalar quantization*. In this case, we have a one-dimensional quantization map q, and the quantization map $Q : C \to C'$ is defined by

L. Velho et al., *Image Processing for Computer Graphics and Vision*,
Texts in Computer Science, DOI 10.1007/978-1-84800-193-0_11,
© Springer-Verlag London Limited 2009

$$Q(c) = (q(c_1), q(c_2), \ldots, q(c_n)).$$

When the quantization is not scalar, it is called *vector quantization* or *block quantization*.

Example 11.1 (Two-level quantization). Consider the problem of quantizing a monochrome color space with 256 gray levels into one with only two levels. One possible method is to map to 0 all colors below the halfway intensity level (that is, up to 127) and to map the remaining colors to the maximum value of 255. This quantization process partitions color space into two sets, one that maps to 0 and one that maps to 255. Figure 11.1 shows the effect of this quantization.

(a)   (b)

Fig. 11.1. (a) 8-bits quantization (b) 1-bit quantization.

Why should one quantize an image? There are two basic reasons: display and compression.

Image Display

For an image to be displayed in a graphics device, the image's color gamut cannot be greater than the gamut of the device (that is, the number of colors in the device's physical color space). Here the quantization space C' is directly linked to the color space of the display graphics device. Image display will be discussed in Chapter 16.

Image Compression

The quantization of an image allows a reduction in the number of bits used to store its color gamut. This reduces the overall amount of memory needed to store the image and the amount of data needed for transmitting the image over a communications channel. We will return to the subject of *image compression* later in Chapter 13.

From the perceptual point of view, 8 bits of quantization suffice for grayscale images. For color images an acceptable quantization for most applications uses 24 bits, 8 for each of the color channels R, G, and B. Some applications (such as images for the movie industry) require up to 12 bits for each color component.

It is very common to find display devices that are capable of displaying only 256 colors (8 bits). For those devices we must quantize the colors to 8 bits.

11.1 Quantization Cells

As we saw in Example 11.1, quantization into a one-bit color space partitions the initial color space into two sets, in each of which the quantization function has one value. More generally, consider a quantization map $q : C \to C'$. To each quantized color $c_i' \in C'$ there corresponds a color subset $C_i \subset C$, consisting of all colors in C that are mapped to c_i':

$$C_i = q^{-1}(c_i') = \{c \in C : q(c) = c_i'\}.$$

The (finite) family of sets C_i forms a partition of the color space C. Each set C_i is called a *quantization cell*. In each of them the quantization function takes on a constant value c_i', called its *quantization level* or *quantization value*.

In the case of one-dimensional quantization, let q_i, for $1 \leq i \leq L$, denote the quantization levels taken on by the map q. The quantization cells in this case are intervals $c_{i-1} < c \leq c_i$, for $1 \leq i \leq L$. Figure 11.2(a) shows three quantization intervals, $[c_1, c_2]$, $[c_2, c_3]$, and $[c_3, c_4]$. In each interval lies the associated quantization level q_i. Figure 11.2(b) shows the graph of the quantization function q, constant within each quantization interval.

In one-dimensional quantization each quantization cell is an interval, and we can control only its length. In n-dimensional quantization the cells are regions of the color space. For scalar quantization each cell is an n-dimensional box. For vector quantization the cell can assume arbitrary shapes.

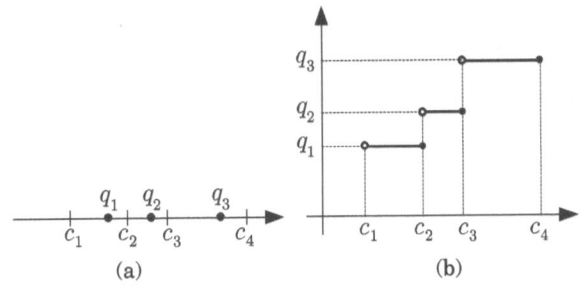

Fig. 11.2. Quantization levels and the graph of the quantization function.

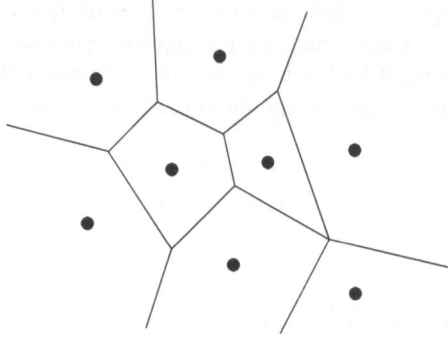

Fig. 11.3. Two-dimensional quantization cells.

In vector quantization the cells are regions of color space that can have more complex shapes. Figure 11.3 shows an example of two-dimensional quantization with eight cells, and therefore eight quantization levels (three bits).

11.2 Quantization and Perception

Consider a monochrome image function $f : U \to C$ whose color space is quantized to L levels $c_1, c_2, \dots, c_L$. This quantization partitions the domain U of the image into subsets

$$U_i = f^{-1}(c_i) = \{(x,y) \in U : f(x,y) = c_i\},$$

each consisting of the pixels of the image whose intensity maps to a certain quantization level. If the number of quantization levels is small, there are few U_i. Moreover, if the image function is well behaved (for example, of class C^1), the boundary separating neighboring regions is a regular curve. Depending on the difference in value between the quantization levels of neighboring regions, the boundary curve may be perceptible to the eye. Such curves are called *quantization contours*.

Figure 11.4 shows the problem of quantization contours. In (a) we have an image with 256 quantization levels, in (b) one with 16 levels, in (c) one with 8 levels, and in (d) the image is quantized to 2 levels. Clearly, as the number of quantization levels decreases, the quantization contours become more perceptible. One aim of the study of quantization is to obtain techniques to avoid the perception of the quantization contours.

Whether or not quantization contours are perceptible depends not only on the number of quantization levels but also on the method of quantization used. In general, for grayscale images, 256 levels (8 bits) are enough to avoid the appearance of quantization contours, regardless of the quantization method

Fig. 11.4. Quantization contours under different numbers of quantization levels: (a) 256; (b) 16; (c) 8; (d) 2.

used. This is illustrated in Figure 11.4(a), which shows an image quantized to eight bits. For color images it is usually enough to use 24 bits of quantization, 8 bits for each color component in RGB space.

However, depending on the image and the quantization method, we can sometimes reduce the number of quantization levels further, without the appearance of contours.

Our perception of quantization contours is worsened by the phenomenon known as *Mach bands*: the human eye magnifies transitions in color intensity, so we can perceive more easily the difference between very similar colors if they are immediately juxtaposed. This is illustrated by the two parts of Figure 11.5. Part (a) shows five vertical stripes of different intensities. The first four stripes are also shown in (b), but slightly separated. One perceives the difference between consecutive stripes as being much more marked in (a) than in (b).

Because the quantization error usually correlates well for neighboring pixels, the boundary between two regions of the domain with distinct but adjacent quantization levels is in general a connected curve and therefore easily picked out by the human eye because of the Mach band phenomenon.

In two-level quantization, contouring is extremely marked, since the final image has only two intensities: black and white. One way to minimize

Fig. 11.5. Mach bands.

quantization contours, even in the more critical case of grayscale images, is *dithering*. This is a filtering process that decreases the correlation in the quantization error between neighboring pixels, thus avoiding that the boundary between quantization levels be a connected curve. Dithering is very important for two-level quantization and therefore has a chapter to itself in this book (Chapter 12).

Connections between quantization and color perception have been extensively exploited in the search for good quantization methods. An example of this is the NTSC system for color television signals, which is based on color coordinates in YUV space, where Y is the luminance and (U, V) are the chrominance coordinates (Chapter 4). In quantizing the components, we can use significantly fewer bits for each chrominance channel than for the luminance information, because the eye is more sensitive to variations in luminance than in chrominance.

11.2.1 Overview of the Quantization Process

From what we have seen, the quantization of a color space, that is, the choice of a quantization map q, involves two parts: determining the quantization cells and determining the quantization level for each cell. Once we know q, the quantization of an image is simple: for each pixel color c in the image, we must identify the quantization cell containing c and replace c by the cell's quantization level $q(c)$.

Existing quantization methods deal in different ways with the two parts of the task of choosing q. One can

- first determine the quantization cells and then choose the quantization level for each cell; or
- first choose the quantization levels and then decide what colors should be mapped to each level; or
- choose quantization cells and levels at the same time, in an interdependent way.

We shall soon see examples of each approach.

11.3 Quantization Error

The optimal determination of the quantization cells, and of the quantization level for each cell, depends on the criterion used to gauge the quantization error and on the distribution of colors in the image. If q is the quantization map and c is a color to be quantized, we can write

$$c = q(c) + e_q,$$

where e_q is the *quantization error* or *quantization noise*.

Distortion is caused when we replace a color c by its quantization value $q(c)$. To measure the distortion we use a distortion measure $d(c, q(c))$. There are many possibilities for choosing a distortion measure d in C. In choosing one, we must take into account perceptual criteria as well as computational efficiency. It is common to use a pseudometric instead of a metric, or even some positive function that, in some sense, gives information about "proximity" in color space. One possible choice is the square of the Euclidean distance, $d(c_1, c_2) = \langle c_2 - c_1, c_2 - c_1 \rangle$.

The quantization of an image implies the quantization of each of its pixels' colors. Thus, a measure of the distortion must take into account not only the quantization distortion of each color from the image color space, but also the frequency of this color on the image. A good measure is given by the mean-square error

$$E((c, q(c)) = \int_C p(c)\, d(c, q(c))\, dc, \tag{11.1}$$

where p is the color probability distribution function in C. The use of this equation to measure the distortion in the quantized image is quite intuitive: it averages the quantization error, taking into account the probability of occurrence of each color in the space being quantized.

11.3.1 Color Frequency Histograms

We have just mentioned the importance of knowing the color distribution of an image in estimating the distortion caused by quantization and therefore in determining optimal quantization cells and levels.

In general, it is very hard to know the color probability distribution of an image. An approximation of this distribution is given by the *color histogram*. In this histogram we associate to each color intensity c present in an image its frequency of occurrence, that is, the number of pixels in the image that have color c. Figure 11.6 shows the image of a house and its histogram. The horizontal axis shows the gradation of the 256 levels of gray (from black to white).

A look at this histogram shows that the image has a great number of pixels with low intensity values, and several pixels have average intensity values.

In the case of color images, one can compute a separate histogram for each color component or create a three-dimensional histogram.

 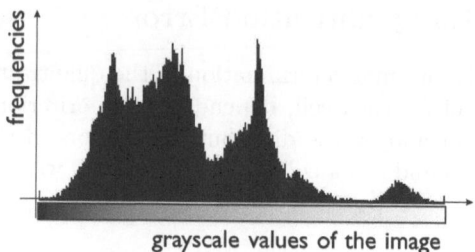

grayscale values of the image

Fig. 11.6. Histogram of a grayscale image.

11.4 Uniform and Adaptive Quantization

We have seen that quantization is straightforward once we have chosen the quantization cells. But how should one determine them? A simple and natural method would seem to be this: divide color space into congruent cells and take the center of each cell as the corresponding quantization level. This is called *uniform quantization*. For L-level one-dimensional quantization, the L quantization cells are intervals $(c_{i-1}, c_i]$ of equal length, $c_i - c_{i-1} = $ constant, and in each cell the quantization value is the average

$$q_i = \frac{c_i + c_{i-1}}{2}, \quad \text{for } 1 \leq i \leq L.$$

If the color space is the RGB cube and we use uniform quantization in each component separately, the quantization cells are little cubes in color space, and the center of each cube defines the quantization value. Figure 11.7 shows two possible cell geometries for uniform quantization in the two-dimensional case.

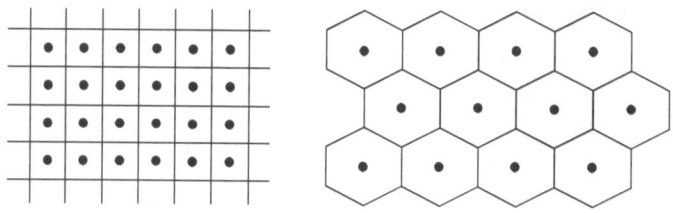

Fig. 11.7. Cells in two-dimensional uniform quantization.

Although uniform quantization is easy to compute, it is not necessarily to be preferred. It has significant shortcomings: for example, some quantization cells may not even contain colors from the image gamut.

Suppose that the image's color distribution is very nonuniform, that is, colors in certain regions of color space occur much more often than those in other regions. If we subdivide the better-represented regions more finely, thus decreasing the size of the cells, the quantization error for pixels having these colors decreases. These colors occur more often, so the overall quantization distortion as defined by (11.1) decreases.

A quantization method that does not partition color space into congruent cells is called *nonuniform*. Nonuniform quantization is called *adaptive* when the geometry of the cells is chosen according to the specific characteristics of the image's color distribution.

11.4.1 Color Map Quantization

It is very common to use a color map to obtain the colors of an image's pixels. More precisely, suppose we have an image $f : U \subset \mathbb{R}^2 \to C$ taking values on some color space C. We define a color map $\varphi : [0,1] \subset \mathbb{R} \to C$, and the image color values are taken as a subset of the color map values $\varphi([0,1]) \subset C$ (see Figure 11.8).

We discretize the unit interval $[0,1]$ into n subintervals defined by some partition $0 = t_1 < t_2 < \cdots < t_n = 1$. By choosing a point $x_i \in [t_i, t_{i+1}]$, for $i = 1, \ldots, n-1$, we obtain a quantization of the set $\varphi([0,1])$ into n levels $\varphi(t_1), \varphi(t_2), \ldots, \varphi(t_n)$. This discretization of the color map is called a *palette*. Quantization of the color map implies quantization of the image.

When we refer to uniform quantization of a color image, we mean uniform quantization of its palette, obtained as above, by subdividing the interval $[0,1]$ into n uniform subintervals.

11.4.2 Test Images

We will use the image shown in Figure 11.9 to compare various quantization methods. The image was initially quantized at 24 bits (8 bits per channel) and therefore does not have perceptible quantization contours.

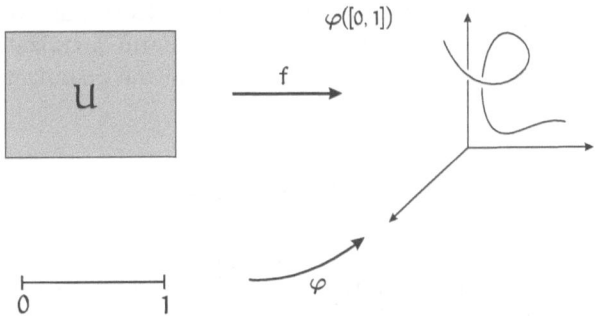

Fig. 11.8. Color map of an image.

Fig. 11.9. Digital color image quantized at 24 bits. See Plate 3 in color insert.

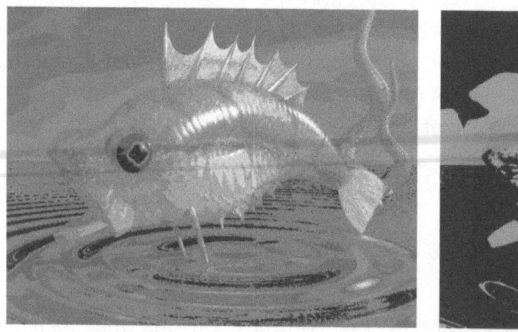 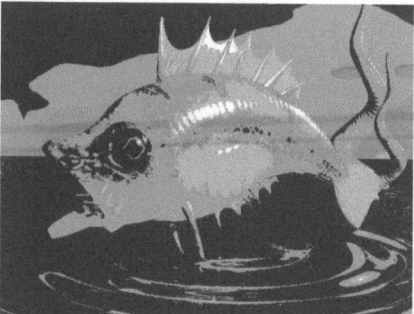

Fig. 11.10. Uniform quantization at eight bits (left) and four bits (right). See Plate 4 in color insert.

For later comparison with other quantization algorithms, we show in the two parts of Figure 11.10 the result of applying uniform quantization to Figure 11.9, using eight and four bits, respectively. Here the quantization contours are quite visible.

11.5 Adaptive Quantization Methods

We now study methods for adaptive color quantization. They consist of two steps: estimating the relevant statistical properties of the image and applying that information to the partitioning of color space. Thus, we start by constructing a color frequency histogram of the image, which approximates the color probability distribution function.

11.5.1 Quantization by Direct Selection

Methods of direct selection start by choosing the quantization levels to be used, based on the statistical properties of the image, and then determine the quantization function in a way that minimizes quantization error. An example is the *populosity algorithm.*

Populosity Algorithm

This method starts by constructing the color frequency histogram of the image. It then chooses as the L quantization values the L colors that occur most often in the image's gamut. The quantization function can be defined as follows: for each color c in the gamut of the image, let $q(c)$ be the quantization value nearest to c, as measured, for example, by the Euclidean metric. (If there is more than one quantization value at minimal distance, we must decide which one to use. One possibility is to choose randomly among the candidates. A more prudent choice would take into account the quantization of neighboring pixels.)

The problem with the populosity algorithm is that it completely ignores colors in regions that are not well represented in color space. Thus, a highlight in the image may disappear entirely in the quantization process, since it only takes up a few pixels. However, the algorithm can be used satisfactorily for images that have an approximately uniform color distribution.

Figure 11.11 shows the result of applying the populosity algorithm to Figure 11.9, at eight bits and four bits. Compare with the uniform quantization shown in Figure 11.10.

11.5.2 Quantization by Recursive Subdivision

Whereas direct selection methods start by choosing quantization levels and then determine the quantization function, *recursive subdivision methods* start

Fig. 11.11. Populosity algorithm: result of quantization at eight bits (left) and four bits (right). See Plate 5 in color insert.

by determining the quantization cells and then compute the quantization function in each cell. As the name implies, such methods work by recursively subdividing color space, in order to choose quantization cells.

We start with a region of color space containing all colors present in the image. At each step of the recursion, this region is subdivided into two or more subregions, based on statistical data on the image's color distribution. The recursion continues until there is just one color of the original image contained in a set of the subdivision, or until the desired number of quantization cells (levels) has been reached. Once the quantization cells have been determined, the quantization function is derived by choosing a quantization value within each cell.

A simple and effective criterion for quantization by recursive subdivision (and adaptive quantization in general) is choosing the quantization levels in such a way that each level occurs in approximately the same number of image pixels. In terms of the image's histogram, this corresponds to carrying out *histogram equalization*, that is, replacing the original histogram by a uniform one, as shown in Figure 11.12. The statistical estimator that allows one to achieve this is the *median*.

The Median of a Set

Given a finite, ordered set of points

$$C = \{c_1 \leq c_2 \leq \cdots \leq c_{n-1} \leq c_n\},$$

the *median* m_C of C is the middle element $c_{(n+1)/2}$ if n is odd and is the mean of the two middle elements if n is even. Thus, the median divides the data set into two equal portions, each having the same number of elements. Unlike the mean of the elements in a set, the median is insensitive to magnitude variations at the extremes of range. If we have an unordered set (possibly with repetitions), the median is defined by first ordering the set and then applying the definition above. Note that one must take into account how many times each value occurs. Thus, the construction of the color frequency histogram of the data set is an important step in the calculation of the median.

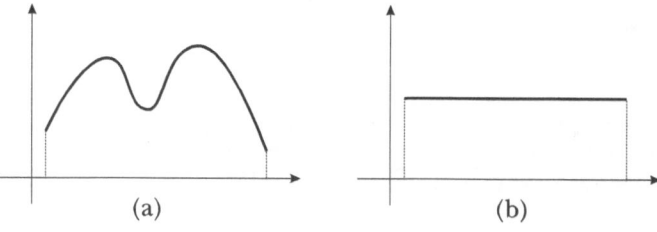

Fig. 11.12. Adaptive quantization. (a) Original histogram. (b) Equalized histogram.

In quantization by histogram equalization for monochrome images, one successively subdivides the range of intensities, using the median of each subrange as the point for the subdivision. The extension of this idea to color images is known as the *median cut algorithm*. It is one of the most popular quantization algorithms in computer graphics, due to its ease of implementation, its computational efficiency, and the good results obtained in converting 24-bit images to 8 bits. Simply put, the algorithm consists in applying histogram equalization recursively to the most spread-out color component of the image's color gamut. We spell this out in more detail, assuming we're working in RGB color space.

Median Cut Algorithm

Let L be the desired number of quantization levels. Take the smallest rectangular box

$$V = \{[r_0, r_1] \times [g_0, g_1] \times [b_0, b_1]\}$$

that contains all the colors present in the image to be quantized. Take the axis of color space in whose direction V is longest—say the green or g-axis. Order the colors present in the image according to their g component, and compute the median m_g of the set of colors based on this ordering. This divides V into two subregions:

$$V_1 = \{(r, g, b) \in C : g \leq m_g\} \quad \text{and} \quad V_2 = \{(r, g, b) \in C : g \geq m_g\}.$$

(One could arbitrarily assign colors with $g = m_g$ evenly between the two boxes, but we will ignore this complication.) Recursively subdivide in the same way each of the regions V_1 and V_2 that contains more than one color, unless the target of L quantization cells has already been reached. For each cell thus created (which is a rectangular box), define the quantization value as the mean of the color values in the box.

Now, to quantize the image, one must find, for each pixel, the cell that contains the pixel's color, then replace that color by the quantization value for that cell. This can be done efficiently by using an appropriate spatial data structure associated with the recursive subdivision of the color space.

Here is an example in two-dimensional color space.

Example 11.2. Consider a two-dimensional set with nine distinct colors, as shown in Figure 11.13(a), with frequencies given by

color	c_1	c_2	c_3	c_4	c_5	c_6	c_7	c_8	c_9
frequency	2	3	2	1	2	1	1	1	2

Suppose we wish to assign these colors to four quantization levels (two bits). The longest side of the color rectangle is vertical. Ordering according to the y-coordinate, we get the following distribution:

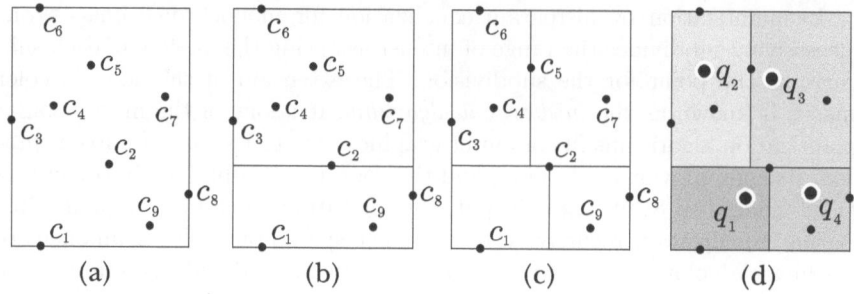

Fig. 11.13. (a), (b), (c): Applying the median cut algorithm. (d) Quantization levels obtained.

color	c_1	c_9	c_8	c_2	c_3	c_4	c_7	c_5	c_6
frequency	2	2	1	3	2	1	1	2	1

More precisely, by taking into account the frequency, we have the ordered set $(c_1, c_1, c_9, c_9, c_8, c_2, \dots)$ with 15 elements. The median is the average of the seventh and eighth elements in the array, which are both c_2. We thus choose c_2 (or rather, its y-coordinate) as the place for the cut, as shown in Figure 11.13(b).

Next we apply subdivision to each subrectangle. For both, the longest side is horizontal. Ordering the colors in each rectangle according to their x-coordinates, we get

color	c_1	c_2	c_9	c_8
frequency	2	3	2	1

and

color	c_3	c_6	c_4	c_5	c_2	c_7
frequency	2	1	1	2	3	1

For the first rectangle, the median color is c_2, and for the second it is c_5. The subsequent subdivisions are shown in Figure 11.13(c).

Now that we have the four quantization cells, we choose the quantization level for each cell by taking the average of the colors in each cell. Colors that are on the edge of a cell should be mapped to the nearest quantization level. Figure 11.13(d) shows the values associated with the four quantization cells. The quantization function q is given by

c (color)	$q(c)$
c_1, c_2	q_1
c_3, c_4, c_6	q_2
c_5, c_7	q_3
c_8, c_9	q_4

Fig. 11.14. Median cut algorithm: result of quantization at eight bits (left) and four bits (right). See Plate 6 in color insert.

The median cut algorithm can be modified in two ways: by changing the method for calculating the quantization levels and by changing the method for choosing a subdivision point. Using the median as the subdivision point causes the histogram of the image to be equalized, so that each quantization level is assigned approximately the same number of pixels in the image. Another possibility is to choose the partition so as to minimize the variance between the colors of each subregion and the corresponding quantization level. For most images, the many possible variations in the algorithm lead to imperceptible differences in the result when we are quantizing a 24-bit image to 8 bits. These variations can therefore be selected largely from the computational efficiency point of view.

Figure 11.14 shows the result of applying the median cut algorithm to the image in Figure 11.9. Compare with the results of uniform quantization (Figure 11.10) and quantization using the populosity algorithm (Figure 11.11).

11.6 Optimization Methods for Quantization

We saw in Section 11.3 that quantization introduces an error that can be estimated by the formula

$$E((c, q(c))) = \int_C p(c) \, d(c, q(c)) \, dc.$$

If we're performing N-level quantization, that is, partitioning color space into N cells $K_1, \ldots, K_N$, with quantization values $q_1, \ldots, q_N$, the preceding equation takes on the form

$$E((c, q(c))) = \sum_{1 \le j \le N} \int_{K_j} p(c) \, d(c, q_j) \, dc.$$

Or, since we have finitely many colors in each cell,

$$E((c, q(c))) = \sum_{1 \le j \le N} \sum_{c \in K_j} p(c)\, d(c, q_j)\, dc. \tag{11.2}$$

Ideally, therefore, we should try to minimize the quantity (11.2) over all possible N-element partitions of color space. The fact that there are an enormous number of such partitions means that the problem is computationally very difficult, so that complete optimization is usually not possible. Existing optimization methods, therefore, tend to solve only a restricted form of the problem, or use heuristic guesses, or return an approximate answer. We now turn to some of these methods.

11.7 Optimal One-Dimensional Quantization

Recall that the notion of optimality depends on the statistics of the color distribution in the image and on the measure of distortion one uses. Suppose, in the one-dimensional case, that the color probability distribution function is $p(c)$ and that we use the mean-square error as our measure of distortion (Section 11.3). Thus we want to minimize

$$D = \int_{-\infty}^{+\infty} p(c)\, (q(c) - c)^2\, dc. \tag{11.3}$$

The quantization cells are intervals $[c_{i-1}, c_i]$, each corresponding to a quantization value q_i. Since $q(c) = q_i$ in the interval $[c_{i-1}, c_i]$, the quantity (11.3) that we wish to minimize can be rewritten as

$$D(c_0, \ldots, c_N, q_1, \ldots, q_{N-1}) = \sum_{i=1}^{N} \int_{c_{i-1}}^{c_i} p(c)\, (q_i - c)^2 dc. \tag{11.4}$$

Necessary conditions for the global quantization error to be minimal are obtained by computing the critical points of the function D. To do this, we take the derivative of (11.4) with respect to c_k and q_k and set it to zero. We obtain

$$\frac{\partial D}{\partial c_k} = (c_k - q_{k-i})^2 p(c_k) - (c_k - q_k)^2 p(c_k) = 0,$$

$$\frac{\partial D}{\partial q_k} = 2 \int_{c_{k-1}}^{c_k} (c - q_k) p(c)\, dc = 0.$$

Solving both sets of equations, we get

$$q_k = \frac{\displaystyle\int_{c_{k-1}}^{c_k} c\, p(c)\, dc}{\displaystyle\int_{c_{k-1}}^{c_k} p(c)\, dc} \quad \text{and} \quad c_k = \frac{q_k + q_{k+1}}{2}, \tag{11.5}$$

where $k = 1, \ldots, N$.

The first of these equations says that the quantization value is the centroid of the probability distribution $p(c)$ in the interval $[c_{k-1}, c_k]$. The second says that the endpoints of quantization intervals are the average between consecutive quantization values. It is easy to check that, when the probability distribution is uniform, that is, $p(c) = 1/(c_N - c_0)$, the optimal solution results in the one-dimensional uniform quantization described earlier.

Note that a solution of (11.5) is a priori just a critical point of D, and not necessarily a minimum. One can show that, for some probability distributions, the solution does minimize D. This is the case, for example, when the probability distribution is uniform or normal (gaussian). As previously mentioned, the solution in the case of uniform distribution corresponds to uniform quantization. For a normal distribution the solution is more complicated and must be computed numerically. The reader interested in details should consult the references section at the end of this chapter.

11.8 Optimal Quantization by Relaxation

The calculations of the preceding section extend to dimensions greater than 1, but in this case the counterpart to (11.5) is more complicated, because it involves integration over the boundary of the quantization cells. A more appropriate quantization method is a relaxation procedure, which attains a solution through successive approximations.

Consider a color vector c, which must be mapped to a vector c', and suppose we have N quantization levels $c'_1, \ldots, c'_N$. Moreover, using the squared Euclidean metric $d(c_1, c_2) = \langle c_1 - c_2, c_1 - c_2 \rangle$, Equation (11.1) implies that the mean-square error is

$$D = \int_{-\infty}^{+\infty} d(c, c')\, p(c)\, dc = \sum_{i=1}^{N} \int_{c \in C_i} d(c, c'_i)\, p(c)\, dc,$$

where p is the probability distribution and C_i is the quantization cell corresponding to the level c'_i. Two conditions must be satisfied in order for an optimal quantization to be obtained:

(a) The quantization map q must map each color to the quantization level nearest to it:

$$q(c) = c'_i \iff d(c, c'_i) \leq d(c, c'_j), \quad \text{for all } 1 \leq j \leq N \text{ with } j \neq i.$$

(b) In each cell C_i we must compute the level c'_i in a way that minimizes the mean error D in C_i.

Condition (a) allows one to calculate C_i from the metric d and the quantization levels. Condition (b) allows one to calculate c'_i from the cell C_i. Just as in one-dimensional optimal quantization, once we know the metric d, the quantization cells and quantization levels are interdependent; if we know one, we can determine the other.

The two conditions can be used to approach an optimal solution by means of a relaxation process, as follows:

- Start with quantization levels $c_i^{(1)}$ chosen in some way.
- From the $c_i^{(1)}$ and the metric d, compute the quantization cells $C_i^{(1)}$ using condition (a).
- From the quantization cells $C_i^{(1)}$ and the metric d, find new quantization levels using (b).
- Repeat the two preceding steps until the total mean-square error no longer changes (to within the desired precision).

This process involves several difficulties:

- We must compute the quantization levels $c_i^{(j)}$ for all possibilities of color in the space, which is computationally expensive.
- The probability distribution of colors is usually not known exactly.
- Conditions (a) and (b) are necessary, but not sufficient, for optimal quantization, so the process may not converge, or it may converge to a nonoptimal solution.

The literature includes many attempts to get around these problems. Additional conditions can be added to ensure that the relaxation process converges and to improve its computational efficiency. See Section 5.6 for references.

11.8.1 Optimal Quantization by Simulated Annealing

Since the optimal quantization depends on the statistical properties of the image, it is natural to try to achieve it using *simulating annealing*, a stochastic relaxation procedure. Such methods work by successive approximations like the preceding one, but they rely on stochastic, rather than deterministic, algorithms to proceed from one approximation to the next. Statistical mechanics shows that the least-energy configurations of certain physical systems are the most probable ones. The energy functionals describing these systems are associated to the so-called Gibbs distribution. The minimum of the system can be obtained, for example, through a stochastic relaxation algorithm. Such an algorithm uses a priori the probability distribution and performs random perturbations in the initial configuration of the system until the least-energy state is achieved. In an entirely analogous way, this theory can be applied to color quantization. See Section 6.3 for references.

11.9 Comments and References

A classic work in the area of quantization is (Heckbert 1982). This is where the median cut algorithm discussed in this chapter was introduced. It has become the algorithm of choice in computer graphics due to its ease of implementation, its computational efficiency, and its good perceptual performance in quantizing 24-bit color images at 8 bits.

The subject of spatial data structures for partitioning space, necessary for the implementation of the median cut algorithm, is well covered in (Samet 1990). Heckbert's original article suggested an implementation using a K-D-tree structure.

The classical reference for optimization methods in quantization is (Lloyd 1957). A quantizer based on the minimization of the mean-square error (11.3) is called a Lloyd–Max quantizer (Max 1960). The problem is extended to vector quantization in (Linde et al. 1980), which includes a demonstration of the convergence of the method and applications to voice and image signals. See also (Heckbert 1982), which includes a fairly comprehensive discussion of the image quantization problem.

For stochastic relaxation methods in the quantization of digital images, see (Fiume and Ouellette 1989).

The original image used in Figures 5.4, 5.12, 5.15, and 5.17 is "Barn and Pond" (KINSA Photo Contest), by Cindy Branham, from the Kodak PhotoCD, Photo Sampler.

References

[Heckbert 1982]Heckbert, P. S. (1982). Color quantization for frame buffer display. *Computer Graphics (SIGGRAPH '82 Proceedings)*, 14(3): 297–307.

[Samet 1990]Samet, H. (1990). *The Design and Analysis of Spatial Data Structures*. Addison–Wesley, Reading, MA.

[Lloyd 1957]Lloyd, S. P. (1957). Least square quantization in pcm's. Technical Memo. Bell Telephone Labs.

[Max 1960]Max, J. (1960). Quantizing for minimum distortion. *IEEE Transactions Inform. Theory*, Vol. IT-6, March 1960, 7–12.

[Linde et al. 1980]Linde, Y., Buzo, A., and Gray, R. M. (1980). An algorithm for vector quantizer design. *IEEE Transactions Comm.*, COM-28(12):84–95.

[Fiume and Ouellette 1989]Fiume, E. and Ouellette, M. (1989). On distributed, probabilistic algorithms for computer graphics. *Proceedings of Graphics Interface '89*, 211–218.

12

Digital Halftoning

There are extreme situations where even with the best quantization algorithms it is hard or impossible to mask quantization contours. This is the case, for example, in two-level quantization, which must be used for display on one-bit graphical output devices. Such output devices are very common: laser printers and phototypesetters are examples of them. In spite of their limitations in what concerns color reconstruction, they can be used to display grayscale images. As we will see in Chapter 16, offset color printing also reduces, via color separation, to the printing of grayscale images on one-bit devices.

In this chapter we will study a filtering process called *dithering*, whose purpose is to minimize the perceptual effects of quantization contours under color discretization. We will devote most of our attention to the problem of two-level (one-bit) quantization, due to the importance of this case in the area of digital publishing. We suggest that you review Chapter 11, on color quantization, which is a prerequisite for this chapter.

12.1 Dithering

When we quantize an image, we create at each pixel a quantization error: the difference between the pixel's original color and its quantized value. Because of the strong correlation among values of neighboring pixels, we obtain a quantization contour that is usually a connected curve, and this makes the passage from one quantization level to the next perceptible.

The basic idea is to replace a sharp boundary between quantization levels by a fuzzy one, where the two levels are so intimately mixed together that the eye is fooled into seeing intermediate levels and perceives the transition as smooth rather than abrupt. (The everyday meaning of the word "dither"—to act nervously or indecisively—is applied metaphorically to the wavering between two levels that appears to take place.)

Analog methods for producing the perception of halftones from a two-level image have been used by the printing industry (in magazines, newspapers, and

L. Velho et al., *Image Processing for Computer Graphics and Vision*,
Texts in Computer Science, DOI 10.1007/978-1-84800-193-0_12,
© Springer-Verlag London Limited 2009

so on) since the end of the nineteenth century. For this reason, dithering algorithms are also known as *digital halftone algorithms*. In fact, some dithering algorithms attempt to implement digital versions of the analog process of halftone generation.

Dithering and Perception

Dithering is effective because of a basic characteristic of human vision: our eye integrates (averages) the stimuli received within a certain solid angle. Thus, we are able to perceive color intensities that are not necessarily present in the image but instead arise from the averaging of the intensities in the neighborhoods of each image element contained in a certain area. From this point of view what matters is the average intensity in a region, not at a pixel.

Physically, perceptual resolution is measured by *visual acuity*, that is, the eye's ability to detect details in observing a scene. The field of vision of the human eye extends for about 150° horizontally and about 120° vertically. Yet the eye cannot distinguish details separated by less than about one minute of arc, that is, one-sixtieth of a degree. This angle is called the *angle of visual acuity*. It depends on the wavelength, the geometry of the eye's optical apparatus, and, above all, on the dimensions and distribution of the eye's photosensitive cells, because the eye can only resolve two objects when the photons they emit reach different cells. Thus, the perception of details in an image depends on three parameters: the distance from the image to the eye, the resolution density of the image, and the eye's aperture.

Distance from the Image to the Eye

This distance is measured along the direction of the optical axis. The greater it is, the fewer the details that can be perceived.

Resolution Density

As we defined in Chapter 5, the *resolution density* of the image is the number of pixels per unit length. The smaller the distance between adjacent pixels, the less able the eye is to resolve individual pixels. The most common unit for the resolution density is *pixels per inch* (ppi), also known as *dots per inch* (dpi). Some devices have different vertical and horizontal densities.

Eye Aperture

When one looks at an image with half-closed eyes, the field of vision becomes narrower, and the angle of visual acuity decreases. One therefore perceives fewer details in the image.

When displaying a digital image, we must look for ways to increase the perceptual resolution of the image, so as to get rid of the artifacts introduced by the color discretization and reconstruction process. We do this by changing the parameters that affect visual acuity. Thus, there are three physical methods that help increase the perceptual resolution of an image:

- displaying the image on a device having higher resolution density;
- observing the image from further away; and
- observing the image with the eyes somewhat closed.

Whichever method is more convenient should be used to improve image perception in any given case.

Dithering Strategy

Given a region $R_k(i, j)$, where $i, j \in \mathbb{Z}$, in the domain of an image f, the average intensity of the image in this region is

$$I_f = \frac{1}{|R_k|} \sum_i \sum_j f(i, j), \tag{12.1}$$

where $|R_k|$ is the number of pixels in the region.

When we quantize the image on the region R, we obtain a bitmap image $\bar{f}$ on R. The *average quantization error* or *local quantization error* is the absolute value $|I_f - I_{\bar{f}}|$ of the difference between the average intensity of the image f on R, and the average intensity on R of the bitmap image $\bar{f}$.

The basic strategy of dithering filters is to distribute the local quantization error over areas of the image, so that the error averages out to zero. There are two possible approaches: statistic and deterministic. *Statistic dithering* attempts to change the elements of the image in such a way that the average quantization error is zero for the image as a whole. *Deterministic dithering* tries to minimize the global error, making the quantization error incurred in one image element be compensated for in the quantization of the neighboring elements.

Dithering and Optimization

We can pose the dithering problem from an optimization point of view: we must find a map $\ell : \mathcal{G} \to \mathcal{G}$ on the space of monochrome images $\mathcal{G}$, and a metric d on $\mathcal{G}$, such that for every grayscale image $f \in \mathcal{G}$ we have

- $\ell(f) \in \mathcal{G}$ is a binary image;
- the distance $d(f, \ell(f))$ is minimized.

This is a viable approach for posing the dithering problem, but additional conditions are necessary in order to obtain good solutions. Indeed, the metric

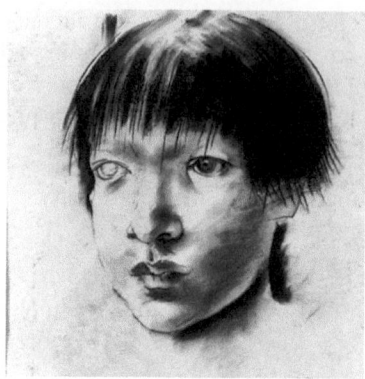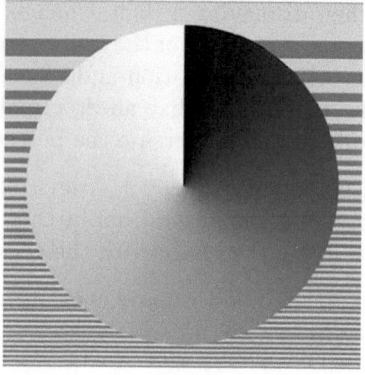

Fig. 12.1. Grayscale images for the comparison of dithering methods.

must have perceptual characteristics, and side conditions must be imposed—for example, the characteristics of the display devices. We will not pursue this approach, but the interested reader can consult the references mentioned in Section 12.5.

Test Images

In this section we will use the images shown in Figure 12.1 in order to illustrate various dithering algorithms. On the left is a reproduction of a charcoal drawing by the Brazilian artist Cândido Portinari; we chose it because it displays subtle variations of shading on the boy's face, as well as a wealth of detail (high-frequency information), as in the hair, for example. The other image is synthetic; it was created so as to contain lots of high-frequency information, as well as areas with mild intensity gradations.

A reader with average visual acuity will not notice any difference between the gray levels in Figure 12.1 and those in a photograph. Both images in Figure 12.1 were reproduced using digital methods to be discussed in this chapter. We have selected the reproduction parameters in a way that minimizes the perceptual difference between the reproduced image and the halftone originals. Ideally, we should have reproduced these images using the traditional analog process for halftone reproduction. We did not do this, because our primary aim in this chapter is not to compare dithering algorithms in terms of their advantages and disadvantages in the quantization process. Our intent, instead, is to give a conceptual exposition of the problem and describe the various algorithms, highlighting, whenever possible, their virtues and limitations.

12.1.1 Dithering by Random Modulation

One basic two-level quantization algorithm for a halftone image was discussed in Chapter 5 (page 294). It consists in choosing a certain constant gray level and using it as a global threshold in deciding whether a pixel should be quantized to 0 or to 1. All pixels whose intensity level lies above the threshold are quantized to 1; all others get a value of 0.

Figure 12.2 shows the result of applying this method to the images in Figure 12.1, with 50% gray as a threshold. Quantization contouring is quite perceptible.

A simple change in this algorithm decreases the correlation in quantization error: namely, changing the intensity level by a random amount before quantization. Thus, instead of comparing the value $f(x, y)$ of the intensity at a given pixel with the threshold, we compare the value $f(x, y) + i$, where i is a random value, chosen independently for each pixel, according to some appropriate probability distribution. This is known as *dithering by random modulation*.

Fig. 12.2. Two-level quantization with 50% threshold.

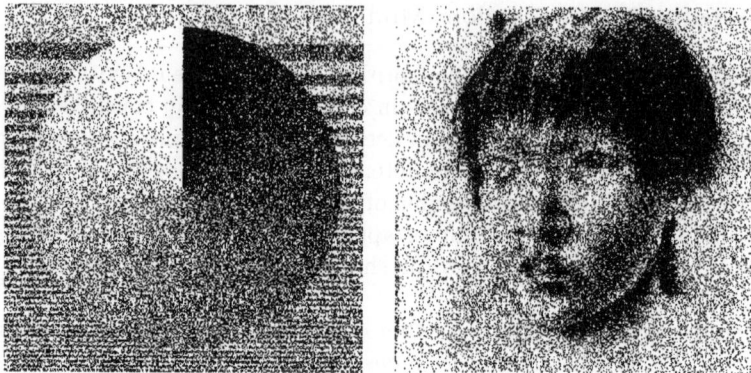

Fig. 12.3. Effect of dithering by random modulation.

The random perturbations cause the intensity of each pixel to be less correlated with that of neighboring pixels. This makes the quantization contour no longer be a connected curve, so that the boundary separating the two quantization regions is not well defined. In Figure 12.3 we show the effect of dithering by random modulation on the images of Figure 12.1, using a 50% threshold and a uniform random variable. Notice how quantization contours, so obvious in Figure 12.2, are absent from Figure 12.3.

The most common distribution for the random perturbation in this algorithm is a uniform distribution over a range of intensities. This process introduces a uniform amount of noise at all frequency ranges, which degrades the quality of the final image. Comparing Figure 12.3 with Figure 12.1, one sees a great loss of high-frequency information. In spite of that, dithering by random modulation can be very effective when used for quantization at more than one bit.

We will see later that there exist digital halftoning algorithms that are far superior to random modulation in terms of perceptual results, and about as easy computationally.

12.1.2 A Classification of Dithering Algorithms

An effective way to analyze and classify dithering algorithms is through the patterns that they produce in areas of constant intensity. These patterns are particularly obvious in two-bit images, where intermediate levels are realized by means of black and white dots. The validity of this criterion is justified by the fact that the configuration of dots is the only difference among the discretizations of an image produced by distinct dithering algorithms when these discretizations have the same average intensity and therefore the same number of black and white dots.

Our eye performs a perceptual characterization of an image using basic low- and high-frequency information. High frequencies delimit regions of interest (contours), whereas low frequencies define the texture information inside each region. Digital halftoning algorithms, therefore, must strive to maintain the high-frequency information and to replace gray shades by binary texture patterns that are perceptually the same. Texture patterns generated by dithering algorithms are classified according to *regularity* and *structure*. Regarding regularity, they can be *periodic* or *nonperiodic*. Regarding structure, the patterns can be *clustered* or *dispersed*.

In general, *periodic patterns* are generated by deterministic processes that try to minimize the local quantization error based on a regular sampling. *Nonperiodic patterns* arise from methods that attempt to minimize quantization error by distributing it globally around the image. *Dispersed configurations* realize gray levels by distributing individual dots as uniformly as possible, whereas *clustered configurations* concentrate dots into small clusters of constant value.

The technique of dispersed dithering is more appropriate for graphical devices that allow precise control over the placement of pixels, as in the case of video monitors. By contrast, the clustering approach is the better alternative for devices that don't reproduce isolated dots well, such as laser printers and phototypesetters.

The above classification will be used to study the different dithering methods discussed in this chapter. It is convenient to use a graphical representation for this classification, as shown in Figure 12.4. In this representation we have a horizontal axis for measuring clustering, and a vertical axis for periodicity. Clustering increases as we move to the right along the horizontal axis; periodicity decreases as we move up the vertical axis.

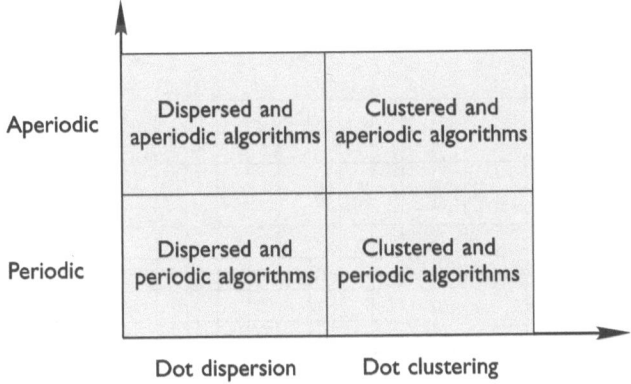

Fig. 12.4. Basic classification of dithering algorithms.

By subdividing the first quadrant of the plane into four areas, we obtain four basic classes of dithering algorithms:

- dispersed and aperiodic algorithms;
- dispersed and periodic algorithms;
- clustered and periodic algorithms;
- clustered and aperiodic algorithms.

Dithering by random modulation, introduced in Section 12.1.1, is an example of a nonperiodic, dispersed dithering filter. In this chapter we will study dithering algorithms for each of the above classes.

12.2 Periodic Dithering

An important family of periodic dithering algorithms consists of *ordered dithering* filters. Here "ordered" is used in contrast with "random": the basic idea of these algorithms is to use a finite, deterministic, and localized threshold function to achieve the goal of decreasing the correlation of quantization error. A detailed description will be given below.

A digital image defines a lattice $R_\Delta = R_{(\Delta x, \Delta y)}$ of the plane. We consider a sublattice $R_{\Delta'}$ of R_Δ, where $\Delta' = (N\Delta x, M\Delta y)$, N and M positive integers. Each "pixel" in the lattice $R_{\Delta'}$ is called a *dithering cell*. Thus, dithering cells are made out of blocks of $N \times M$ pixels from R_Δ. $R_{\Delta'}$ is called the *dithering lattice*. The shaded region in Figure 12.5 shows a dithering cell of order 4×4.

For each dithering cell we define a threshold function $t(x, y)$. Thus, two-level quantization inside the cell is very easy to accomplish. For each pixel $p = (x, y)$ in the cell, if $t(x, y) \le f(x, y)$, we quantize $f(p)$ to 1; if not, we quantize the pixel $f(p)$ value to 0.

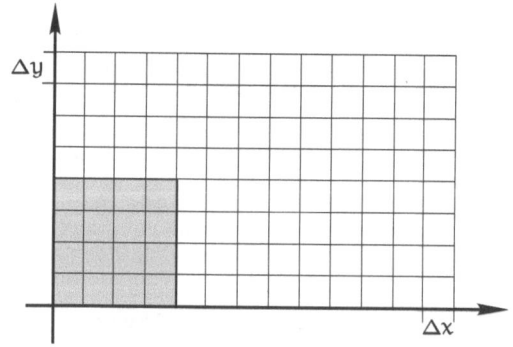

Fig. 12.5. Dithering cell.

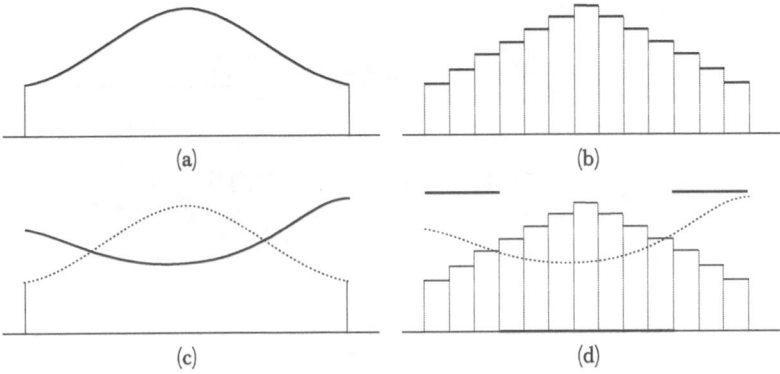

Fig. 12.6. Dithering cell and threshold quantization function.

In the discrete domain the threshold function is discretized to obtain a matrix of thresholds of order $N \times M$, called the *dithering matrix*. Figure 12.6 illustrates, in dimension 1, the quantization function (a), its discretization (b), the image values inside the cell (c), and the quantized image inside the cell (d).

Several variations are possible in the scheme just described: the threshold function may vary from one cell to another, or it can be defined to produce either clustered or dispersed dithering.

Dithering Patterning

In the above description of dithering we clearly have a trade-off between spatial and tonal resolution: pixels are grouped into cells, and inside each cell we are able to render more tonal values (the average intensities of the bitmap patterns inside the cell). For an $N \times N$ cell we are able to display from 1 to N^2 black pixels, corresponding to N^2 intensity levels. Considering the additional intensity of no black pixels inside the cell, we have a total of $N^2 + 1$ tonal values in the dithering cell. Note that we are not considering the many different arrangements of the pixels inside the cell for each of the intensity levels; they depend on the threshold function.

By using a constant image with intensity varying from 0 to $N^2 + 1$, we can have a glimpse on the different "shapes" of dot patterning for each intensity level. As an example, for the dithering matrix

9	5	6
4	1	2
8	3	7

we obtain the 10 patterns shown in Figure 12.7.

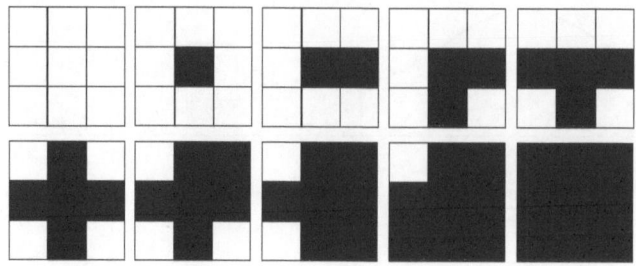

Fig. 12.7. Patterns of intensity levels in a dithering cell.

12.2.1 Clustered Ordered Dithering

Clustered dithering attempts to simulate on the computer the traditional analog photographic method for obtaining image halftoning. This analog halftoning process consists of using a special camera to record, on high-contrast film, a superimposition of the image with a screen. Thus, the light coming from the image is modulated by the screen lattice before hitting the film. Each small opening in the screen works as a lens, focusing the light coming from a small region of the image onto a dot on the film. The size of the dot depends on the luminance of the region that is being sampled: bright areas produce small dots, whereas dark areas produce large dots that generally overlap with their neighbors. The geometry of the lattice obtained depends on several factors, such as the characteristics of the screen, the time of exposure, and so on.

In ordered dithering by clustering, a similar effect is achieved digitally by arranging the threshold function so that black pixels are clustered together.

Consider the problem of quantizing an image of constant intensity, $f(x,y) = k$, inside a dithering cell. Parts (a) and (b) of Figure 12.8 illustrate the quantization by plotting the threshold function and the image value k.

For a sufficiently regular threshold function t, the inverse image of the level, $t^{-1}(k) = \{(x,y) : t(x,y) = k\}$, defines a curve γ in the dithering cell. We choose t so that γ is a closed, nonself-intersecting curve, as shown in Figure 12.8(c). The quantization is easily done: all pixels inside γ will be quantized to 1, and the pixels outside it will be quantized to 0. This is shown, in discrete form, in Figure 12.8(d). It is clear from this picture that the geometry of the black dots clustering is controlled by the shape of the threshold function.

For a grayscale image of constant intensity, the dithering pattern is repeated periodically in the quantized image. In the general case of an image with variable gray values, the threshold function is repeated periodically along the dithering lattice, but the percentage of dark and light regions within each dithering cell changes according to the intensities of the pixel in the original image. The final result is a collection of pixel clusters of varying sizes: an

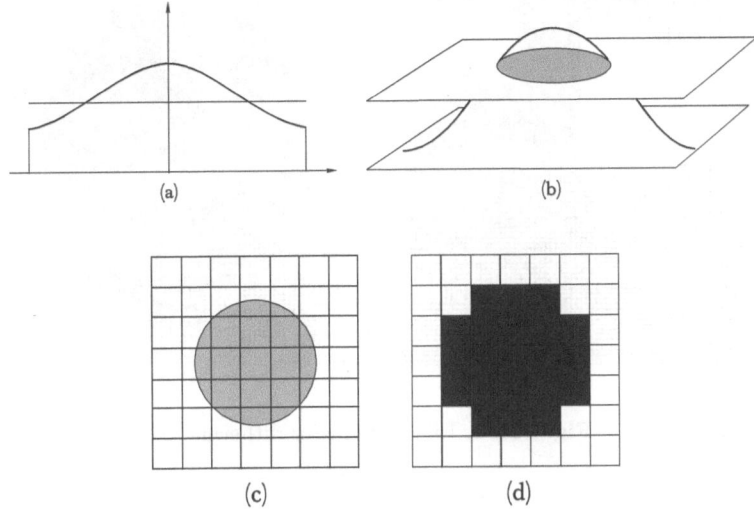

Fig. 12.8. Threshold function and constant image.

effect very similar to the one obtained with traditional, optical, halftoning techniques.

Figure 12.9(a) shows a 6×6 ordered dithering matrix that performs clustered dithering. Note that the threshold function of this matrix is shaped like the function in Figure 12.8(b). The numbers in the dithering matrix indicate the ordering of the threshold values, not their intensity in absolute terms; one can regard the intensities as being $\frac{1}{37}$, $\frac{2}{37}$, etc., since there are $6 \times 6 = 36$ thresholds and therefore 37 intervals of intensity can be distinguished. In order to apply the algorithm, the image intensities are normalized to the interval $[0, 1]$.

35	30	18	22	31	36
29	15	10	17	21	32
14	9	5	6	16	20
13	4	1	2	11	19
28	8	3	7	24	25
34	27	12	23	26	33

(a)

35	30	18	22	31	36
29	15	10	17	21	32
14	9	5	6	16	20
13	4	1	2	11	19
28	8	3	7	24	25
34	27	12	23	26	33

(b)

Fig. 12.9. Clustered dithering matrix.

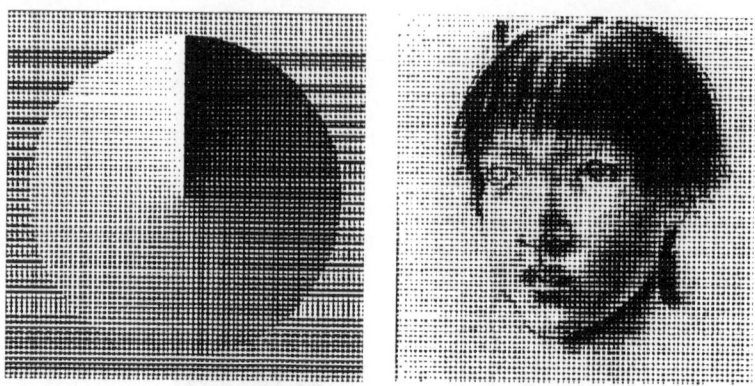

Fig. 12.10. Effect of cluster dithering.

Figure 12.9(b) shows, in gray, pixels with intensity threshold above 50%, and in white those with threshold below 50%. Now consider an image with constant intensity, equal to 50%. After two-level quantization and dithering with the filter defined by this matrix, the image would be a periodic pattern obtained by repeating Figure 12.9(b). If we take a constant-gray image with intensity higher than 50%, the white region in Figure 12.9(b) is reduced; if the intensity is less than 50%, the white region is enlarged.

Figure 12.10 shows our test images after one-bit quantization using the ordered dithering filter defined by the matrix of Figure 12.9(a).

Diagonal Patterning

Perceptual studies show that the human eye's sensitivity to artifacts produced by a periodic pattern changes according to the angle of the pattern and is least when the periodicity axis makes an angle of about 45° or −45° with the horizontal direction. This perceptual fact can be exploited to create better cluster dithering algorithms. Toward this goal, we try to define a threshold function that forces the clusters to be aligned along the 45° diagonals.

The easiest way to achieve this uses diamond-shaped cells: we subdivide the square dithering cells into four pieces and construct a square using the four midpoints (see Figure 12.11(a)). The threshold function is defined in such a way that, for a 50% intensity value, the black and white quantized values produce the patterns shown in Figure 12.11(b).

An explicit example is given by the dithering matrix in Figure 12.12(a). This cell has 32 pixels, so it serves to distinguish 33 levels of quantization. Figure 12.12(b) shows in gray all those pixels with intensity threshold above 50%, and in white those with threshold below 50%.

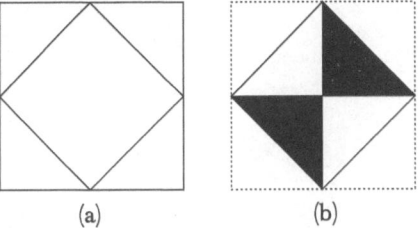

Fig. 12.11. Diamond patterning for diagonal clustering.

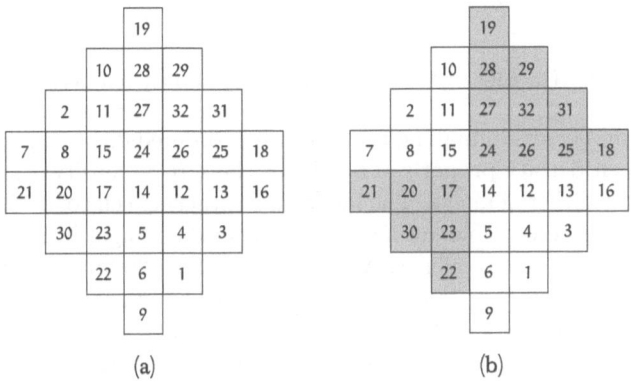

Fig. 12.12. Cluster dithering matrix with pixel clusters along the diagonal.

Figure 12.13 shows the result of applying this dithering matrix to the test images in Figure 12.1. Notice how the periodicity pattern now has a diagonal bias of 45°.

Practical Hints

As we have seen, ordered dithering increases tonal resolution (number of realizable intensity levels) at the cost of spatial resolution. The effective spatial resolution of the dithered image is not that of the display device, but that of the dithering lattice. The former is what matters in terms of the perceptual quality of the dithered image. Of course, the device resolution density also matters, in that it gives an upper bound for the number of elements in the dithering cells.

By analogy with the traditional analog halftoning process, the dithering lattice is often called a *screen*, and the density of dithering cells is called the *screen density* (or, more precisely, *screen resolution density*). It is measured in *lines per inch* (lpi).

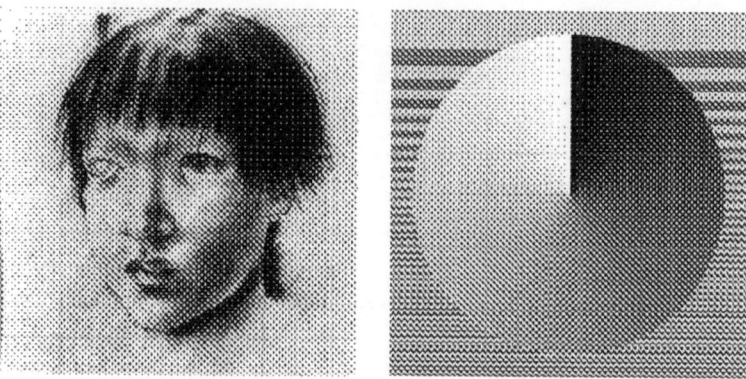

Fig. 12.13. Effect of cluster dithering with diagonal periodicity.

One of the secrets in making the best use of ordered dithering lies in choosing the screen density properly. The higher the screen density, the better the quality of the dithered image. The individual dithering cells of a 150-lpi screen are virtually imperceptible, even at close range; the images in Figure 12.1 are printed at that screen frequency. A screen density of 120 lpi can still give good results. By contrast, the screen density in Figure 12.13 is about 35 lpi, and in Figure 12.10 it is about 30 lpi. See also Figure 12.14.

A high screen density requires that the device resolution density be even higher, by a factor equal to the order of the dithering cell. Thus, in Figure 12.1, the screen density of 150 lpi is still many times less than the resolution density of the phototypesetter used to produce this book. This means the dithering cell is large (has lots of entries), so that many levels of gray can be distinguished, but the display resolution of the device does not allow us to perceive the dithering cells. (The ratio between the device resolution density and the screen density is the order of the dithering cell, and the number of intensity levels that can be distinguished is roughly the square of the order of the dithering cell.)

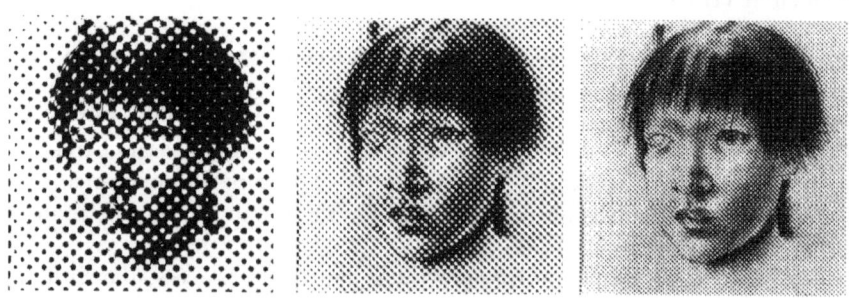

Fig. 12.14. Ordered dithering at screen densities of 15 lpi, 30 lpi, and 60 lpi.

Artistic Screens

As observed before, by changing the shape of the threshold function we change its level curves and consequently the cluster geometry of the dithered image. This fact can be used to create various effects in the dithered image, as illustrated in Figure 12.15.

12.2.2 Dot Dispersion Ordered Dithering

Ordered dithering by dot dispersion is the preferred method in the display of images on devices with fine control over pixel placement, such as video monitors. As in the case of cluster dithering, the crux of the algorithm lies in the choice of a dithering matrix.

While cluster dithering tries to imitate the lattice obtained by traditional photographic processes, dithering by dot dispersion tries to arrange the quantization thresholds in the dithering matrix in such a way that the texture created has the same frequency distribution as the original texture. In particular, in regions of constant intensity, the resulting dithered texture should avoid low-frequency components as much as possible. This is done by distributing quantization thresholds as homogeneously as possible among the entries of the dithering matrix.

Ordered dithering by dot dispersion is also known in the literature as *Bayer dithering*, because it was B. Bayer who determined the dithering matrices of various orders that maximize the homogeneity of the distribution of quantization thresholds. The 2×2 Bayer dithering matrix is

$$\begin{array}{|c|c|}\hline 2 & 3 \\ \hline 4 & 1 \\ \hline \end{array}.$$

The distribution of intensities using this matrix is shown in Figure 12.16. For a region with 50% intensity, the resulting texture is the checkerboard pattern obtained by periodic replication of Figure 12.16(c).

Fig. 12.15. Changes in the dithering matrix lead to different geometries for the pixel clusters.

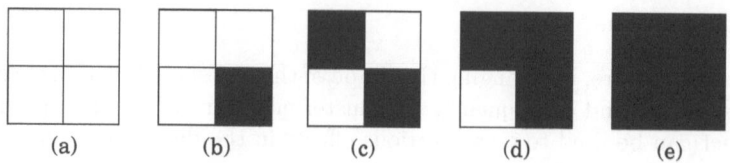

Fig. 12.16. Dithering cells filled at various intensities, for order-2 Bayer dithering.

The 4 × 4 Bayer matrix is

2	16	3	13
10	6	11	7
4	14	1	15
12	8	9	5

Notice that the each group of four threshold levels (1 to 4, 5 to 8, 9 to 12, and 13 to 16) is dispersed in the same arrangement as levels 1 to 4 in the 2 × 2 Bayer matrix. This stems from the recursiveness of the algorithm used in the generation of Bayer matrices or arbitrary order. For a description of this algorithm, and further details on Bayer dithering, see the references given in Section 12.5.

Figure 12.17 shows the result of applying order-4 Bayer dithering to the test images in Figure 12.1.

As already mentioned, dispersed dot dithering algorithms do not perform well with display devices where pixel placement and size cannot be controlled very well. This is because a small variation in the dot size can lead to significant changes in average intensity over the dithering cell. Cluster dithering

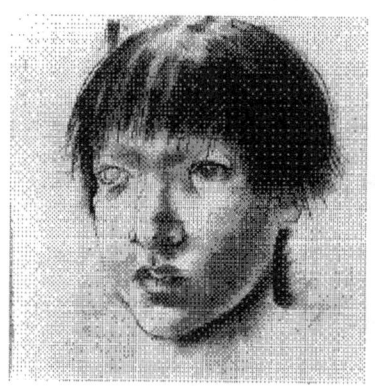 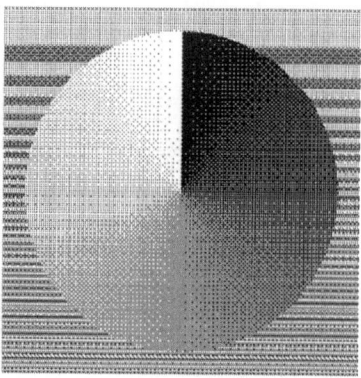

Fig. 12.17. Effect of order-4 Bayer dithering.

is not so sensitive to this problem because pixels tend to coalesce anyway, so variations in pixel size affect only the edges of the cluster. If pixel placement and size *can* be controlled well, it is advantageous to use dot dispersed dithering, because the perceived resolution density is, in effect, higher than the density of the dithering lattice.

The upshot of this is that dot dispersion algorithms—in particular, Bayer dithering—are used very commonly for dithering on video monitors that have insufficient "frame buffer depth" (that is, simultaneously addressable bits of color resolution) for the desired purpose.

For a given device resolution, Bayer dithering has a much better performance, from the perceptual point of view, than dithering by random modulation, studied in Section 12.1.1, and its computational cost is no higher. For this reason dithering by random modulation is seldom used nowadays; its interest is primarily historical and academic.

12.3 Pattern Dithering

Ordered dithering is commonly confused with a technique called *pattern dithering*. In this technique we subdivide the image into dithering cells of size $N \times N$, similar to what we did in ordered dithering. Each cell defines $N^2 + 1$ quantization levels.

The algorithm proceeds as follows: we use the $N^2 + 1$ quantization levels to address a table of dithering patterns; for each cell, we compute the average intensity of the image using Equation (12.1) and quantize it to one of the $N^2 + 1$ quantization levels. The quantized value is used to address the table of dithering patterns.

Notice that if we take the table of dithering patterns as the $N^2 + 1$ patterns for ordered dithering as computed in Section 12.2, the pattern dithering algorithm performs almost, but not quite, as well as the ordered dithering algorithm. In fact, for each average intensity, pattern dithering addresses the same pattern on the table, whereas ordered dithering may associate different bitmap patterns depending on the distribution of the image intensity values inside the dithering cell.

Note that the dithering pattern table is quite arbitrary. Indeed, pattern dithering can be used to create several special rendering effects on the dithered image. A well-known and classical example of this technique is the images created on line printers, like the one shown in Figure 12.18. Such effects are produced by creating a dithering table that associates to each quantization level a well-chosen set of overprinted characters.

But even in this arena, dithering with threshold functions, or ordered dithering, is much more flexible than pattern dithering. We remind the reader that the geometry of the cluster produced by ordered dithering is controlled by the geometry of the level curves of the threshold function, which can be quite arbitrary.

Fig. 12.18. Pattern dithering for line printers.

12.4 Nonperiodic Dithering

Among nonperiodic dithering algorithms, one of the earliest was dithering by random modulation, which, as observed, has now been largely supplanted by superior algorithms. A more recent method, already a classic and very popular, is the Floyd–Steinberg algorithm, which was originally introduced for grayscale images but is easily adapted to color images.

12.4.1 The Floyd–Steinberg Algorithm

The Floyd–Steinberg algorithm computes the effective quantization error at each pixel and compensates for it in neighboring pixels. In this way, the overall error can be minimized. More precisely, suppose we want to perform one-bit quantization on an image defined by the image function $f(x, y)$. Starting with a pixel of coordinates (x, y), we replace $f(x, y)$ by its quantized value $\bar{f}(x, y)$ equals 1 or 0, depending on whether this value is above or below the desired threshold (usually 50%). Then take the quantization error $\delta = f(x, y) - \bar{f}(x, y)$

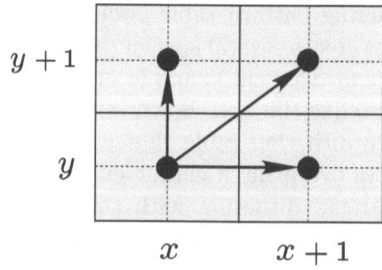

Fig. 12.19. Propagation of the quantization error in the Floyd–Steinberg algorithm.

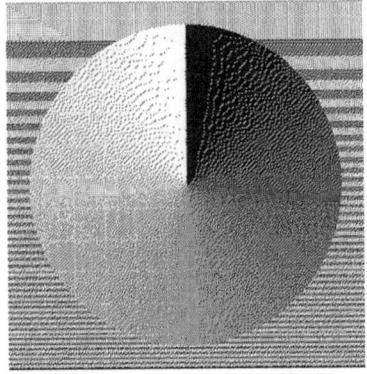

Fig. 12.20. Effect of Floyd–Steinberg dithering.

and proceed as follows (see Figure 12.19): add $\frac{3}{8}\delta$ to $f(x+1, y)$, add $\frac{3}{8}\delta$ to $f(x, y+1)$, and add $\frac{1}{4}\delta$ to $f(x+1, y+1)$. Move to the next pixel on the same row, $(x+1, y)$, and repeat the process with the modified value $f(x+1, y) + \frac{3}{8}\delta$. Continue in this way to the end of the row, then move on to the next row, and so on. (At the last pixel of a row, do not apply the horizontal and diagonal corrections, and on the last row, do not apply the vertical and diagonal corrections.)

The problem with this dithering method is that it propagates the error along the diagonal, which causes a certain directionality effect in the resulting image. See Figure 12.20, which shows our test images after processing with the Floyd–Steinberg algorithm. The effect of diagonal error propagation can be clearly seen. There are several variations of this algorithm in the literature aimed at avoiding the directional artifacts. For more information, consult Section 12.5.

Dithering of Color Images

Although we have been focusing primarily on the use of dithering for one-bit quantization of grayscale images, the same techniques can be used in color quantization, to avoid or minimize quantization contours. For example, the image in Figure 12.21, left, was quantized from 24 to 8 bits without dithering. Figure 12.21, right, shows the same image, quantized to the same number of bits, but using the Floyd–Steinberg dithering algorithm. Observe how the dithered image is devoid of perceptible quantization contours.

12.4.2 Dithering with Space-Filling Curves

Before proceeding, we look back at our classification scheme for dithering algorithms (Figure 12.4) in order to locate the algorithms studied so far. The result is shown in Figure 12.22.

Fig. 12.21. Quantization from 24 to 8 bits, without dithering (left) and with Floyd–Steinberg dithering (right). See Plate 7 in color insert.

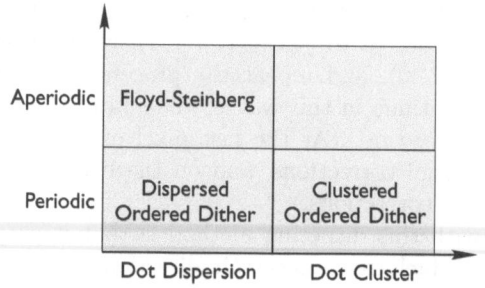

Fig. 12.22. Classification of dithering algorithms encountered so far.

Looking at the figure, we observe that we haven't discussed any nonperiodic clustered dithering algorithm. In principle, such an algorithm should give good results, since it would emulate the action of photographic emulsion. (In black-and-white photography, the film is covered with an emulsion of small, light-sensitive grains, distributed randomly, and therefore nonperiodically. Each grain has finite area, so the effect is that of a cluster being turned on or off by light.)

The missing slot in Figure 12.22 will be occupied by an important class of nonperiodic digital halftoning algorithms, which uses fractal curves generically known as *Peano curves* or *space-filling curves*. Such algorithms possess several advantages over those already studied:

- they use a combination of dot clustering with dot dispersion, by diffusing the average quantization error of each dithering cell;
- the diffusion of the average quantization error that arises does not have a preferred directionality as in the case of the Floyd–Steinberg algorithm;

- they are nonperiodic, so the texture of the final image does not have spurious regular patterns;
- they enable us to change the size of the dithering cell during execution time, thus achieving a better rendition of image details.

In the digital publishing literature clustering algorithms are called *amplitude modulation* (AM) algorithms, because they modulate the size of the dot clustering. Dot dispersion algorithms are called *frequency modulation* (FM) algorithms, because they disperse the dots and control the frequency of their distribution. The space-filling curve algorithm to be described here performs in a range from AM to FM dithering techniques, and so can be tuned to work with a wide range of devices with different degrees of point placement precision.

Pixel Enumeration

A *pixel enumeration* of an image $f : U \subset \mathbb{R}^2 \to C$ is a one-to-one path $c : I \subset \mathbb{R} \to U$, defined on a subset I of the real line and such that the image $c(I)$ contains all the pixels in the image's domain. Thus, an enumeration c associates with each pixel, having coordinates (i, j), say, a unique $k \in \mathbb{R}$ satisfying $c(k) = (i, j)$. Intuitively, an enumeration is a method to run through all the pixels of an image exactly once, in an orderly way.

When we use images encoded in matrix format, the standard enumeration consists in running through the pixels row by row, one after another, as in Figure 12.23.

This enumeration, although convenient and efficient, has two drawbacks regarding its use in dithering algorithms: it has a horizontal directionality; and it has discontinuities, when the curve jumps from one row to the next.

The second drawback is not present in the *boustrophedonic* enumeration of Figure 12.24, but the first one, the directionality of the enumeration, still is. This directionality is responsible, in particular, for the appearance of diagonal patterns in the Floyd–Steinberg algorithm. We can eliminate both drawbacks by using an enumeration curve that approximates a space-filling curve.

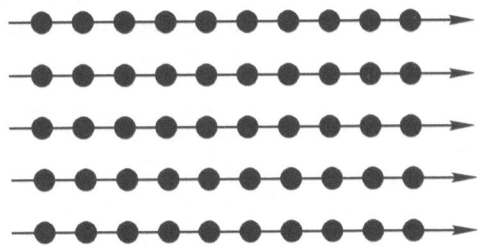

Fig. 12.23. Standard enumeration of the pixels in an image matrix.

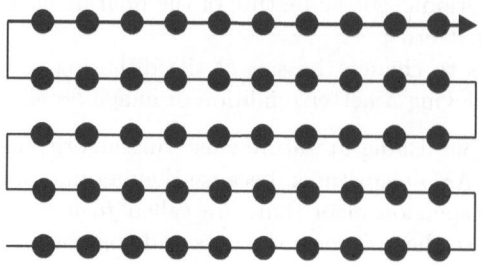

Fig. 12.24. Boustrophedonic enumeration.

Enumeration with Peano Curves

A *space-filling curve* is a map $\varphi : [0, 1] \rightarrow [0, 1] \times [0, 1]$ that is continuous and surjective. Such a curve visits, in a continuous way, every point in the unit square. By means of a scaling transformation, we obtain space-filling curves for an arbitrary rectangle of the plane. Our intent is to use a one-to-one approximation to some space-filling curve in order to define an enumeration for the pixels in a digital image.

Space-filling curves were discovered by the mathematician Giuseppe Peano in the late nineteenth century. Peano constructed explicitly a continuous curve that visits every point in the unit square; this curve became known as the *Peano curve*. In honor of this mathematician, all space-filling curves are commonly called Peano curves as well, and we adopt this nomenclature.

Figure 12.25 shows four steps in the recursive construction of a space-filling curve known as the *Hilbert curve*.

It is interesting to observe that each step of the iteration in the construction of the Hilbert curve gives a polygonal curve that uniquely visits all vertices of a regular lattice. Thus, by iterating successively, we obtain a method to enumerate the pixels of an image with arbitrary resolution. The enumeration thus obtained has no discontinuities and no preferred direction. Figure 12.26 shows the enumeration with the Hilbert curve for a 4×4 image.

At first sight an enumeration with the Hilbert space-filling curve works only for images with resolution $2^p \times 2^p$, for p a positive integer. Nevertheless, the uniform lattice associated with an image of resolution $m \times n$ can be

Fig. 12.25. Recursive construction of the Hilbert curve.

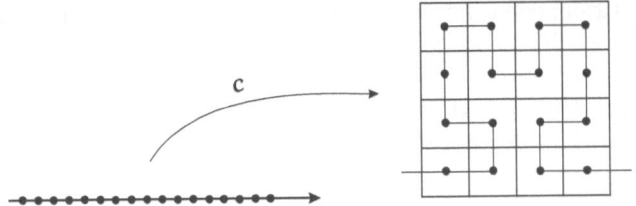

Fig. 12.26. An enumeration of a 4×4 block of pixels.

embedded into a power-of-two square lattice, and it is easy to induce an enumeration for the embedded lattice from an enumeration of the square lattice. Working in this way is more effective than implementing an enumeration for rectangular space-filling curves.

Several other space-filling curves can be recursively constructed using polygonal curves and used to define an enumeration of the pixels of a discrete image. We will use the Hilbert curve for illustration.

Partition by Peano Curves

In order to describe the Peano curve halftoning algorithm, it is necessary to understand the partition that such curves determine in the image's domain. Consider an enumeration $c : I \to U \subset \mathbb{R}^2$ of a digital image by a Peano curve. Let $I_1 \cup I_2 \cup \cdots \cup I_n$ be a partition of the interval I into n subintervals. Since the enumeration is one to one, this partition determines a partition $c(I_1) \cup c(I_2) \cup \cdots \cup c(I_n)$ of the image's domain. Note that the restriction $c|I_j : I_j \to R_j = c(I_j)$ is an enumeration of the pixels of the image contained in the region R_j. Figure 12.27 shows a block of 4×4 pixels partitioned into three regions of 5, 4, and 7 pixels (each region is shaded with a different intensity).

We stress that the partition obtained in the image has a natural ordering: when we describe the Peano curve once, we visit every set of the partition, each exactly once, in an orderly way.

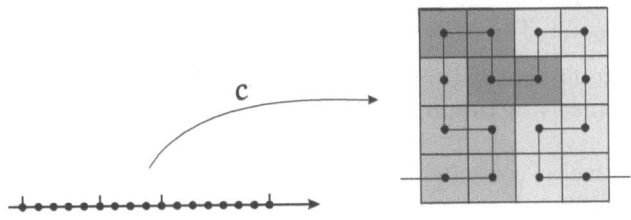

Fig. 12.27. Three partition cells of the Hilbert curve.

The partition determined by a Peano curve is essential for the understanding of the dithering algorithm about to be introduced. Each set in this partition will play a role similar to that of the dithering cells in the ordered dithering algorithm. For this reason each set of the partition is also called a *dithering cell*, or simply a *cell*.

Adaptive Partitioning

We have already mentioned that nonperiodic clustered dithering algorithms provide a rough emulation of the analog process of halftoning by photographic emulsion. This emulation gets better if we are able to change the dithering cell size; this corresponds to a variation in the film grain size.

Each pixel of the image belongs to some cell, but this correspondence is not one to one. A *pixel size function* is a function that associates to each pixel the size of the cell to which it belongs (that is, the number of pixels in the cell).

In order to get a better rendition of image details without compromising tonal reproduction in the dithering process, it would be interesting to vary the size of the cell associated with a pixel P based on how fast image intensities vary in a neighborhood of P. If the intensity varies quickly (high frequencies), the cell size should be small; otherwise, a bigger cell is enough. This variation can be measured by using the absolute value of the directional derivative at P in the direction of the curve being used as an approximation to the space-filling curve.

In fact, the correct relationship between the cell size and the directional derivative values is based on the fact that the eye's response to intensity changes obeys a logarithmic law. For this reason the cell size function is the exponential of the directional derivative magnitude. This rule maintains a linear relationship between the perceptual intensity inside each cell and the directional slope of the image intensity.

Figure 12.28 shows a grayscale image and an adaptive partition associated with an enumeration using the Hilbert space-filling curve.

Computing the Partition

Using the size function s, we are able to compute the adaptive partition of the image domain. The partitioning algorithm proceeds as follows:

- take the (user-specified) maximal cell size as the current cell size s_0;
- scan the pixels of the image using the enumeration of the space-filling curve;
- for each pixel p, compute its cell size $s(p)$;
- compare $s(p)$ with the current cell size s_0.
  If $s(p) < s_0$, make $s(p)$ the current cell size, and continue with the next pixel $p + 1$.

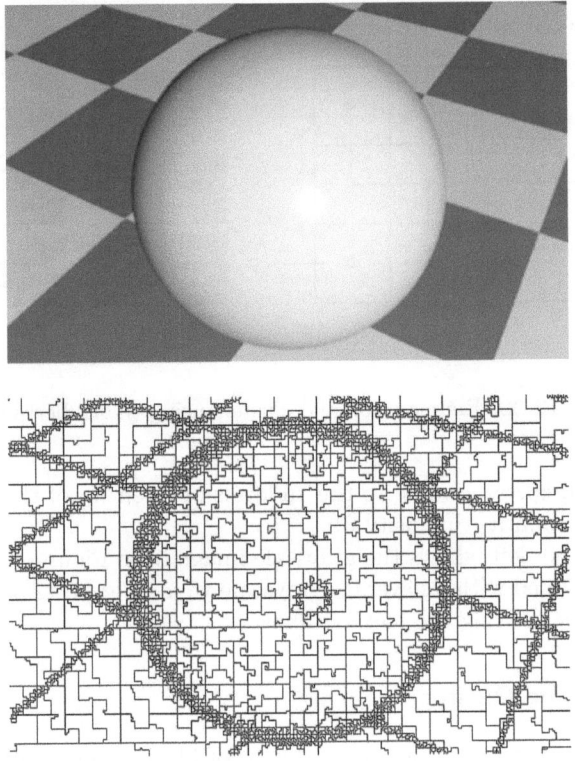

Fig. 12.28. Grayscale image and adaptive partition.

If $s(p) \geq s_0$, p is the last pixel of the cell. Then start a new cell of the partition by repeating the procedure for the next pixel $p + 1$.

Cell Quantization

To perform the quantization of each dithering cell, we use the previously defined enumeration of the cell's pixels, as shown in Figure 12.29. More precisely, we work as follows: the average image intensity inside the cell is computed using Equation (12.1). We turn on and off pixels inside the cell so that the average intensity of the bitmap pattern obtained is approximately equal to the average intensity of the cell. The error is called the *quantization error* of the cell. The bitmap pattern of on and off pixels is created in such a way that the black pixels are clustered together inside the cell.

Figure 12.29(a) shows a 16-pixel cell in the domain of the image, enumerated using a Hilbert curve. Figure 12.29(b) shows the clusters constructed by

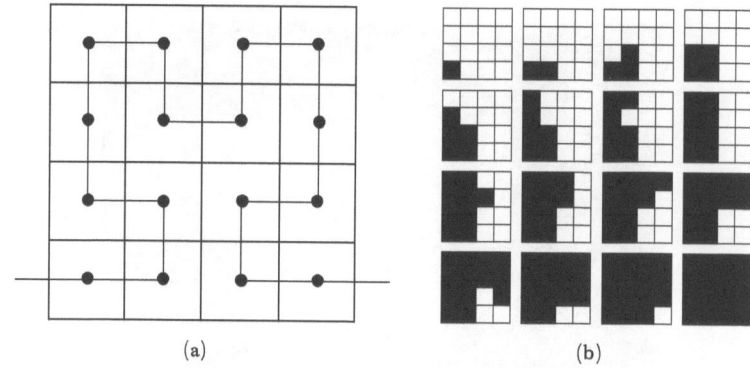

(a)   (b)

Fig. 12.29. Enumeration and clusters using the Hilbert curve.

the algorithm, corresponding to the 16 levels of intensity of the region in the part (a) (not counting intensity 0).

Figure 12.30, left, shows the grayscale image from Figure 12.28 with each dithering cell filled with its average grayscale value. Figure 12.30, right, shows the dithered cells using the method above.

Positioning the Cluster

When quantizing the cell, we get a cluster of black pixels inside it. We have one degree of freedom in positioning this cluster of black pixels inside the cell, by sliding it along the space-filling curve. A good choice consists in aligning the central pixel of the cluster with the central pixel of the cell. This centering process adds a certain degree of randomness to the dithered image and perceptually improves the quality of the final image. This operation is illustrated in Figure 12.31: in (b) we locate the pixel with darkest intensity in the 4×4

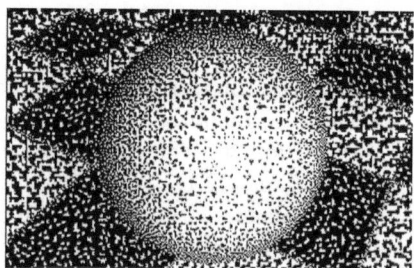

Fig. 12.30. Left: average values inside each cell. Right: two-level quantization of each cell.

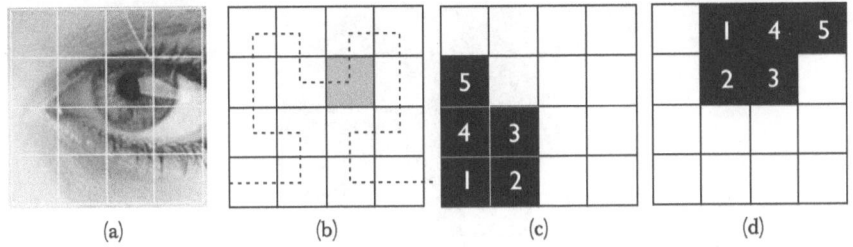

Fig. 12.31. Centering the cluster in the dithering cell.

cell shown in (a); in (c) we show the cluster of five pixels; and in (d) we show the centered cluster.

The Dithering Algorithm

In brief, the idea of dithering with Peano curves is quite simple. After choosing an appropriate Peano curve to enumerate the pixels, we

- determine the partition of the image into dithering cells, as explained above;
- calculate the average intensity of the image in each region of the partition, using Equation (12.1);
- compute the cluster for each cell so that the bitmap pattern obtained has approximately the same intensity as the average image intensity inside the cell;
- position the cluster inside the cell;
- move on to the next cell along the Peano curve, and repeat the steps above, adding the quantization error of the previous cell when computing the average intensity.

This algorithm can be implemented very efficiently; the partition can be computed simultaneously with the average intensities for each cell.

Figure 12.32 shows our test images processed with the Peano curve dithering algorithm just described. The curve used here is the Hilbert curve of Figure 12.25, with a fixed cell size of seven pixels. Figure 12.33 shows the effect of the same algorithm but using adaptive dithering cells with a maximum value of seven pixels.

Observe that this algorithm diffuses the quantization error along the Peano curve. If each cell of the partition has a single pixel, the algorithm performs dot error diffusion along the Peano curve; if the cells are bigger, the algorithm performs clustering within each cell, with error diffusion from one cell to the next.

In ordered dithering, the pixel clustering pattern is defined by the dithering matrix. Once the size of the dithering cell has been chosen, this matrix is fixed.

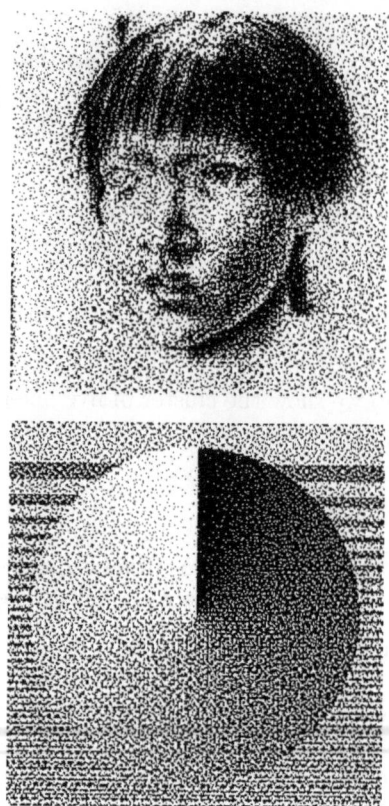

Fig. 12.32. Effect of Peano curve dithering with cells of seven pixels.

In the Peano curve algorithm the size of the cluster depends only on the chosen partition of the domain and can be modified at execution time.

Figure 12.34 is part of a cartoon, containing line drawings and areas of constant gray level. The shadow on the wall in the background is rendered as a regular pattern of dots simulating a standard halftone screen. The image was processed with both constant and adaptive Peano curve dither algorithms. The effect of adaptiveness is striking. The improvement obtained is mainly due to the treatment of high frequencies in the image. In this case, the algorithm was capable of matching exactly the edges of the drawings, at the same time reproducing with uniform dot patterns the different gray shades.

12.5 Comments and References

A fairly comprehensive study of the techniques of digital halftoning, including an analysis of the algorithms in the frequency domain, can be found in

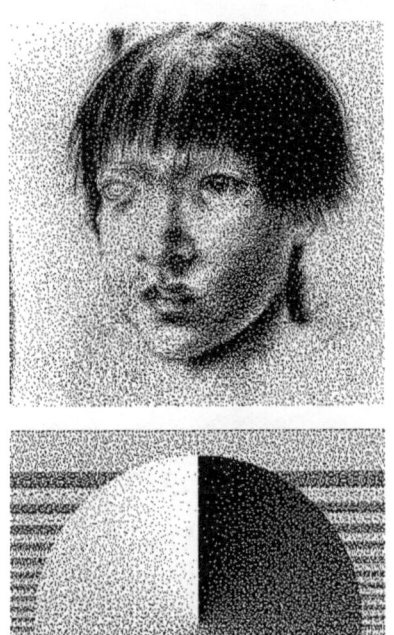

Fig. 12.33. Adaptive dithering with maximum cell of seven pixels.

(Ulichney 1987). The book also introduces a class of dot dispersion dithering algorithms called "blue noise" dithering.

The classic article (Jarvis et al. 1976), although out of date, should be read by anyone interested in dithering techniques; it describes several dithering algorithms very well.

Dithering by random modulation, one of the earliest methods, was introduced (in the context of image quantization for broadcasting) in (Roberts 1962).

As we have mentioned, ordered dithering is generally confused with pattern dithering; (Knowlton 1972) is a pioneer work on pattern dithering. For a discussion of the use of pattern dithering in the creation of "halftones" on line printers, see (Hamill 1977) or (Perry and Mendelsohn 1964).

The idea of using a fixed table of thresholds for determining quantization levels was introduced in (Limb 1969), in the context of color quantization of television images. The first use of a dot dispersion dithering algorithm for

Fig. 12.34. Line-drawing cartoon dithered with (a) constant cluster size of 27 pixels and with (b) variable cluster size.

displaying images on two-level devices was in (Lippel and Kurland 1971). Bayer's article (Bayer 1973) was responsible for definitively associating his name with dot dispersion algorithms, since there he constructed dithering matrices that minimize the occurrence of low frequencies in regions of constant intensity.

The ordered dithering matrices we used in this chapter's examples were taken from (Ulichney 1987). That book uses a beautiful recursive algorithm to generate dot dispersion dithering matrices, based on planar subdivision. The matrices thus obtained coincide with Bayer's.

The Floyd–Steinberg algorithm was introduced in (Floyd and Steinberg 1975). Several variations and extensions of this algorithm can be found in the literature. The interested reader should consult (Ulichney 1987). One algorithm that attempts to minimize the directionality problems that characterize the Floyd–Steinberg algorithm is the method of *dot diffusion* presented in (Knuth 1987). Of course, space-filling curve dithering with a cell size of one pixel is a dot dispersion, nonclustered dithering method and has no directionality artifacts. The use of the Floyd–Steinberg algorithm to minimize quantization artifacts in color images first appeared in the literature in (Heckbert 1982).

The use of space-filling curves in dithering algorithms was introduced in (Witten and Neal 1982). The algorithm was generalized to the clustering case in (Velho and Gomes 1991) and later extended to use an adaptive cell size (Velho and Gomes 1995). Another digital halftoning algorithm using space-filling curves can be found in (Cole 1991).

Optimization-based dithering algorithms are covered in (Geist et al. 1993). A different optimization dithering algorithm, which is appropriate for dithering sequences of images, can be found in (Gotsman 1993).

The flexibility of the ordered dithering technique in changing the shape of the cluster is nicely exploited to create artistic screening effects in (Ostromoukhov and Hersch 1995).

A comprehensive coverage of clustered dithering with a detailed study of screening, dot shapes, and moiré problems can be found in (Peter 1992).

The Indian boy used as a running example in this chapter is a charcoal drawing by the twentieth-century Brazilian painter Cândido Portinari (see, for example, (MOMA 1940) or (Fabris 1990)). We thank Projeto Portinari for permission to use it.

The original of Figure 12.18 is titled *Portrait of Fabiana* and was produced by the late Brazilian artist Waldemar Cordeiro in 1969 (Belluzzo et al. 1986).

References

[Bayer 1973]Bayer, B. E. (1973). An optimum method for two-level rendition of continuous-tone pictures. In *International Conference on Communications, Conference Record*, pp. 26–11 to 26–15.

[Belluzzo et al. 1986]Belluzzo, A. M. et al. (1986). Waldemar Cordeiro: Uma aventura da razão. Museu de Arte Comtemporânea, Universidade de São Paulo, São Paulo.

[Cole 1991]Cole, A. J. (1991). Halftoning without dither or edge enhancement. *The Visual Computer*, 7:232–246.

[Fabris 1990]Fabris, A. (1990). *Portinari, pintor social.* Editora Perspectiva and Editora da Universidade de São Paulo, São Paulo.

[Floyd and Steinberg 1975]Floyd, R. W. and Steinberg, L. (1975). An adaptive algorithm for spatial gray scale. In *SID 75, Intl. Symp. Dig. Tech. Papers*, 36.

[Geist et al. 1993]Geist, R., Reynolds, R., and Suggs, D. (1993). A markovian framework for digital halftoning. *ACM Transactions on Graphics*, 12(2):136–159.

[Gotsman 1993]Gotsman, C. (1993). Halftoning of image sequences. *The Visual Computer*, 9:255–266.

[Hamill 1977]Hamill, P. (1977). Line printer modification for better grey level pictures. *Computer Graphics and Image Processing*, 6:485–491.

[Heckbert 1982]Heckbert, P. (1982). Color image quantization for frame buffer display. *Computer Graphics (SIGGRAPH '82 Proceedings)*, 14(3): 297–307.

[Jarvis et al. 1976]Jarvis, J. F., Judice, C. N., and Ninke, W. H. (1976). A survey of techniques for the display of continuous tone pictures on bilevel displays. *Computer Graphics and Image Processing*, 5:13–40.

[Knowlton 1972]Knowlton, K. (1972). Computer-produced greyscales. *Computer Graphics and Image Processing*, 1:1–20.

[Knuth 1987]Knuth, D. E. (1987). Digital halftones by dot diffusion. *ACM Transactions on Graphics*, 6(4):245–273.

[Limb 1969]Limb, J. O. (1969). Design of dither waveforms for quantized visual signals. *Bell System Tech. J.*, 2555–2582.

[Lippel and Kurland 1971]Lippel, B. and Kurland, M. (1971). The effect of dither on luminance quantization of pictures. *IEEE Transaction of Communication Technology*, 6:879–888.

[MOMA 1940]MOMA (Museum of Modern Art) (1940). *Portinari of Brazil*. The Museum of Modern Art, New York.

[Ostromoukhov and Hersch 1995]Ostromoukhov, V. and Hersch, R. (1995). Artistic Screening. *Computer Graphics (SIGGRAPH '95 Proceedings)*, 219–228.

[Perry and Mendelsohn 1964]Perry, B. and Mendelsohn, M. L. (1964). Picture generation with a standard line printer. *Comm. of the ACM*, 7(5): 311–313.

[Peter 1992]Peter, F. (1992). *Postscript Screening: Adobe Accurate Screens*. Adobe Press, Mountain View, CA.

[Roberts 1962]Roberts, L. G. (1962). Picture coding using pseudo-random noise. *IRE Trans. Infor. Theory*, IT-8:145–154.

[Ulichney 1987]Ulichney, R. (1987). *Digital Halftoning*. MIT Press, Cambridge, MA.

[Velho and Gomes 1991]Velho, L. and Gomes, J. (1991). Digital halftoning with space filling curves. *Computer Graphics (SIGGRAPH '91 Proceedings)*, 25(4):81–90.

[Velho and Gomes 1995]Velho, L. and Gomes, J. (1995). Stochastic screening dithering with adaptive clustering. *Computer Graphics (SIGGRAPH '95 Proceedings)*, 273–276.

[Witten and Neal 1982]Witten, I. H. and Neal, M. (1982). Using Peano curves for bilevel display of continuous-tone images. *IEEE Computer Graphics and Applications*, 47–52.

Image Compression

In this chapter we are going to discuss the problem of image encoding and compression. We will present the classical methods for image compression based on transformation to the frequency domain (i.e. Discrete Cosine Transform) and exploiting multiresolution decomposition (i.e., the Wavelet Transform). These methods are employed respectively on the JPEG compression and JPEG 2000 compression standards.

13.1 Image Encoding

As we saw in Section 6.1 (Figure 6.1), there are three abstraction levels for the specification of an image in the computer: the continuous image, the digital image (image after discretization of domain and range), and the *encoded image*. In an encoded image, the digital image is transformed into a set of symbols, which are organized according to some data structure, and encoded using strings of bits.

When we encode an image, we usually try to store the image as compactly as possible. Thus, encoding is directly linked to image compression.

Matrix Encoding

As an example, consider a monochrome image with geometric resolution $m \times n$, and quantized at 8 bits (256 color levels). One of the possible encodings for this image is a matrix format arising directly from the matrix representation introduced earlier. In this representation an image might be encoded by means of the following data:

- a descriptor indicating the geometric resolution $(m \times n)$;
- the number of components per pixel (in this case 1, since the image is monochrome), and the number of quantization bits per component (8 in this case);

L. Velho et al., *Image Processing for Computer Graphics and Vision*,
Texts in Computer Science, DOI 10.1007/978-1-84800-193-0_13,
© Springer-Verlag London Limited 2009

- an ordered list of the $m \times n$ elements of the matrix of pixels.

Even for such an extremely simple encoding scheme, the data structure used must allow the information to be extracted correctly. Thus, details such as whether the matrix entries are listed by row or by column must be stated in the specification of the matrix encoding.

Run-Length Encoding

The classical encoding method for images takes advantage of the fact that neighboring pixels often have the same color. Consider, for instance, each image scanline. Instead of storing the information $f(x_i, y_j)$ for each pixel (x_i, y_j), we encode, for each row of the image, the length of the intervals where the image function is constant, together with the constant value in that interval. This is called *run-length encoding.*

It is intuitively clear that run-length encoding leads to good compression only if the intervals along which the image function is constant are relatively long with respect to the image's horizontal resolution.

Instead of using color correlation along scanlines of the image, we can devise a version of the algorithm to exploit spatial correlation, using images that are piecewise constant as functions of two variables. In this case the sets in the partition have a more complicated geometry and topology, and we must devise better spatial data structures to represent them. Two good choices are quadtrees or BSP trees (binary space partition trees).

Pulse Code Modulation

Instead of encoding the image based on regions of the domain where there is no color variation, a more flexible and efficient procedure encodes the color variation itself. Many encoding techniques, known as *pulse-code modulation* methods, use this strategy.

Entropy Encoding

In order to compare two encoding schemes with respect to the compactness achieved in the encoding process, we must devise a way to measure the amount of information carried by the image. This problem is studied in information theory, and it is important to get acquainted with the basic principles of this theory.

The techniques of this area apply to the encoding of any information and not only to images. Generally, we can regard the information to be encoded as a "message" to be stored or transmitted. The message is encoded using symbols from a fixed, finite alphabet

$$\mathcal{A} = \{f_0, f_1, \ldots, f_{L-1}\},$$

with L elements. It is natural to associate the alphabet a probability measure P. The value of P at each symbol $f_i \in A$ gives the probability of occurrence of the symbol in a message. The message itself is modeled as a random function F defined over the alphabet.

The probability of the symbol f_k occurring in a message is

$$P(F = f_k) = p_k, \quad \text{for } k = 0, 1, \ldots, L - 1.$$

In the case of images, the symbols are the color quantization levels of the image color space. That is, an image can be considered as a source of independent pixels; each pixel has its own color. The set of all pixel colors is the message to be encoded. The probability can be approximated by using the image histogram.

Intuitively, the amount of information when a symbol f_k is present in the message is related to the inverse of p_k: if the probability is high, the symbol will almost certainly occur; therefore, the information it carries is low. On the other hand, if the probability is low, and the symbol does occur, it must carry a great amount of new information.

In accordance with the above intuition, the amount of information I is measured by

$$I(f_k) = \log\left(\frac{1}{p_k}\right).$$

The base of the logarithm is irrelevant from the mathematical point of view. In applications, the choice of a base depends on the data we wish to represent. A widely used standard base is two. In this case, the unit of information is called a *bit*. In this case, if the alphabet has two symbols $\{f_0, f_1\}$ and $p_0 = p_1 = \frac{1}{2}$, then $I(f_k) = 1$. Thus, one bit is the amount of information we gain when we have the choice of two symbols that can occur with equal probability.

One should be careful not to confuse the use of the word "bit" here with the notion of a binary digit when dealing with numbers represented in base two. One bit of information is not necessarily encoded on the computer using one bit (binary digit) for its storage.

The amount of information carried by the message is the sum of the amount of information carried by each symbol in the message. That is,

$$\mathcal{E} = \sum_{k=0}^{L-1} p_k I(f_k) = \sum_{k=1}^{L} p_k \log_2 \frac{1}{p_k} = -\sum_{k=1}^{L} p_k \log_2 p_k. \tag{13.1}$$

The number E is called the *entropy* of the message.

Suppose we have L symbols in an alphabet (in the case of an image, this means L quantization levels). Using the convexity property of the logarithmic function, it is possible to show that

$$\mathcal{E} \le \log_2 L. \tag{13.2}$$

That is, the entropy is bounded above by $\log_2 L$.

Now that we have devised a concept for the amount of information carried by a message, we are able to discuss the binary encoding of the message, that is, the binary encoding of the alphabet symbols the message uses. This is done by associating a string of bits (binary digits) to each symbol in the alphabet. This string of bits is called a *codeword*.

The simplest method to accomplish this association is to use codewords with the same number of bits. For example, say we have four symbols in our alphabet (four levels of quantization in the case of an image), namely f_0, f_1, f_2, and f_3. Using uniform two-bit codewords, we are able to obtain the following encoding for the symbols:

symbol	f_0	f_1	f_2	f_3
code	00	01	10	11

The number of bits used to encode a codeword is called the *length of the codeword*. When all of the codewords have the same length, we say that the encoding is *uniform*, or that the encoding has a constant *bit rate*. In the above example, we have uniform encoding with bit rate 2.

Usually, the various symbols in a message occur with different probabilities. If we know ahead of time the probability distribution of these symbols, we can use this knowledge to design a nonuniform encoding such that the more frequent levels use fewer bits than rarer ones. We call such an encoding *adaptive*. Adaptive encoding leads to a reduction in the total number of bits needed to encode the image.

Suppose that for each symbol f_k we use a codeword of length l_k. If we have L symbols in the alphabet, the average codeword length M is given by

$$M = \sum_{k=0}^{L-1} p_k l_k,$$

where p_k is the probability of the occurrence of the symbol f_k. The number M represents the average number of bits (binary digits) used per symbol in the encoding process. Thus, for uniform encoding M coincides with the number of bits to encode each symbol. The *source coding theorem* by Shannon (see the references) says that $\mathcal{E} \leq M$. That is, the entropy is the lower bound for the average number of binary digits used in the binary encoding. For this reason, the entropy is called the *fundamental encoding limit*.

Two-Symbol Encoding

Consider the case where the alphabet has two symbols ($L = 2$) with probabilities $p_1 = p$ and $p_2 = 1 - p$, $0 \leq p \leq 1$. The entropy will be

$$\mathcal{E}(p) = -p \log_2 p - (1 - p) \log_2(1 - p).$$

The graph of $\mathcal{E}$ is shown in Figure 13.1.

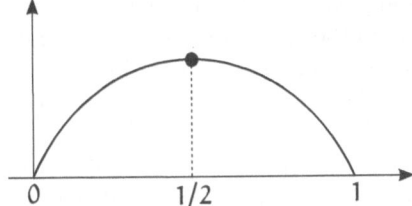

Fig. 13.1. Graph of the entropy of two-symbol alphabet encoding.

The maximum entropy is 1, and corresponds to $p = \frac{1}{2}$: each symbol has the same probability of occurring in the message. Shannon's source coding theorem says that we must use one bit to encode the message. When $p \to 0$ or $p \to 1$, the entropy decreases and from the source coding theorem it is possible to encode the message using less than 1 bit, on the average.

Huffman Encoding

One of the most common types of nonuniform encoding is called *Huffman encoding*. It uses the probability distribution of the image's quantization levels (obtained from the frequency histogram) in order to perform an adaptive encoding whose average bit rate is fairly close to the fundamental limit of the entropy. Moreover, Huffman encoding is relatively easy to implement. For details, see the relevant references given in Section 5.6.

13.2 Image Compression

The goal of compression is to obtain an image representation while reducing as much as possible the amount of memory needed to encode the image. Image compression is possible because images, in general, are highly coherent (nonrandom), which means that there is redundant information. Visual data, like other meaningful data, are usually structured, and this structure means that the data over different parts of an image are interrelated.

For example, consider an image in matrix format. If we take an arbitrary pixel, its color will likely be close to that of neighboring pixels, since they are more likely than not to belong to the same object. Or perhaps the colors aren't close, but some more complex relationship applies: for instance, the pixel might be on the boundary of two objects, or it may be part of a texture pattern. In any case, there are usually some redundant data because of the image's structure. Image compression methods try to eliminate some of this redundancy to produce a more compact code that preserves the essential information contained in the image.

Depending on the application, the information we wish to preserve may be objective or subjective. In the first case, compression must allow the data to be recovered exactly in its original form when desired. This is known as *reversible*, or *lossless*, compression. When the original data cannot be recovered exactly, we have *irreversible*, or *lossy*, compression.

Typical applications of compression involve the storage and transmission of an image, or a sequence of images (as in the case of television). Whether or not reversibility is needed depends directly on the application. In medicine, for example, it may be necessary to preserve the contents of a digital X-ray exactly. On the other hand, when sending a fax, one usually cares only that the recipient be able to read the message.

In order to understand the problem of image compression, we return to the three levels of image representation shown in Figure 6.1. In the continuous level the image is represented by some model describing the image function f : $U \subset \mathbb{R}^2 \to \mathbb{R}^3$. In the discrete level the image is given by some representation (e.g., matrix representation resulting from spatial point sampling). There are essentially two methods of encoding an image: encode a model of the image function or encode some representation of the image function.

13.2.1 Compression by Image Model

This class of compression algorithms includes the methods that encode a model of the image function. In general, these methods are resolution independent, because from the image model we can sample the image at any resolution.

For a simple example of this method, take a grayscale "ramp" given by the linear function $f(x, y) = ax + by + c$. The linear equation is the model of the image function, and the image is compressed by encoding the coefficients a, b, and c.

Another example is a procedural description of a scene in a computer graphics system (containing geometric models for the objects, space transformations, illumination information, and so on). Such a description may actually be longer than a single image generated from it, so no compression results. However, if we are dealing with an animation, the synthetic description is usually much more compact than the resulting sequence of images. The main drawback of such a description is that the decoding time—the time needed to generate the actual images—may be very long.

Compression by Approximation

This type of compression, by transformation of the model, is commonly used. We compute an image model that approximates the original image function within a specified tolerance. The approximation techniques involved usually consist of interpolation methods, and the encoding of the approximated model data (in general, control points and interpolation basis) should take less space than encoding the data in the original image.

A simple example of compression by approximation uses a piecewise linear approximation of the image on each scanline. This is illustrated in Figure 13.2, where the dashed curves represent the tolerance used for the approximation (uniform metric).

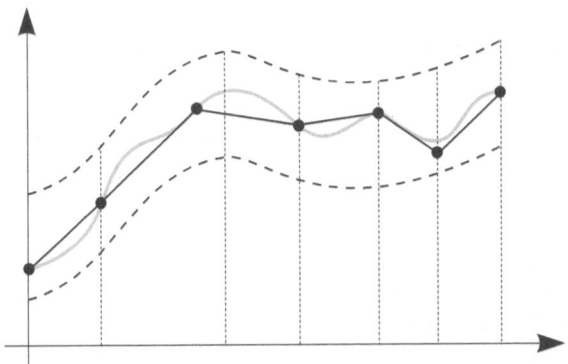

Fig. 13.2. Piecewise linear approximation of an image.

Note that, once the linear approximation has been obtained, encoding the image reduces to encoding the equation $y = ax + b$ on each linear part of the approximating function. This is a simple task and uses less storage space than storing data for all pixels.

Fractal Image Compression

For a given image function f, a mapping $T : \mathcal{I} \to \mathcal{I}$ of the image space is constructed such that f is obtained as the limit set of the sequence of images

$$T(f_0), T^2(f_0), T^3(f_0), \dots,$$

where f_0 is an arbitrary image. In this case the image function is modeled by the dynamical system associated to the map T. The image is encoded by encoding this dynamical system. The image is decoded by a relaxation procedure that consists in iterating the transformation T.

13.2.2 Compression by Image Representation

These are the methods that encode some representation of the image function.

Compression by Discretization

Methods of compression by discretization use a functional model of the image and achieve compression by reducing the information in the range (color resolution) or in the domain (spatial resolution).

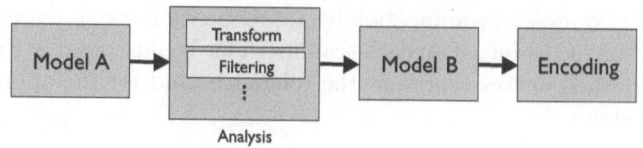

Fig. 13.3. Compression by transformation of the model.

A simple method is *compression by subsampling*, which resamples the image at a lower rate. The decoding process consists in reconstructing the original image from the subsamples. If low frequencies predominate on the image, the original image can be reconstructed with little or no loss, by the Shannon–Whittaker sampling theorem.

Compression by domain discretization is closely related to decomposition schemes of the image domain. There are several compression techniques in this direction. These techniques use spatial data structures, such as quadtrees or BSP trees, to encode the domain decomposition efficiently.

Compression by Transformation

Compression by transformation of the model involves analyzing the image using filters, transforms, and so on, with the goal of finding a better representation of the image for the purposes of encoding. This new model should be pursued in such a way that its encoding occupies less storage than the original image representation, thus achieving compression. Figure 13.3 illustrates the pipeline of this method.

Many classical transforms can be used for the analysis of an image: the Fourier transform, the cosine transform, the Hadamard transform, and so on. More recently, the wavelet transform has been added to this arsenal. A detailed study of the use of such transforms is well outside the scope of this work. The interested reader should consult the references in the next section.

13.3 Compression and Multiscale Analysis

In this section we will apply our study of image filtering to two very important areas: image compression and multiscale representation of images.

The use of filters in image compression stems from the idea that one can take advantage of the spectral characteristics of the image, that is, of information about the various frequencies present in the image. Roughly speaking, one can run an image f through n bandpass filters $h_1, h_2, \ldots, h_n$, obtaining n components $f_1, f_2, \ldots, f_n$ of the image with distinct spectral characteristics. Each of these components is called a *spectral band* of the image and is characterized by the presence of frequencies in a given region of the spectrum.

One then applies compression methods adapted to each band separately. This is called *subband encoding*. An important particular case is that of two-band compression, also known in the literature as *two-channel encoding*.

13.3.1 Two-Channel Encoding

This type of encoding uses a highpass and a lowpass filter to split the image signal f into two components: a high-frequency component H and a low-frequency component L. The original image is the sum $f = H + L$.

Since L contains only low frequencies, one can sample it at a low rate without substantial loss of information. On the other hand, the high-frequency component can be compressed using quantization with a small number of bits, again without significant loss of information. Figure 13.4 shows one possible implementation of this process. In this figure, the original image f goes through a lowpass filter, and the output is resampled at a lower rate, yielding a low-frequency component with lower spatial resolution. This component is then quantized with an appropriate number of bits, resulting in a compressed version of the low-frequency portion of the image. We will denote by $\bar{L}$ this subsampled low-frequency component.

The encoding of the high-frequency portion is obtained by subtracting the low frequency component from the original image, that is,

$$H(x,y) = f(x,y) - L(x,y).$$

After this, we quantize H, to obtain an encoded version of the high-frequency portion of the image.

The low-frequency component L is obtained from the subsampled version $\bar{L}$ by using an interpolation filer I. Therefore, we can write

$$f(x,y) = L(x,y) + H(x,y) = I(\bar{L}(x,y)) + H(x,y).$$

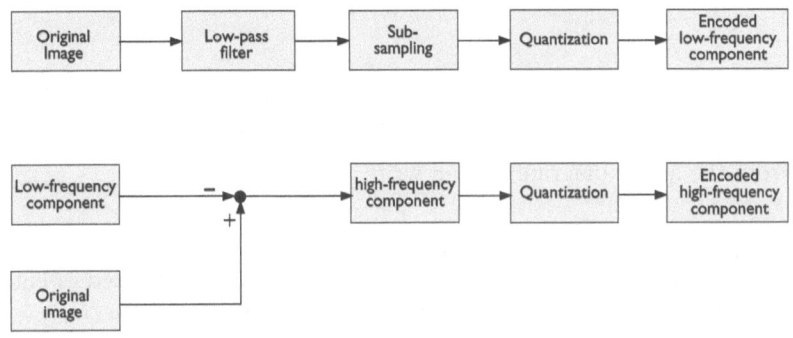

Fig. 13.4. Two-channel encoding: low and high frequencies.

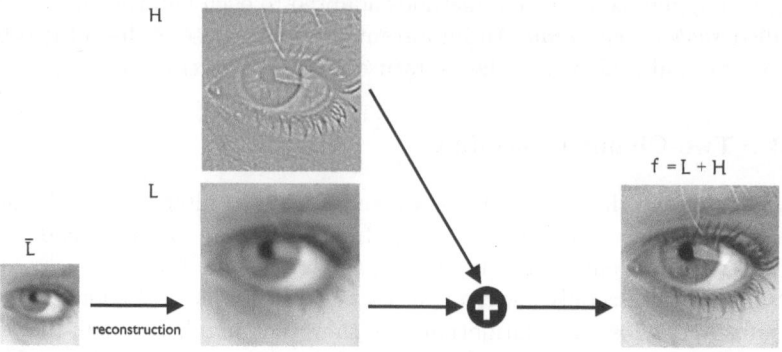

Fig. 13.5. Two-channel decoding.

In general the reconstruction (interpolation) is not exact and we have

$$f(x,y) \approx I(\bar{L}(x,y)) + H(x,y).$$

The whole process is illustrated in Figure 13.5.

We can iterate this process, repeatedly applying the method of two-channel encoding to the low-frequency component $\bar{L}$ of the image. After M steps, we get three sequences of M images, denoted $(\bar{L}_j)$, (L_j), and (H_j) (for $j = 1, \ldots, M$), where $L_1 = L$, $\bar{L}_1 = \bar{L}$, and $H_1 = H$. As j varies from 1 to M, each L_j represents an image with increasingly coarser spatial resolution, representing the low-frequency component of f at that resolution. Each element H_j has the same spatial resolution of L_j and represents the high-frequency component of the image f at that resolution.

To summarize, the sequences (L_j) and (H_j), for $j = 1, \ldots, m$, encode the low- and high-frequency information for f at a particular spatial resolution. This is illustrated by the decomposition diagram below:

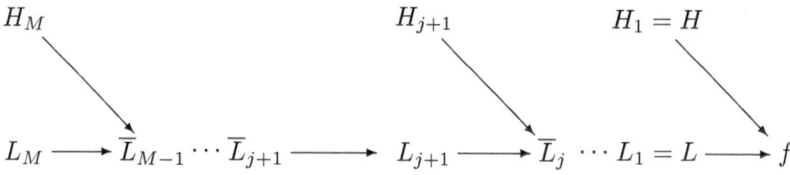

From the above diagram we can write

$$H_{j+1} = \bar{L}_j - L_{j+1},$$

Since L_{j+1} is a blurred copy of $\bar{L}_j$ we conclude that H_j is obtained essentially by applying a laplacian filter to $\bar{L}_j$. For this reason, the sequence $(H_j)_{j=1,\ldots,m}$ is known as the *laplacian pyramid*.

On the other hand, each L_j is obtained from the previous one by the application of a lowpass filter, followed by subsampling. The lowpass filter can be considered an approximation to a gaussian filter. For this reason, the sequence $(L_j)_{j=1,\ldots,m}$ is known as the *gaussian pyramid*.

Figure 13.6 shows some steps in the construction of the gaussian (on the right) and the laplacian (on the left) pyramids, for the image of an eye. In this figure, each image L_i on the right is obtained by doubling the size of the image L_{i+1} above it, and adding to the result the image on the left, H_i.

To encode an image f, we construct the gaussian and laplacian pyramids, and keep the low-frequency, blurred and subsampled image L_m, and the associated laplacian pyramid. For decoding we must start with the coarsest image L_m of the gaussian pyramid and the whole sequence of H_j's (the laplacian pyramid). We then repeatedly apply the reconstruction procedure described above. Indeed, this reconstruction procedure gives

$$H_1 + L_1 = f,$$
$$H_2 + L_2 = L_1,$$
$$\vdots$$
$$H_{m-1} + L_{m-1} = L_{m-2},$$
$$H_m + L_m = L_{m-1}.$$

Adding these equations term by term and simplifying, we obtain

$$H_1 + H_2 + \cdots + H_m + L_m = f. \tag{13.3}$$

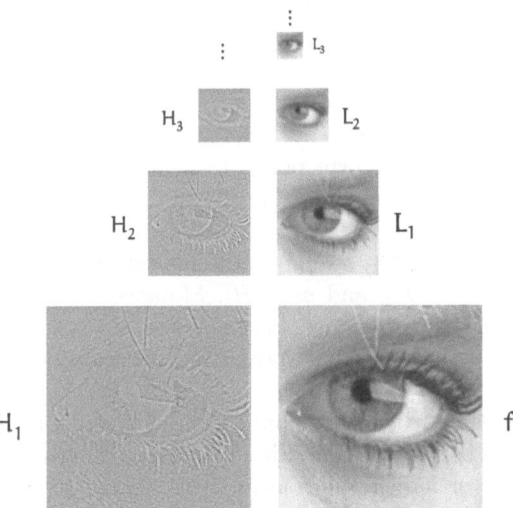

Fig. 13.6. Gaussian and laplacian pyramids. Starting from the image on the top of the gaussian pyramid, and applying the reconstruction operation recursively, we recover the original image f.

Perceptually, the image L_m is a blurred version of f, with very low spatial resolution. The decoding process reconstructs the original image f by stages; at the end of each stage we get an image with higher spatial resolution and more details (high frequencies).

The above encoding method allows the gradual reconstruction of the original image during the decoding process. We have at each stage a coarse (but increasingly finer) version of the final image. This enables us to get information about the final image before having a complete reconstruction. This property is important in some applications. In image transmission we can stop the process, if desired, before receiving the whole image.

Subband encoding and decoding is also important historically; it introduced the use of multiscale analysis in image processing. We will outline this development in the next section.

13.3.2 Multiscale Representation of an Image

Let $\mathcal{F}$ be a space of signals, and $S_1 \supset S_2 \supset \cdots \supset S_{m-1} \supset S_m$ a sequence of subspaces of $\mathcal{F}$. Given a signal $f \in \mathcal{F}$, a multiscale decomposition of f is a sum

$$f = g_1 + \cdots + g_{m-1} + g_m + f_m, \qquad (13.4)$$

where

$$g_i \in S_{i+1} - S_i. \qquad (13.5)$$

Intuitively, each subspace S_i contains signals with more detail (information about variations in frequency) than the space S_{i+1}. Therefore, each component signal g_i in equation (13.4) measures the variation in details (frequencies) between two consecutive levels. We conclude that the sum in (13.4) represents the signal f as a low-frequency component $f_m \in S_m$ plus variations g_i between consecutive levels, as expressed by (13.5). For a digital image, f_m contains the low frequencies and so corresponds to a blurred image; as we add each component g_i, we get increasingly sharper and more detailed versions of the image f.

Note that equation (13.3) resembles the multiscale decomposition in equation (13.4), with $f_m = L_m$ and $g_i = H_i$. Moreover, since $H_j = L_{j+1} - L_j$, we see that H_j is the difference between two images of different resolutions and therefore that (13.3) is indeed a multiscale decomposition of the image f.

Besides presenting a beautiful application of linear filtering theory to image processing, we included this section because it describes some basic ideas behind the theory of multiresolution analysis. These techniques have found enormously wide application in recent years. This is the natural setup for the theory of wavelet transforms, which enables us to construct different multiresolution decompositions of the space of images. This is quite beyond the scope of this book; see the next section for references.

13.4 Comments and References

In our section on image compression, we tried to establish a framework that would allow a general understanding of the many available methods, without getting into the details of any one of them. For certain specific classes of images, one can find more efficient algorithms than the generic ones. Thus, the compression of bilevel images, of great importance in telecommunications (bitmaps are used for text image transmissions, for example), has been intensively studied and heavily optimized. For details, see (Jain 1989).

The chapter about image encoding in (Lim 1990) is very complete and has a good classification of compression techniques, along with an abundant bibliography about this topic.

A recent new approach to compression, introduced in (Barnsley et al. 1988), uses *iterated functions systems;* such methods are also loosely known as fractal compression. Details can be found in (Barnsley and Hurd 1993). For an approach to fractals from the viewpoint of computational computer graphics, see (Peitgen and Saupe 1988).

The JPEG group (Joint Photographic Experts Group) was established in the 1980s to develop a standard format for the compression of digital images. The standard it developed, likewise known as JPEG, has been adopted by the ISO (International Standards Organization) and is well described in (Pennebaker and Mitchell 1993). Likewise, the MPEG standard was developed by the Motion Photographic Experts Group for the compression of sequences of images (animations); it has also been adopted by the ISO. The MPEG compression format is always evolving to account for the requirements from the different segments of the video and television industries.

The paper (Velho and Alvarenga 1990) describes an image compression algorithm that employs piecewise linear approximation along the rows of an image, as discussed in the text. This algorithm is very well suited to synthetic images.

Huffman encoding was originally introduced in (Huffman 1962). For a full treatment, see (Rosenfeld and Kak 1976). The very popular Lempel–Ziv–Welch (LZW) compression algorithm, which performs even better than Huffman's, can be found in (Welch 1984). For an information-theoretic discussion of run-length encoding, see (Jain 1989).

The use of transforms in image compression is covered in many books. We mention (Pratt 1978) and (Jain 1989) for a simple introduction, and (Clark 1990) for a deeper and more complete treatment.

Reconstructing an image can be regarded as a problem of interpolation of scattered data. A solution from this point of view is proposed in (Chen et al. 1994).

For an appealing and intuitive treatment of information theory, and in particular of the origin and evolution of the concept of entropy, see (Resnikoff 1987). An introductory and well-written book covering this topic is (Haykin

1988). In this reference the reader will find a proof of inequality (13.2). The classical reference on the subject is Shannon's seminal paper (Shannon 1949), which laid down the mathematical base of information theory. Here the reader will find a proof of the source coding theorem.

Different data structures are used in commercial and academic software to encode the image on the computer. These structures give rise to dozens of different formats for image storage. A detailed description of these formats can be found in (Murray and Ryper 1994).

References

[Barnsley and Hurd 1993]Barnsley, M. and Hurd, L. P. (1993). *Image Compression with Fractals.* A. K. Peters, Wellesley, MA.

[Barnsley et al. 1988]Barnsley, M. F., Jacquin, A., Malassenet, F., Reuter, L., and Sloan, A. D. (1988). Harnessing chaos for image synthesis. *Computer Graphics (SIGGRAPH '88 Proceedings)*, 22(4):131–140.

[Chen et al. 1994]Chen, Y.-S., Yen, H.-T., and Hsu, W.-H. (1994). Compression of color images via the technique of surface fitting. *CVGIP: Graphical Models and Image Processing*, 56(3):272–279.

[Clark 1990]Clark, R. J. (1990). *Transform Coding of Images.* Academic Press, London.

[Haykin 1988]Haykin, S. (1988). *Digital Communications.* John Wiley and Sons, New York.

[Huffman 1962]Huffman, D. A. (1962). A method for the construction of minimum redundancy codes. In *Proceedings of IRE*, 40:1098–1101.

[Jain 1989]Jain, A. K. (1989). *Fundamentals of Digital Image Processing.* Prentice-Hall, Englewood Cliffs, NJ.

[Lim 1990]Lim, J. S. (1990). *Two-Dimensional Signal and Image Processing.* Prentice-Hall, Englewood Cliffs, NJ.

[Murray and Ryper 1994]Murray, J. D. and Ryper, W. V. (1994). *Encyclopedia of Graphics File Formats.* O'Reilly and Associates, Sebastopol, CA.

[Peitgen and Saupe 1988]Peitgen, H. O. and Saupe, D. (1988). *The Science of Fractal Images.* Springer-Verlag, New York.

[Pennebaker and Mitchell 1993]Pennebaker, W. and Mitchell, J. (1993). *JPEG: Still Image Data Compression Standard.* The Color Resource, San Francisco.

[Pratt 1978]Pratt, W. (1978). *Digital Image Processing.* Wiley–Interscience, New York.

[Resnikoff 1987]Resnikoff, H. L. (1987). *The Illusion of Reality.* Springer-Verlag, New York.

[Shannon 1949]Shannon, C. E. (1949). *The Mathematical Theory of Communication.* University of Illinois Press, Urbana, IL.

[Velho and Alvarenga 1990]Velho, L. and Alvarenga, C. (1990). Image compression by first-order approximation. In *Proceedings IMAGE'COM, First International Conference on Image Chains*, 387–391.

[Welch 1984]Welch, T. A. (1984). A technique for high-performance data compression. *Computer*, (6):8–19.

14

Combining Images

Techniques for combining digital images have been used extensively to obtain special effects in movies and television. (In fact, the motion picture industry has been using *analog* methods for the same purposes since the 1920s.) In this chapter we study a number of such techniques.

14.1 Preliminaries

The purpose of combining images is to create a new image that contains features from each of the components. This allows great flexibility in the process of image generation. Thus, if we want a single image containing elements created by different processes—for example, a live shot and a computer-generated scene—we must combine the two images using an appropriate method. Even in the case of purely synthetic scenes, it is often advantageous to synthesize different elements of the scene separately and then combine them into a single image. The interaction between windows in a graphical user interface is also an example of image combination.

Several analog methods for combining images have been used by the movie industry, either to lower production costs or to achieve special effects. For example, live action can be filmed at a studio and superimposed on footage of an outdoor scene. A related technique is to use small-scale models for the background and superimpose the live action in such a way that it appears to take place in front of a full-scale background scene.

With the advances in computer graphics and image processing, and with the increase in computational power, analog techniques have increasingly given way to digital ones, in both the television and the movie industries.

Our mathematical approach to the problem of combining images will consist in defining operations *comp* that associate with two images f and g an image $h = \mathrm{comp}(f, g)$. For some applications we will need operations on more than two images, $\mathrm{comp}(f_1, \ldots, f_n)$; but in general we can regard the process of combining n images as $n - 1$ successive binary operations.

L. Velho et al., *Image Processing for Computer Graphics and Vision*,
Texts in Computer Science, DOI 10.1007/978-1-84800-193-0_14,
© Springer-Verlag London Limited 2009

Note that the operation $\mathrm{comp}(f, g)$ is well defined only if f and g belong to the same image space: in particular, they must have the same support set and take values in the same color space.

We will start by developing the theory in the continuous domain, and then we will examine the problems arising from discretization.

14.2 Combining Images Algebraically

We saw in Chapter 7 that an image space $\mathcal{I} = \{f : U \subset \mathbb{R}^2 \to \mathcal{C}\}$ is a vector space. Thus, given a set f_i, where $i = 1, \ldots, n$, of images in $\mathcal{I}$, and real numbers c_i, where $i = 1, \ldots, n$, we can obtain an image f using a linear combination

$$f = c_1 f_1 + c_2 f_2 + \cdots + c_n f_n. \tag{14.1}$$

This operation can be generalized as follows: instead of n real numbers c_i, take n real-valued functions $\alpha_i : U \subset \mathbb{R}^2 \to \mathbb{R}$, defined in the support set U of the image space. Equation (14.1) is then replaced by

$$f(x, y) = \alpha_1(x, y)f_1(x, y) + \cdots + \alpha_n(x, y)f_n(x, y). \tag{14.2}$$

In order for f in this equation to be well defined, we must suppose that $\alpha_i(x, y) \geq 0$. Moreover, the effect of (14.2) should normally be a weighted average of intensities at each point, so that each pixel has an intensity that is an average of those from the component images, avoiding problems of color overflow. Algebraically, we should require that

$$\alpha_1(x, y) + \cdots + \alpha_n(x, y) = 1, \quad \text{for all } (x, y) \in U.$$

Obviously, if we start from functions α_i that do not satisfy this constraint, we can obtain new functions β_i that do, by setting $C(x, y) = \sum \alpha_i(x, y)$ and applying the normalization $\beta_i = \alpha_i / C$.

The two conditions we have imposed on the α_i make this family of functions into a *partition of unity*. A (finite) partition of unity for a subset $U \subset \mathbb{R}^n$ is a family of functions $\alpha_i : U \to \mathbb{R}$, where $i = 1, \ldots, n$, satisfying the following conditions for every $(x, y) \in U$:

(i) $\alpha_i(x, y) \geq 0$ for $i = 1, \ldots, n$;
(ii) $\alpha_1(x, y) + \alpha_2(x, y) + \cdots + \alpha_n(x, y) = 1$.

Two important properties follow from this definition:

- For every $(x, y) \in U$, we have $0 \leq \alpha_i(x, y) \leq 1$, by (i) and (ii).
- For every $(x, y) \in U$, there exists at least one i such that $\alpha_i(x, y) > 0$, by (ii).

If the functions α_i are continuous, we say that the partition of unity is continuous. Figure 14.1(a) shows a two-element partition of unity $\{\alpha_1, \alpha_2\}$

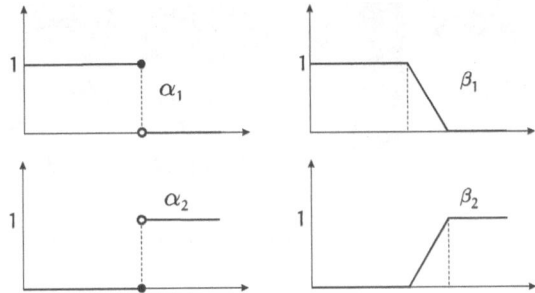

Fig. 14.1. Two partitions of unity of an interval. The partition in (b) is continuous, but not the one in (a).

on an interval of the real line. Figure 14.1(b) shows another such partition, $\{\beta_1, \beta_2\}$, this time continuous.

A discontinuous partition of unity α_i can be turned into a continuous one—in fact, one of any desired class C^k, for $k \geq 0$—by taking the convolution $\alpha_i * \varphi$ of each function in the partition with a convolution kernel φ of class C^{k+1}, satisfying

$$\varphi(x, y) \geq 0 \quad \text{and} \quad \int_U \varphi = 1.$$

The partition of unity illustrated in Figure 14.1(b) can be obtained from the one in Figure 14.1(a) by convolving with a one-dimensional box filter.

14.2.1 Mixing Images

An important particular case of algebraic combination of images occurs when the functions α_i are constant over the domain U of the image space, that is, $\alpha_i(x, y) = c_i$ for every $(x, y) \in U$. This reduces to the linear combination defined earlier in (14.1) and is called *mixing* the images. A classical example is *cross-dissolving*: given two images f and g, cross-dissolving (or simply dissolving) f to g means taking a sequence of images

$$h_t = \text{dissolve}_t(f, g) = (1 - t)f + tg,$$

where the number t ranges from 0 to 1. We have $\text{dissolve}_0(f, g) = f$ at $t = 0$, and $\text{dissolve}_1(f, g) = g$ at $t = 1$. For other values of t, the intermediate image h_t displays, at each point, a weighted average of the colors at that point in f and g; moreover, the weights are the same everywhere, since the functions in our partition of unity are the constants $\alpha_1(x, y) = 1 - t$ and $\alpha_2(x, y) = t$.

Figure 14.2(c) shows a dissolve of the images in (a) and (b).

The more general equation (14.2) for combining images is useful in applications because it allows for an adaptive choice of weights: the contribution of each image to the end result in this case depends on the pixel.

Fig. 14.2. The images f_1 and f_2 shown in (a) and (b) are mixed in (c), with parameter $\alpha_1 = 0.4$ and $\alpha_2 = 0.6$.

14.3 Combining Images by Decomposing the Domain

It is common in computer graphics to obtain an image by combining specific elements from different images belonging to the same image space. This is illustrated in Figure 14.3, where we superimpose an element from image (a) on top of image (b), to obtain the image in (c).

To describe this type of superimposition using the formulation of (9.2), we need to decompose the domain. Denote by f_1 and f_2 the images in Figure 14.3(a) and (b), and let their domain be U. The image f of Figure 14.3(c) is given by

$$f(x,y) = \begin{cases} f_1(x,y) & \text{if } (x,y) \in A, \\ f_2(x,y) & \text{if } (x,y) \in B, \end{cases} \tag{14.3}$$

where the sets A and B form the partition (decomposition) of U shown in Figure 14.4 (B has two connected components).

In general, given n images $f_1, \ldots, f_n$ in the same image space, with domain U, and a decomposition of U into m sets $U_1, \ldots, U_m$, one can define an operator $\text{comp}(f_1, \ldots, f_n)$ piecewise, by giving a different rule for each U_i. As a more elaborate example, parts (a) and (b) of Figure 14.5 show images f_1 and f_2, and part (c) shows a decomposition of the domain U into sets A, B, and C. Part (d) shows the result of combining f_1 and f_2 according to the piecewise rule

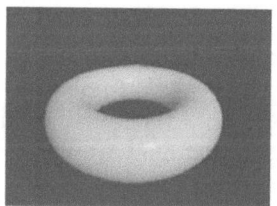  

Fig. 14.3. Superimposing an element of an image onto another image.

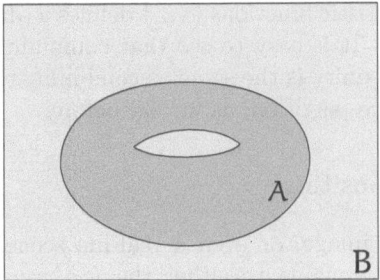

Fig. 14.4. Decomposition of the domain.

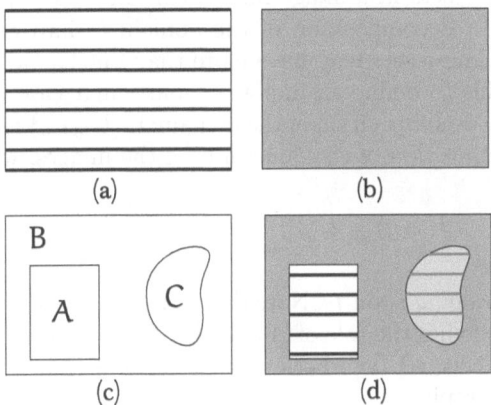

Fig. 14.5. Combining images using piecewise constant weight functions.

$$f(x,y) = \begin{cases} f_1(x,y) & \text{if } (x,y) \in A, \\ f_2(x,y) & \text{if } (x,y) \in B, \\ \text{dissolve}_{0.5}(f_1, f_2) & \text{if } (x,y) \in C. \end{cases}$$

14.3.1 Partitions of Unity and Decompositions

The reader may have realized that, of the two general methods for combining images that we've seen so far (using a partition of unity, in Section 14.2, and using a decomposition of the domain, in Section 14.3), the second is in fact a particular case of the first. Indeed, given a decomposition of U into sets U_i, for $i = 1, \ldots, n$, we associate to each U_i its *characteristic function* χ_{U_i}, defined by

$$\chi_{U_i}(x,y) = \begin{cases} 1 & \text{if } (x,y) \in U_i, \\ 0 & \text{if } (x,y) \notin U_i. \end{cases}$$

The family of characteristic functions $\{\chi_{U_i}\}$ defines a (discontinuous) partition of unity, of the set U. It is easy to see that combining images algebraically using this partition of unity is the same as combining them by using different rules for each set of the partition, as we did before.

14.3.2 Image Compositing

When we observe an image, or even a real-life scene, we often classify its objects mentally as belonging to either the *foreground* or the *background*. The foreground consists of those elements, generally toward the front, where attention is concentrated. The background has the role of "visual support" for the scene's foreground objects.

Dividing the objects in a image into foreground and background has the effect of defining a decomposition of the domain U into two corresponding sets, U_f and U_b. These sets may have more than one connected component.

We can use this to define an important particular case of combining two images f and g by decomposition of the domain: if $\{U_f, U_b\}$ is the background–foreground decomposition of the domain U of the images, we define

$$f(x,y) = \begin{cases} f_1(x,y) & \text{if } (x,y) \in U_f, \\ f_2(x,y) & \text{if } (x,y) \in U_b. \end{cases} \tag{14.4}$$

This is called *overlaying f_1 on f_2*. Note that (14.4) is a particular case of (14.3), with U_f and U_b playing the role of A and B. Figure 14.6 shows the overlaying of a computer-generated (synthetic) image on a digital image obtained by scanning a photograph.

Overlaying is clearly not a commutative operation, since we're picking the elements of one image to define the foreground and the background regions.

We now define the more general notion of *image compositing*, a very common way of combining images in computer graphics. Consider images $f_1, f_2, \ldots, f_n$ in the same image space, with domain U. Suppose each image

Fig. 14.6. Superimposing a synthetic image on a photograph. See Plate 8 in color insert.

f_i defines a background–foreground decomposition $\mathcal{U}^i = \{U_f^i, U_b^i\}$ of U. The intersection of the partitions $\mathcal{U}^i$ defines a partition $\mathcal{V}^k$ of the domain U. In each set $V^k \subset \mathcal{V}^k$ of this partition, compositing is defined by mixing the images $f_1, f_2, \ldots, f_n$ with certain weights. Figure 14.3, as we have seen, shows the compositing of two images. By contrast, Figure 14.5 is not an example of compositing, although the images are being combined using a decomposition of the domain: the decomposition is not a background–foreground decomposition associated to each image.

Besides a single overlaying operation, probably the most important particular case of image compositing is a succession of overlays. That is, given n images $f_1, f_2, \ldots, f_n$, one recursively sets $g_1 = \mathrm{over}(f_1, f_2)$, $g_k = \mathrm{over}(g_{k-1}, f_{k+1})$ for $2 \le k \le n - 1$, and outputs the image g_{n-1}.

Example 14.1 (Depth-of-scene merging). Suppose we have two images f_1 and f_2, in which to each point (x, y) we have associated not only the color information (R, G, B) but also the depth Z (distance to observer) of the object visible at that point. Write

$$f_1(x, y) = (R_1(x, y), G_1(x, y), B_1(x, y), Z_1(x, y)),$$
$$f_2(x, y) = (R_2(x, y), G_2(x, y), B_2(x, y), Z_2(x, y)).$$

We can define a merging operation $h = \mathrm{comp}(f_1, f_2)$ as follows:

$$f(x, y) = \begin{cases} f_1(x, y) & \text{if } Z_1(x, y) \le Z_2(x, y), \\ f_2(x, y) & \text{if } Z_1(x, y) > Z_2(x, y). \end{cases}$$

Thus, the merged image gets at each point the value of whichever object is closer to the observer; when there is a tie, we choose f_1 arbitrarily.

We can reformulate this problem in terms of a decomposition of the domain, by setting

$$U_1 = \{(x, y) \in U : Z_1(x, y) \le Z_2(x, y)\},$$
$$U_2 = \{(x, y) \in U : Z_1(x, y) > Z_2(x, y)\}.$$

We then have $f = f_1$ in U_1 and $f = f_2$ in U_2.

The background–foreground decomposition $\{U_f, U_b\}$ corresponds to the partition of unity given by the two functions $\alpha_f = \chi_{U_f}$ and $\alpha_b = (1 - \alpha_f)$. Clearly, $1 - \alpha_f$ is the characteristic function of the background region U_b.

Thus, the overlaying operator of the previous section can be defined by a partition of unity with two functions α_f and α_b, where $\alpha_f = 1$ at the pixels of f_1 belonging to foreground elements (those that should be overlaid on the background), and $\alpha_f = 0$ at the remaining pixels. We therefore have

$$f(p) = \alpha_f(p) f_1(p) + (1 - \alpha_f)(p) f_2(p). \tag{14.5}$$

14.4 Combining Images in the Discrete Domain

In this section we study the operators for combining images in the discrete domain. We will sometimes work with one-dimensional signals, for ease of illustration. We can regard a one-dimensional signal as an image scanline.

Recall that the images to be combined must lie in the same image space, and in particular must have the same domain and same color space. When discretizing two images we should use the same grid, and we must quantize the color information to the same number of bits for both images. These assumptions will be implicit during the rest of this chapter.

When we mix or cross-dissolve two images, the partition functions are constant over the domain. Therefore, assuming the images were correctly sampled, there is no problem working in the discrete domain: we just work pixel by pixel, combining the sampled values from the two images.

In the case of compositing, we work with the restriction of the images to the sets of decomposition of the domain. Thus, when discretizing the domain, we must take into account how the sets of the decomposition behave inside each pixel.

Consider the overlay operator $f = \text{over}(f_1, f_2)$ of (14.5), based on the background–foreground decomposition of the domain. Working in the continuous domain, we had two options for the value of $f(p)$ at a point p: it might equal $f_1(p)$ or $f_2(p)$, depending on whether p was in the foreground or the background. In the discrete case, we have another option for pixels that straddle the boundary between background and foreground (Figure 14.7). Here we are dealing with sampling problems. Indeed, the discontinuities of the α function on the boundaries of the decomposition gives rise to high frequencies, and this causes trouble when sampling pixels that contain boundary points. We will discuss the different sampling options in more detail later in this chapter.

14.4.1 The Opacity Function

The overlaying operation can be thought of as placing the first image, f_1, "in front" of the second, f_2, except that only parts of f_1—namely, where $\alpha_f = 1$

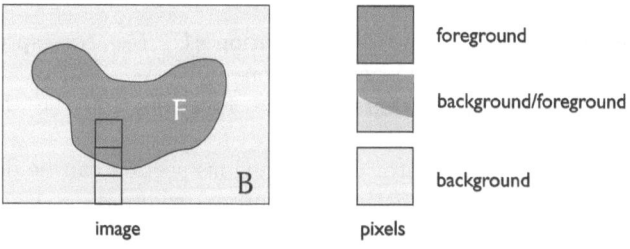

image   pixels

foreground

background/foreground

background

Fig. 14.7. Types of pixels in the background–foreground decomposition of an image in the discrete domain.

in (14.5)—actually cover f_2: the rest of f_1 is discarded. It is useful to think of α_f as a measure of the degree of opacity of the image in front. When $\alpha_f = 1$, we think of f_1 as being opaque at that point, and the image in the back does not appear. Where $\alpha_f = 0$, the image in front is transparent, and the result has the same color as the back image at that point. Finally, as we mentioned, it is useful to allow values $0 < \alpha_f < 1$ after discretization, corresponding to a partially transparent pixel: we can see the color of the image in back, but only in part.

The notion of opacity can be extended to other operators and turns out to lend great versatility to techniques for combining images. For example, we can have a partially transparent overlay, where the objects in front are assumed not to be fully opaque in the first place but to have an intrinsic degree of opaqueness, perhaps varying from point to point. The background of the figure in front, as before, is considered fully transparent and has no influence on the final result. Then the operation of partially transparent overlaying is given by

$$f(p) = \alpha(p)f_1(p) + (1 - \alpha)(p)f_2(p), \tag{14.6}$$

which is exactly the same as (14.5), except that α_f is replaced by a more general opacity function α, which is no longer the characteristic function of the foreground objects but instead that characteristic function times the intrinsic opacity of the objects.

Thus, images that are to be overlaid on others can naturally be thought of as having an additional *channel* (dimension in the range), encoding the opacity. If we're representing color in RGB space, then each pixel actually has four components (r, g, b, α), the last one being the opacity. For efficiency, one often stores $(\alpha r, \alpha g, \alpha b, \alpha)$ instead of (r, g, b, α).

As explained earlier, values of alpha between 0 and 1 can be useful even in the case of a fully opaque overlay, at the boundary between background and foreground. A pixel entirely contained in the opaque region defined by the foreground objects has an opacity value $\alpha = 1$, as in Figure 14.8(a). One that is entirely contained in the background has $\alpha = 0$, as in part (c) of the same figure. In the intermediate case, shown in part (b), the pixel's opacity is defined as the fraction of the pixel area that lies in the opaque region.

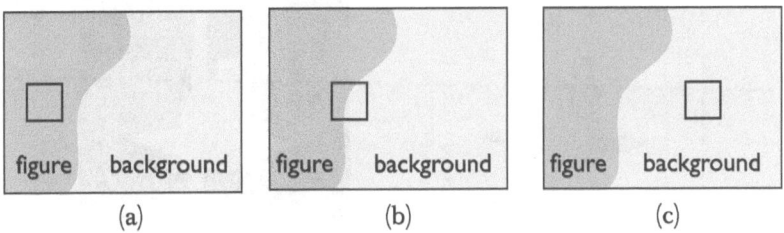

Fig. 14.8. Alpha channel and pixel geometry.

14.4.2 Discretization and Opacity Function

Formally, the discussion of the preceding paragraphs describes the discretization of opacity information by *area sampling*. There is another common way to handle opacity, based on *supersampling* (compare Section 8.3.1). In this section we formalize the two approaches.

Taking the case of an opaque overlay to fix ideas, we recall that in the continuous domain the opacity α is the characteristic function of a subset of the domain U. As a real function $\alpha : U \to [0,1]$, it can be regarded as a monochrome image: a signal with two-dimensional domain and one-dimensional range. In discretizing such an image, we should try to minimize aliasing by using one of the techniques of Section 8.3.1, such as area sampling or supersampling. *Area sampling* discretizes α by averaging its values over each pixel. *Supersampling* consists in subdividing each boundary pixel and computing the value of α in each subpixel. We consider each method in turn.

Area-Sampling Discretization

Area sampling preserves only one aspect of the pixel geometry, namely, the fraction of the pixel that lies in the opaque region. This amounts to using a smoothing (lowpass) filter before sampling the function α. The resulting monochrome digital image is stored together with the color information, constituting the *alpha channel* of the overall, four-channel, digital image.

Figure 14.9 shows a synthetic image on the left, and its alpha channel on the right. Notice how the alpha channel is the digital counterpart of the traditional masks long employed by the movie and television industries in the analog process of compositing.

Together with the opacity function α, the function $1 - \alpha$ forms the partition of unity associated with the background–foreground decomposition of an image. When we discretize $1 - \alpha$ and represent it as a monochrome image, we get the *countermask* of the opacity mask, as shown in Figure 14.10.

Fig. 14.9. A digital image and its alpha channel.

Fig. 14.10. Countermask of the opacity mask.

Figure 14.11, left, is an enlargement of the alpha-channel representation of a detail from Figure 14.9: the rightmost corner of the table. Note the different shades of gray along the edges. Figure 14.11, right, shows the same detail but with one-bit point sampling: the value of α is 1 if the center of the pixel belongs to the opaque region, and 0 otherwise.

Supersampling Discretization

The boundary between the opaque and transparent regions is rich in high frequencies, since it encodes an abrupt variation in the opacity function. The alpha-channel technique minimizes aliasing along this boundary in the process of compositing, but at the cost of discarding most of the information about the pixel geometry, since it uses area sampling. To preserve more of this information, we can instead perform discretization using supersampling.

To do this, we subdivide boundary pixels into subpixels, and in each subpixel we sample the partition function. We use one bit of storage for each subpixel, with value 1 or 0 depending on whether the sample lies in the foreground or in the background, respectively. Thus, instead of a monochrome image, we obtain a bitmask with information about the pixel geometry. Figure 14.12 shows one pixel on the background–foreground boundary, and the corresponding bitmask.

 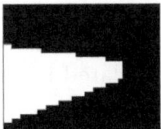

Fig. 14.11. Detail from the alpha channel of an image.

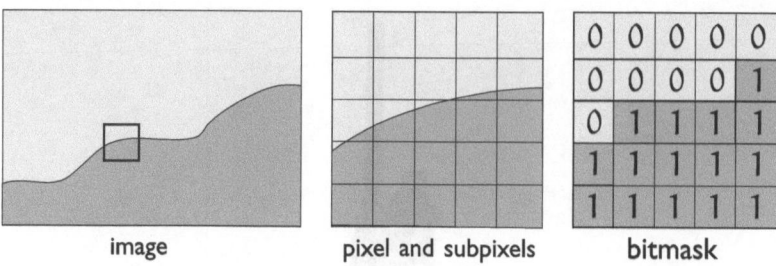

image   pixel and subpixels   bitmask

Fig. 14.12. Bitmask for a pixel on the boundary of a background–foreground decomposition.

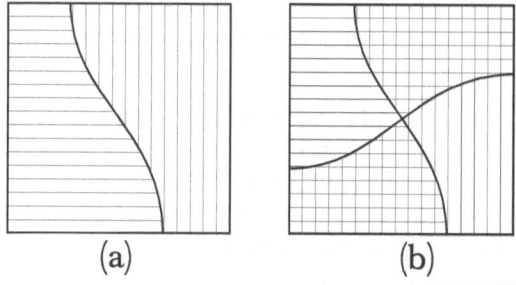

(a)   (b)

Fig. 14.13. Possible pixel geometries.

As remarked before, the bitmask carries information about the pixel geometry. This is very important in order to avoid or minimizing sampling artifacts. Indeed, a pixel might be divided by the background–foreground boundary, as in Figure 14.13(a). But it might at the same time be divided by another boundary relevant to the compositing operation in question, in which case we get more than two regions, as in Figure 14.13(b). When we discretize, it is important to preserve a certain amount of information about the *pixel geometry*, that is, the geometry of the decomposition inside the pixel.

14.5 Computation of the Opacity Function

We now turn to the problem of computing the opacity function of an image and look at the various methods that can be used in the computation. This problem is directly related to the method used for creating the image. There are three main cases:

- synthetic two-dimensional images, created using a program for drawing or painting, for example;

- synthetic "three-dimensional" images, generated from a three-dimensional scene using an image synthesis system; and
- real-life images created using a video camera, scanner, or other raster input device.

Synthetic Two-Dimensional Images

These images are typically produced using programs that model the physical process of painting or drawing. Modern paint programs offer the user control over the shape, size, color, and opacity of the brush. Each brush stroke is treated as an image that is overlaid on what is already on the screen; the opacity information is taken into account in performing the overlay and stored by the program together with the color information. Drawing programs often allow only opaque strokes, but they may use partial opacity internally, to avoid aliasing, as discussed earlier. Since they keep a structured description of the geometry of the scene, they can precisely determine the geometry of each pixel and therefore its opacity.

Synthetic Three-Dimensional Images

Images of this type are produced by visualization programs that have access to the scene's geometric data and that therefore can generate the opacity channel accurately. Three-dimensional objects are projected onto the virtual camera plane and colored according to the lighting model used. Since the geometry of each pixel can be determined accurately, so can the opacity function. The discrete representation of the pixel geometry depends on the visualization method used and may rely on either bitmasks or alpha-channel information. More details can be found in textbooks on three-dimensional image synthesis.

Scanned Images

When we scan a photograph or capture an image using a video camera, we have no information beyond the color at each pixel. If opacity information is desired, it must be deduced from the color information. Typically, what is desired is a background–foreground decomposition of the domain, and the associated opacity function, so the image can later be combined with others. In general, such a decomposition constitutes a hard problem in image analysis. In practice, the geometry of the decomposition is often simple and can be obtained using fairly robust methods. Once the background–foreground decomposition is known, the opacity function can be defined pixelwise, using *area-fill algorithms*; see the references in Section 14.8. Clearly, in this case there is no possibility of generating a bitmask unless one reduces the resolution of the image by grouping pixels.

Fig. 14.14. Blue screen and alpha channel. See Plate 9 in color insert.

Example 14.2 (Blue screen). A special case that makes the calculation of the opacity function easy is when the background has been created with a uniform color. This trick, known as "blue screen" or "chroma key," has been widely used for decades in analog compositing. Figure 14.14 shows a torus against a uniform blue background, then the image's opacity channel, obtained by defining the background to be the set of pixels whose color is that hue of blue. The figure also shows the overlay of the torus with a different background.

14.6 Compositing in the Discrete Domain

The two methods for discretizing the opacity function give rise to two ways of compositing images in the discrete domain: alpha-channel compositing and bitmask compositing. We study each in turn.

14.6.1 Compositing Using the Alpha Channel

As we have seen, the alpha channel is obtained by spatial discretization of the opacity function using the same grid as the color discretization. Normally, the same number of bits is used for the alpha channel as for each color channel. By treating the alpha channel as if it were another color channel, for the purposes of storage, one achieves a fairly homogeneous representation for the image.

We will see now how we can use the opacity function to define other operations for combining images, beyond overlaying.

When we combine two images f and g, we must also combine their alpha channels into an alpha channel for the resulting image, which may be needed later—for example, if the image is to be combined further. Now, the alpha channel of the input images does not remember the pixel geometry, only the percentage of pixel area that comes from an opaque region. Thus, when we combine these two percentages, we have to make assumptions about the pixel geometry from which they come. Several heuristics are possible here.

Specifically, Figure 14.15 shows three possible configurations for a pixel in f and g, prior to discretization. In part (a), the opaque regions don't intersect inside the pixel. In part (b), the opaque region of f contains that of g, always restricted to the pixel in question. Finally, part (c) shows the *general position*, the one most likely to occur in general: the two opaque regions overlap partially. We will always assume that we are in the case of Figure 14.15(c). We then have a subdivision of the pixel into four sub-regions (see Figure 14.16): $\bar{f} \cap \bar{g}$, $f \cap \bar{g}$, $\bar{f} \cap g$, and $f \cap g$, where the bar indicates complement with respect to the full pixel.

We will also assume that, if α_f and α_g are the values of the opacity functions of f and g at a pixel, the fraction of the area of the intersection of the two opaque regions in the pixel is the product $\alpha_f \alpha_g$. This is equivalent to saying that the event that a point in the pixel belongs to the opaque region of f is independent of its belonging to the opaque region of g, so the probabilities of

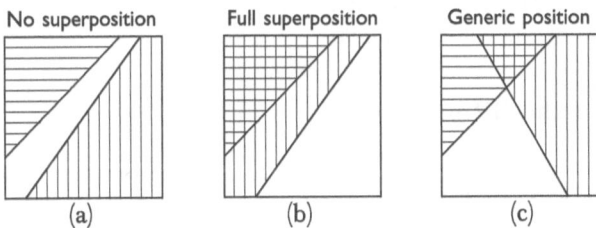

Fig. 14.15. Configurations of pixel geometry.

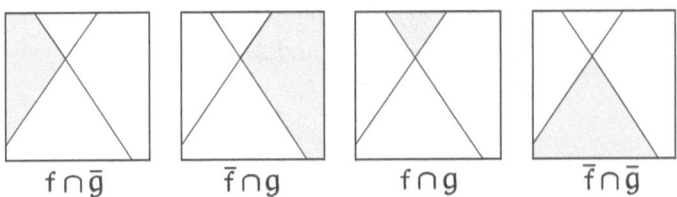

Fig. 14.16. The four subregions of a pixel.

these two events can be multiplied to give the probability of the joint event. It is also equivalent to saying that the fraction of the opaque region of g that lies in the opaque region of f is α_f.

Clearly, this assumption may lead to errors in the estimate of the fraction of the pixel that lies in the intersection. However, these errors are irrelevant in most cases.

The same assumption yields the fraction of the pixel covered by the other three subregions of the partition. The area covered by f but not g is $f - (f \cap g)$, so the corresponding fraction is $\alpha_f(1 - \alpha_g)$. Analogously, the fraction covered by g but not f is $\alpha_g(1 - \alpha_f)$, and the fraction neither covers is $(1 - \alpha_f)(1 - \alpha_g)$.

From these considerations we will deduce in Section 14.7 the weights A_f and A_g with which the colors and opacity of f and g should be combined at each pixel, so that for each compositing operation the resulting image is $A_f f + A_g g$, with A_f and A_g appropriately computed.

14.6.2 Compositing Using Bitmasks

When the opacity function is discretized by supersampling, instead of by area sampling, more information about the pixel geometry is preserved, and the areas of the pixel subregions in Figure 14.16 can be estimated more accurately. Thus, the bitmask technique illustrated in Figure 14.12 is superior from the viewpoint of compositing images. Using this technique, compositing is carried out in three steps:

- compositing of the bitmasks,
- calculation of the alpha channel, and
- calculation of the color channels.

Compositing the Bitmasks

In the first step, the bitmasks of the same pixel from the two images are combined using bitwise Boolean operators, to determine the decomposition of the pixel into subregions. Denoting by M_f and M_g the bitmasks of the pixel associated to the images f and g, we can write

$$\bar{f} \cap \bar{g} = (\text{not } M_f) \text{ and } (\text{not } M_g),$$
$$f \cap \bar{g} = M_f \text{ and } (\text{not } M_g),$$
$$\bar{f} \cap g = (\text{not } M_f) \text{ and } M_g,$$
$$f \cap g = M_f \text{ and } M_g.$$

Calculation of the Alpha Channel

Once we have the bitmask of a pixel in the composite image, we can use it to compute that pixel's opacity, which is the fraction

$$A = \frac{\text{number of subpixels with bit value 1}}{\text{total number of subpixels}}.$$

Note that here no supposition needs to be made regarding the relative position of the opaque sets within the pixel, since we have this information stored (to within the resolution of the supersampling grid).

Calculation of the Color Channels

In this step we use the opacity information just computed to derive the color values in the composite image. This is done in the same way as in the alpha-channel method: the composite image is

$$A_f f + A_g g.$$

14.7 Compositing Operations

Several image compositing operations are obtained from decompositions of the domain. The compositing operation consists of two steps: determining the sets that define the decomposition, and assigning color to each pixel in the sets of the decomposition.

The color assignment comes from the colors in the pixels of the images f and g that are being composited. We stress that, in addition to the colors of f and g, we can also assign a pixel the color 0, that is, eliminate all color information in a particular region of the decomposition. Here are the possible choices of color for each region:

Region	Possible Colors
$\bar{f} \cap \bar{g}$	0
$f \cap \bar{g}$	0 or f
$\bar{f} \cap g$	0 or g
$f \cap g$	0, f, or g

This gives rise to 12 possibilities ($2 \times 2 \times 3$), so there are 12 image compositing operators. We briefly describe each in turn. In each case we denote by A_f the fraction of the pixel area occupied by a color coming from image f, and likewise for A_g. Thus, the color at the corresponding pixel of the composite image is given by

$$A_f f + A_g g. \tag{14.7}$$

Analogously, the alpha channel value in the composite image is

$$A_f \alpha_f + A_g \alpha_g. \tag{14.8}$$

Fig. 14.17. Images used to illustrate compositing operators. Left: f and its alpha channel. Right: g and its alpha channel.

Each compositing operator will also be analyzed using the bitmask method. Here we must compute, for each operator, the bitmask M_r of the composite image, as well as the fractions A_f and A_g.

To illustrate the operators, we use the images of Figure 14.17: f is a torus and g a checkerboard floor. Both images are synthetic. Figure 14.17 also shows the alpha channel of these images.

In Figure 14.18 we show in different shades of gray the regions $\bar{f} \cap \bar{g}$, $f \cap \bar{g}$, $\bar{f} \cap g$, and $f \cap g$. Black indicates $\bar{f} \cap \bar{g}$, the background of both f and g. Note that $\bar{f} \cap g$ is not connected.

14.7.1 The Overlay Operator

The overlay operator, often called simply "over," was introduced in Section 14.3.2. The color of the elements of the front image f always predominates over that of the back image g. Figure 14.19, top, defines this operator by giving the table of color assignments in each region of the pixel partition. Here the area fraction occupied by f is $A_f = 1$, and that occupied by g is $A_g = 1 - \alpha_f$. Clearly, this operator is noncommutative.

In bitmask formulation, we have

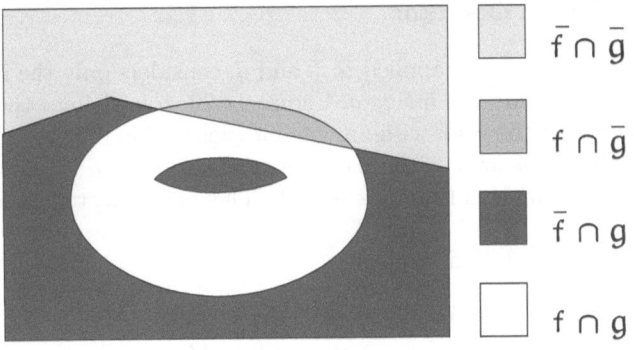

$\bar{f} \cap \bar{g}$

$f \cap \bar{g}$

$\bar{f} \cap g$

$f \cap g$

Fig. 14.18. Decomposition of the domain of f and g.

$$M_r = M_f,$$
$$A_f = 1,$$
$$A_g = \frac{\text{number of bits } M_g}{\text{number of bits } [(\text{not } M_f) \text{ and } M_g]}.$$

Figure 14.19, bottom, shows the effect of the over operator on the images f and g, and the alpha channel of the resulting image, obtained according to (14.7) and (14.8).

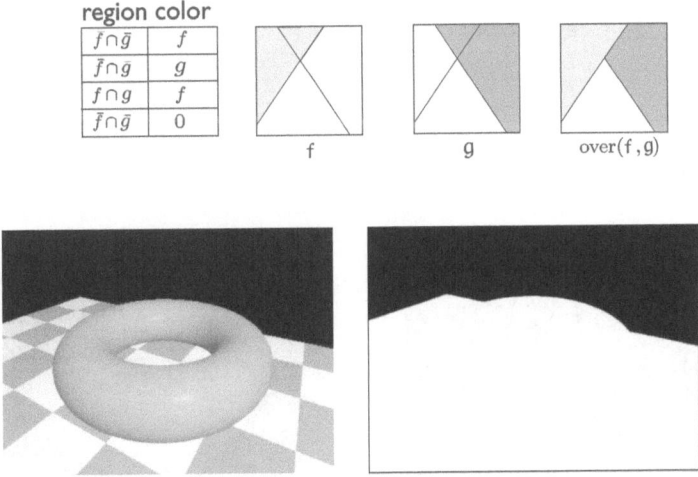

Fig. 14.19. Pixel geometry and effect of the over operator.

14.7.2 The Inside Operator

The inside operator, when applied to f and g, considers only the information from f that lies inside the image g. Figure 14.20, top, defines the operation precisely by giving the color assignments for each region of the partition. The fraction of the pixel area with color information from f is $A_f = \alpha_g$. The fraction with information from g is $A_g = 0$. This operator, too, is noncommutative.

In bitmask formulation, we have

$$M_r = M_f \text{ and } M_g,$$
$$A_f = \frac{\text{number of bits } M_r}{\text{number of bits } M_f},$$
$$A_g = 0.$$

Figure 14.20, bottom, shows the effect of the inside operator on the images f and g, including the resulting alpha channel.

14.7.3 The Outside Operator

The result of the operation outside(f, g) is to preserve that part of image f that lies outside the region delimited by the elements of image g. Figure 14.21 gives the color assignments for each region and illustrates the effect of the operator. The fraction of a pixel with color coming from f is $A_f = 1 - \alpha_g$; the fraction coming from g is 0. This operator is again noncommutative.

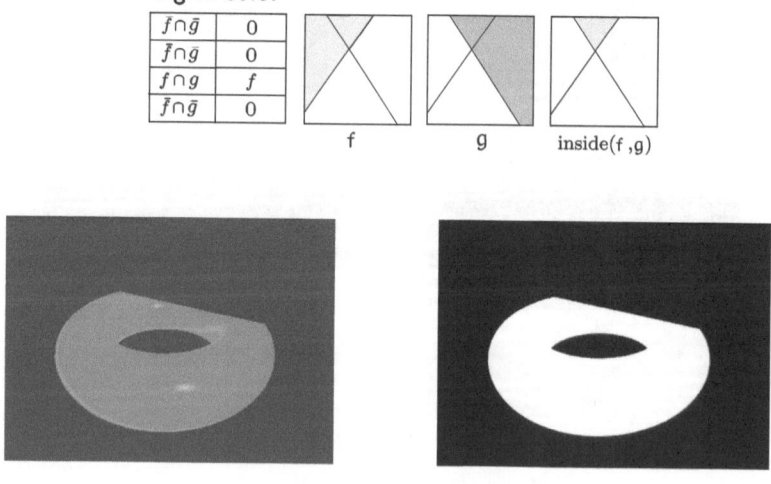

Fig. 14.20. Pixel geometry and effect of the inside operator.

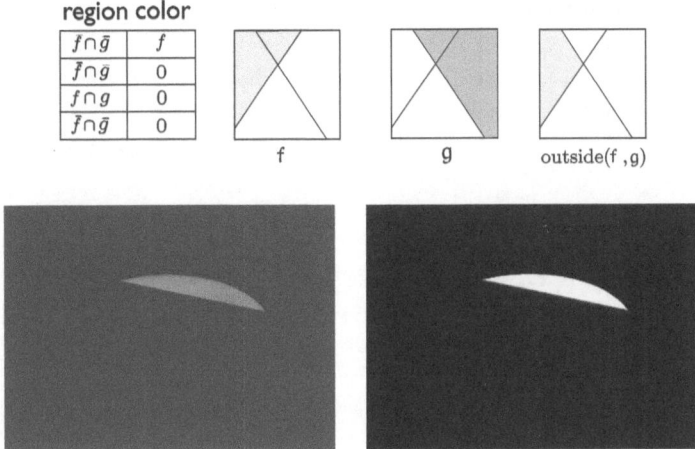

region color

$f \cap \bar{g}$	f
$\bar{f} \cap g$	0
$f \cap g$	0
$\bar{f} \cap \bar{g}$	0

f   g   outside(f,g)

Fig. 14.21. Pixel geometry and effect of the outside operator.

In bitmask formulation, we have

$$M_r = M_f \text{ and (not } M_g),$$
$$A_f = \frac{\text{number of bits } M_f}{\text{number of bits } M_f},$$
$$A_g = 0.$$

14.7.4 The Atop Operator

The result of the operation $\text{atop}(f, g)$ is to superimpose the colors from image f onto regions where there are elements of image g, but not elsewhere. Figure 14.22 gives the color assignments for each region and illustrates the effect of the operator. The fraction of a pixel with color coming from f is $A_f = \alpha_g$; the fraction coming from g is $A_g = 1 - \alpha_f$. The alpha channel of the result coincides with that of g. This operator is again noncommutative.

In bitmask formulation, we have

$$M_r = M_g,$$
$$A_f = \frac{\text{number of bits } (M_f \text{ and } M_g)}{\text{number of bits } M_f},$$
$$A_g = \frac{\text{number of bits } [M_f \text{ and (not } M_g)]}{\text{number of bits } M_g}.$$

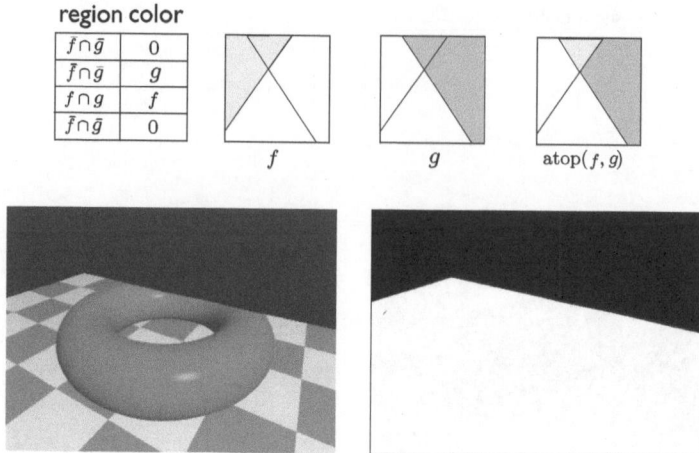

region color

$\bar{f} \cap \bar{g}$	0
$\bar{f} \cap g$	g
$f \cap g$	f
$\bar{f} \cap \bar{g}$	0

f ⠀⠀ g ⠀⠀ atop(f, g)

Fig. 14.22. Pixel geometry and effect of the atop operator.

14.7.5 The Xor Operator

The result of the operation $\text{xor}(f, g)$ is the symmetric difference, also known as exclusive-or, between the images f and g: $\text{xor}(f, g) = (f - g) \cup (g - f)$. Figure 14.23 defines the operator by giving the color assignments for each region and illustrates its effect. The fraction of a pixel with color coming from f is $A_f = 1 - \alpha_g$; the fraction coming from g is $A_g = 1 - \alpha_f$. Unlike the preceding ones, this operator is commutative.

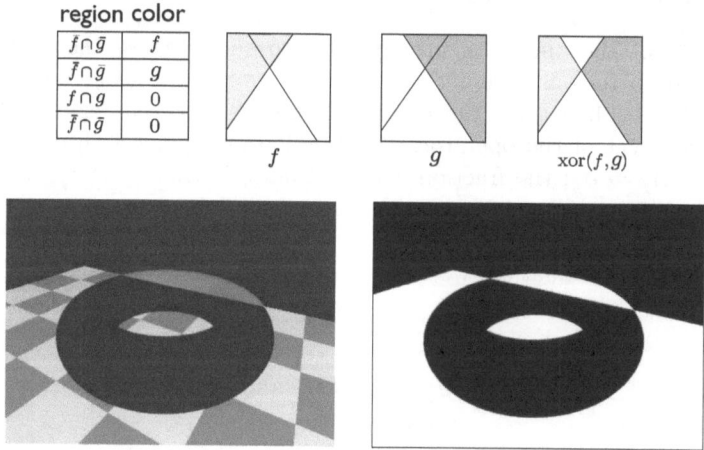

region color

$f \cap \bar{g}$	f
$\bar{f} \cap g$	g
$f \cap g$	0
$\bar{f} \cap \bar{g}$	0

f ⠀⠀ g ⠀⠀ xor(f, g)

Fig. 14.23. Pixel geometry and effect of the xor operator.

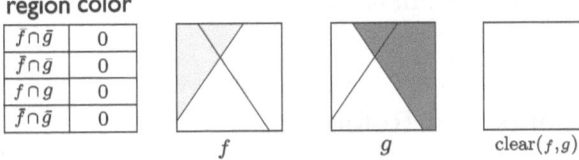

Fig. 14.24. Pixel geometry of the clear operator.

In bitmask formulation, we have

$$M_r = [M_f \text{ and } (\text{not } M_g)] \text{ or } [(\text{not } M_f) \text{ and } M_g],$$

$$A_f = \frac{\text{number of bits } [M_f \text{ and } (\text{not } M_g)]}{\text{number of bits } M_f},$$

$$A_g = \frac{\text{number of bits } [(\text{not } M_f) \text{ and } M_g]}{\text{number of bits } M_g}.$$

The compositing operators introduced above can be used in succession in order to obtain other combinations of images. In this context it is convenient to define, using the same procedure above, two additional operators on images: *clear* and *set*.

14.7.6 The Clear Operator

The clear operator assigns the zero color (background color) to each pixel, independently of its geometry, and makes the pixel totally transparent. Therefore, $A_f = A_g = 0$. Note that there is a difference between a pixel $(0,0,0,\alpha)$, where $\alpha > 0$, and a pixel $(0,0,0,0)$: the form has zero color but is partly opaque, while the latter has zero color and is transparent. Clearing means making the pixel value $(0,0,0,0)$. See Figure 14.24.

In bitmask formulation, we have $A_f = 0$, $A_g = 0$, and $M_r = 0$.

14.7.7 The Set Operator

The result of the operation $\text{set}(f, g)$ is f: the information from image g is discarded. In symbols, $A_f = 1$ and $A_g = 0$. See Figure 14.25. This operator

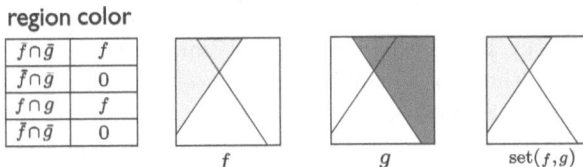

Fig. 14.25. Pixel geometry of the set operator.

is clearly noncommutative. In bitmask formulation, we have $A_f = 1$, $A_g = 0$, and $M_r = M_f$.

14.8 Comments and References

A brief but useful exposition of compositing techniques can be found in (Porter and Duff 1984), the paper that introduced compositing using the alpha channel. See also (Duff 1985), which discusses the problem of compositing synthetic images. For such images, besides opacity information, one can store depth-of-scene information (Example 14.1). The problem of discretization, which we studied in detail for the opacity function, can be posed for the depth function as well, and similar considerations apply.

The bitmask method was developed simultaneously in (Fiume et al. 1983) and (Carpenter 1984) to solve the problem of sampling em three-dimensional image synthesis systems.

The computation of the opacity function from color information is discussed in (Bloomenthal 1983) and (Fishkin and Barsky 1984).

An important problem in compositing images created by distinct processes is to make the lighting compatible. The article (Nakamae et al. 1986) gives a method to superimpose computer-generated images with photographs of real scenes that takes into account the luminance and chrominance of the real image, and even atmospheric effects.

The cross-dissolve image in Figure 14.2 was created by Lucia Darsa and Bruno Costa.

The original image used in Figures 14.6 and 14.9 is the "alias foyer", from the Alias Sketch tutorial.

References

[Bloomenthal 1983]Bloomenthal, J. (1983). Edge inference with applications to antialiasing. *Computer Graphics (SIGGRAPH '83 Proceedings)*, 17(3):157–162.

[Carpenter 1984]Carpenter, L. (1984). The a-buffer, an antialiased hidden surface method. *Computer Graphics (SIGGRAPH '84 Proceedings)*, 18(3):103–108.

[Duff 1985]Duff, T. (1985). Compositing 3D rendered images. *Computer Graphics (SIGGRAPH '85 Proceedings)*, 19(3):41–44.

[Fishkin and Barsky 1984]Fishkin, K. P. and Barsky, B. A. (1984). A family of new algorithms for soft filling. *Computer Graphics (SIGGRAPH '84 Proceedings)*, 18:235–244.

[Fiume et al. 1983]Fiume, E., Fournier, A., and Rudolph, L. (1983). A parallel scan conversion algorithm with anti-aliasing for a general purpose ultracomputer. *Computer Graphics (SIGGRAPH '83 Proceedings)*, 17(3):141–150.

[Nakamae et al. 1986]Nakamae, E., Harada, K., Ishizaki, T., and Sancha, T. L. (1986). A montage method: The overlaying of computer generated images onto a background photograph. *Computer Graphics (SIGGRAPH '86 Proceedings)*, 20(4):207–214.

[Porter and Duff 1984]Porter, T. and Duff, T. (1984). Compositing digital images. *Computer Graphics (SIGGRAPH '84 Proceedings)*, 18(3): 253–259.

15

Warping and Morphing

This chapter studies topological filters designed to change the shape of the objects of an image. This process is called *deformation* or *warping*, and therefore we talk about *warping filters*. Together with amplitude filters, which change the image's color information, warping filters can be used to create a transition between images of different objects, in a technique known as *morphing*. Warping and morphing filters are important in many applications, from the correction of preexisting image distortions to the creation of special effects in the entertainment industry.

15.1 Warping Filters

The study of warping filters involves three basic problems: specification, computation, and implementation.

Warping specification is related to the user interface. This is a very difficult and delicate topic. The user should be given tools to specify a transformation using the smallest possible number of parameters. For some mappings this is an easy task; for example, a rotation requires only the center of rotation and the angle of rotation. But, in general, specifying warpings based on predetermined goals requires a lot of work. In general, the warp is specified by the user in the discrete universe and it must be reconstructed when doing resampling. We will not cover the topic of warping specification here; the interested reader should look for references in the final section of the chapter.

Also, as in the rest of the book, we will not get involved in implementation issues. Again, for this topic we direct the reader to the references in Section 15.8.

In this chapter we will study different techniques involved in the computation of the warping map. Knowledge of these techniques is of great importance in implementing robust and efficient warping filters. Moreover, they constitute a beautiful application of the theory of sampling and reconstruction studied in previous chapters.

L. Velho et al., *Image Processing for Computer Graphics and Vision*,
Texts in Computer Science, DOI 10.1007/978-1-84800-193-0_15,
© Springer-Verlag London Limited 2009

15.2 Warping in the Continuous Domain

Given an image $f : U \subset \mathbb{R}^2 \to \mathcal{C}$, a *warping filter* is defined by a map $h : U \to V \subset \mathbb{R}^2$. Such a map acts on the image, giving rise to another image that can be regarded as a deformation of the original one. Figure 15.1 shows the effect of a particular warping filter.

In general, we make several assumptions about the warping map h. One natural requirement is that h should be injective and continuous. Injectivity says that there is no superposition of points in the deformation, so it is possible to define an inverse map h^{-1}. Continuity says that there are no tears or rips.

From the mathematical point of view, is it natural to require that h be bijective and that both it and its inverse h^{-1} be continuous. A map with these properties is called a *homeomorphism*. The warped image is given by $g = f \circ h^{-1} : V \to \mathcal{C}$; equivalently, $f = g \circ h$:

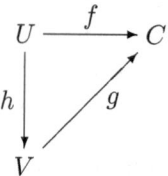

Intuitively, a homeomorphic warping can be imagined as distorting an image stamped on a rubber sheet, without tearing or ripping it.

In fact, we will usually assume that h and h^{-1} have continuous partial derivatives and that the Jacobian determinant of h be nowhere zero. In this case h is called a *diffeomorphism*. These conditions are important when we need to apply techniques from analysis and differential topology; otherwise

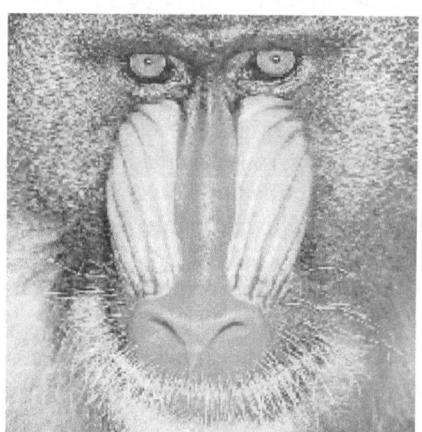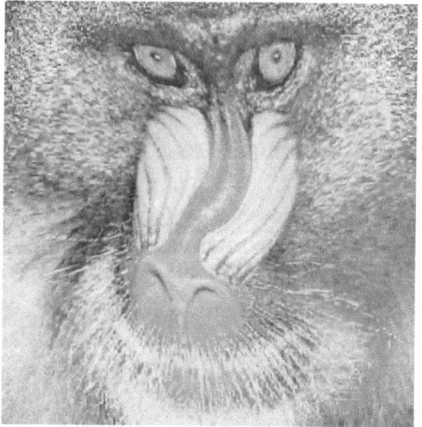

Fig. 15.1. An image before and after warping.

homeomorphic warpings are sufficient to work with. Also, when working in the representation universe, we use discrete images and at most we have a bijective map between the discrete set of pixels and their colors.

A warp map can also be regarded as a coordinate change. One must be careful when taking this point of view: changes in coordinate systems can lead to confusion, because (for example) turning the coordinate axes *counterclockwise* has the same effect on equations as turning the image's objects *clockwise* while the axes stay fixed. (To read the next paragraph, you can move your eyes down or slide the page up.) This is also why we get the warped image g by composing with h^{-1} rather than with h.

We will always suppose that the original coordinate system is (x, y) and the warped coordinate system is (u, v). That is, if the coordinates of a feature in the original image were (x, y), the coordinates of the same feature in the warped image would be

$$h(x, y) = (u, v), \qquad \text{with } u = u(x, y) \text{ and } v = v(x, y).$$

When we apply a warping transformation h to an image f, obtaining an image g, it is very common to call f the *source image* and g the *target image*.

15.2.1 Expansions and Contractions

A map $T : U \subset \mathbb{R}^2 \to \mathbb{R}^2$ is an *expanding transformation*, or an *expansion*, if there is $\lambda > 1$ such that

$$|T(X) - T(Y)| \geq \lambda |X - Y| \quad \text{for all } X, Y \in U.$$

Thus, T always increases distances. Similarly, T is a *contracting transformation*, or a *contraction*, if there is $\lambda \in (0, 1)$ such that

$$|T(X) - T(Y)| \leq \lambda |X - Y| \quad \text{for all } X, Y \in U.$$

Thus, T always decreases distances.

Example 15.1 (Homotheties and scaling maps). A *homothety* is a map of the form $T(X) = \lambda X$, where the real number $\lambda > 0$ is the *scaling factor*. Clearly, this is an expansion for $\lambda > 1$ and a contraction for $\lambda < 1$. A homothety is also called an *isotropic* or *proportional scaling map*; the adjectives mean that the scaling factor is the same in all directions. One can have nonproportional scaling maps, but usually when we say scaling we have proportional scaling in mind.

A transformation that preserves distances is called an *isometry*:

$$|T(X) - T(Y)| = |X - Y| \quad \text{for all } X, Y \in U.$$

Fig. 15.2. Another example of warping.

In general, a warp transformation is neither a contraction, nor an expansion, nor an isometry. Most commonly, it expands certain regions and contracts others, so we can speak of the map being locally expanding or contracting. The map may also expand in one direction while contracting in another. The reader may be amused by trying to identify the contraction and expansion regions in Figure 15.2.

Expanding a signal in the time domain causes a contraction of the signal spectrum support in the frequency domain. Conversely, a contraction in the time domain increases the signal's frequencies. See Figure 15.3. This should be kept in mind when we study certain problems associated with warping.

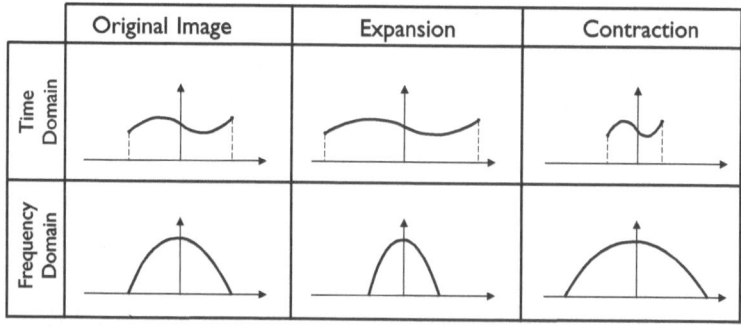

Fig. 15.3. Contractions and expansions in the time and frequency domains.

15.3 Warping in the Discrete Domain

In the computations, we use the discrete-continuous model: images with discretized domains, and colors encoded as floating-point numbers. In general, the warp map takes points of the original lattice to points not necessarily on a lattice. Therefore, we cannot transform the source image by simply copying the color from the source lattice points to the lattice points of the target image. This would most likely lead to pixels in the target image without a value or to pixels with multiple values.

We can illustrate this by using a one-dimensional array of pixels, which, as usual, can be thought of as a single row from an image (image scanline). Figure 15.4 shows the graph of a warp transformation h, which is seen to be roughly expanding between pixels 4 and 5, roughly contracting between pixels 5 and 6, and roughly isometric between pixels 1 and 4. Obviously, this map suffers from the problems we have been discussing: points with integer coordinate are not mapped to points with integer coordinate in the target image, and although h is bijective in the continuous domain, it doesn't give rise to a bijection in the discrete domain.

Following the arrows in the graph, we see that pixel 5 in the target image does not correspond to any pixels in the source, so the naive method outlined above would leave this pixel without a color value. Likewise, pixel 6 of the target is matched with two pixels in the source, so its color assignment is ambiguous—it corresponds to the overlapping of two pixels of the source image.

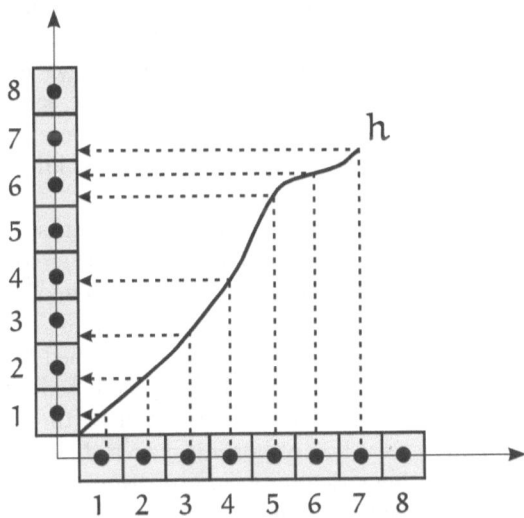

Fig. 15.4. Effect of expansion and contraction in the discrete domain.

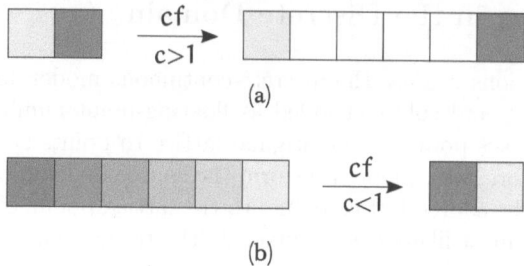

(a)

(b)

Fig. 15.5. Problems arise in the discrete domain under either an expansion (a) or a contraction (b).

To summarize, in the case of an expansion, we have the problem of supplying values for pixels that are "left out," as shown in Figure 15.5(a). In the case of a contraction, we have the problem of several pixels collapsing into one, as shown in Figure 15.5(b), and we must have some method for choosing the final color value. Both problems can be solved in a unified way, by reconstructing the source image in order to work in the continuous universe. This will be explained in next section.

15.3.1 Resampling

When we work in the continuous domain, the warping is either a homeomorphism or a diffeomorphism and we have at our disposal many mathematical tools and results to use. On the other hand, when moving to the discrete universe, we are faced with maps that might not behave well.

The obvious solution is to use reconstruction techniques in order to be able to work in the continuous domain. We reconstruct the source image, apply the warping filter, and finally sample the warped continuous image in order to obtain the target digital image. The whole process is known as *resampling* and is illustrated in Figure 15.6.

This method completely solves the two problems discussed earlier: working in the continuous domain we have a bijective warping; therefore, blanking pixels in the warped image does not occur. Also, the problem of pixel overlapping disappears.

We should point out that the warping map is specified by the user in the discrete universe. Therefore, it must also be reconstructed in the resampling process. Depending on the reconstruction method used the warping mapping in the continuous domain is a homeomorphism or a diffeomorphism and we therefore have several mathematical tools at our disposal. Of course, we must be careful when reconstructing the image and when sampling it after the warp.

In a region where h is expanding, we reconstruct the image and sample at the lattice pixels. The fact that the transformed image has lower frequencies

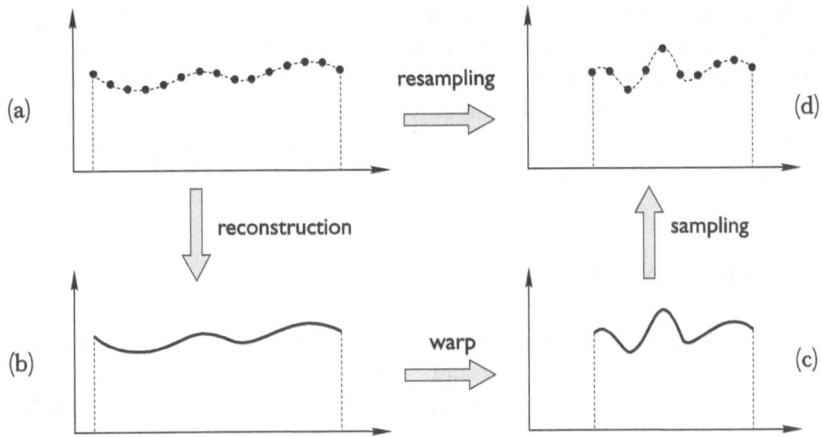

Fig. 15.6. Warping by resampling: (a) original digital image; (b) continuous image obtained by reconstruction; (c) warped continuous image; (d) target image, obtained by sampling (c).

than the original image works in our favor; reconstruction and sampling errors are minimized. By contrast, in a region where h is contracting, frequencies become higher. Since the target image is usually sampled at the same rate as the source image, this can lead to severe aliasing problems in the resampled image. One way to avoid this is to use a smoothing filter prior to the final sampling step. The whole process, then, is composed of the steps shown in Figure 15.7.

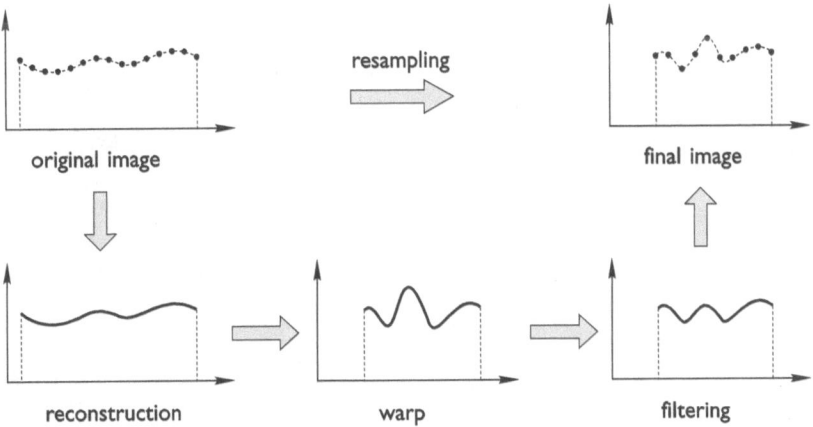

Fig. 15.7. Warping by resampling and filtering.

The filtering of the image after the warping is very delicate. In general, as clarified by the one-dimensional example of Figure 15.4, the warping transformation expands and contracts at different rates in different regions of the image domain. Therefore, the use of a spatially invariant filter is not recommended; rather, we should use a filter that adapts to the warping distortions. Such filters are called *adaptive*.

15.4 Some Examples

A simple example of a warping filter is the flipping of a square digital image with respect to its diagonal. The geometric transformation is defined by $(x, y) \mapsto (y, x)$, so that $u(x, y) = y$ and $v(x, y) = x$. This gives $f(x, y) = g(h(x, y)) = g(u, v) = g(y, x)$.

Another simple example is rotating the image, say counterclockwise by $90°$. The transformation is then given by

$$h \begin{pmatrix} x \\ y \end{pmatrix} \begin{pmatrix} \cos 90° & -\sin 90° \\ \sin 90° & \cos 90° \end{pmatrix} \begin{pmatrix} x \\ y \end{pmatrix} = \begin{pmatrix} 0 & -1 \\ 1 & 0 \end{pmatrix} \begin{pmatrix} x \\ y \end{pmatrix} = \begin{pmatrix} -y \\ x \end{pmatrix}.$$

Figure 15.8 shows the effect of this transformation.

Both of these examples have a special property that makes their implementation very easy. Reflection and rotation are rigid motions, so there is no change in the relationships among pixels as a result of warping. Moreover, under this particular reflection and this particular rotation, the resulting pixel lattice is perfectly aligned with the original pixel lattice. This is the best possible situation.

Fig. 15.8. Effect of a $90°$ counterclockwise rotation.

15.4.1 Zooming

Zooming is the special case of warping with a scaling map (Example 15.1). *Zooming in* refers to an expansion, and *zooming out* to a contraction. Two properties of scaling maps make the implementation of zooming easier than that of a general warp transformation:

- although pixel distances are not preserved, the expansion or contraction ratio is the same everywhere; and
- if the scaling ratio is an integer, the pixel lattice of the target image is a subset or superset of the lattice of the original image.

The first property allows the use of spatially invariant filters, and the second avoids the need for resampling when the scaling ratio is an integer.

Zoom-in with the Box Filter

When we zoom in by a factor of two, there is a simple method to supply the pixels of the enlargement: replace every row by two identical rows, and likewise for columns. This is known as *zooming by pixel replication.*

In formal terms, this is achieved by applying a box interpolation filter. More precisely, we start by inserting a row of zeros at the top and one after each row of the original image, thus obtaining an image with $2n+1$ rows if the original had n rows. We do likewise for columns. Thus, if the original image was

$$\vdots \qquad\qquad \vdots$$
$$\cdots \quad f(i,j) \qquad f(i,j+1) \quad \cdots$$
$$\cdots \quad f(i+1,j) \; f(i+1,j+1) \; \cdots$$
$$\vdots \qquad\qquad \vdots$$

this step leads to

$$\vdots$$
$$f(i,j) \quad 0 \quad f(i,j+1)$$
$$\cdots \quad 0 \quad\; 0 \quad\; 0 \qquad \cdots$$
$$f(i+1,j) \; 0 \; f(i+1,j+1)$$
$$\vdots$$

We then convolve with the box filter of order two, which has mask

$$\begin{array}{|c|c|} \hline 1 & 1 \\ \hline 1 & 1 \\ \hline \end{array}.$$

This leads to an image with $2n$ rows, looking like this:

$$\vdots$$
$$f(i,j) \qquad f(i,j) \qquad f(i,j+1)$$
$$\cdots \quad f(i,j) \qquad f(i,j) \qquad f(i,j+1) \quad \cdots$$
$$f(i+1,j) \; f(i+1,j) \; f(i+1,j+1)$$
$$\vdots$$

Note that this convolution kernel has total mass four instead of one; this makes the average intensity of the target image four times greater than that of the image enlarged with zeros and therefore equal to the average intensity of the original image.

This method can be applied repeatedly to obtain enlargements by factors 2^2, 2^3, and so on. More generally, we can zoom in by any integer factor r by interlacing the original image with $r - 1$ rows of zeros per row, then applying an $r \times r$ box filter. This also works with nonproportional scaling maps of the form $h(x,y) = (rx, sy)$, for r and s integers.

Figure 15.9 shows the effect of two consecutive enlargements by a factor of two. Notice the effect of the box filter's discontinuous character, in the form of clearly defined sharp edges between groups of pixels (compare to Figure 6.4). With each zoom-in, this reconstruction defect inherent in the box filter becomes more perceptible, and the enlarged image appears coarser. Nonetheless, zooming by replication is very popular, due to its ease of implementation and computational efficiency. Its use is especially common in hardware implementations.

Zoom-in with the Bartlett Filter

Carrying out the same zoom-in process using the Bartlett (triangular) filter instead of the box filter, one obtains smoother images, as shown in Figure 15.10. This is because the Bartlett filter has a less spread-out transfer function, and therefore its use does not amplify the high frequencies to nearly the same extent as the box filter.

Formally, one adds rows and columns of zeros exactly as in Section 15.4.1 and then applies the filter, which has mask

$\frac{1}{4}$	$\frac{1}{2}$	$\frac{1}{4}$
$\frac{1}{2}$	1	$\frac{1}{2}$
$\frac{1}{4}$	$\frac{1}{2}$	$\frac{1}{4}$

Fig. 15.9. Zooming in by factors of two and four, using the box filter.

in the case of a zoom-in by a factor of two. The result is an image with twice as many rows minus one, and twice as many columns minus one. Computationally, the effect is to insert between each two consecutive rows of the original image a new row that is the average of the two, and likewise for columns. In other words, we perform linear interpolation between rows and between columns (independently). For this reason the application of this filter is also known as *zoom-in with bilinear interpolation*.

This technique is easily generalized to a scaling map

$$h(x, y) = (rx, sy),$$

with r and s integers. It is almost as computationally efficient and easy to implement as the method of the preceding section. Its main drawback is the fuzziness of the resulting image, which can sometimes be corrected by the subsequent use of other filters. In a great many applications, and especially in television, where spatial and color resolution are relatively low, zoom-ins using bilinear interpolation yield very satisfactory results.

Fig. 15.10. Zooming in by factors of two and four, using the Bartlett filter.

Zoom-out

As we have seen, zooming out increases the incidence of high frequencies and causes several pixels to collapse into one. We need to apply a smoothing filter in order to minimize aliasing problems in the target image. Since the scaling factor is constant, we can use a spatially invariant filter. For a contraction factor of two, each 2×2 block of the original image, say

$$f(i,j) \qquad f(i+1, j)$$
$$f(i, j+1) \; f(i+1, j+1)$$

is mapped to one pixel of the target image. Using a box filter, with mask

$$\frac{1}{4}\begin{array}{|c|c|} \hline 1 & 1 \\ \hline 1 & 1 \\ \hline \end{array}$$

Fig. 15.11. Zooming out by factors of two and four, using the box filter.

we get the value

$$\tfrac{1}{4}\big(f(i,j) + f(i+1,\,j) + f(i,\,j+1) + f(i+1,\,j+1)\big)$$

for the pixel. Figure 15.11 shows the effect of two consecutive applications of this method.

More effective smoothing filters can also be used, in order to obtain a final image that is better in certain respects. Such filters minimize the introduction of high frequencies during reconstruction, in exchange for a loss of definition (sharpness) in the final image.

15.5 Warping in Practice

When working with warping filters we must decide on the best process for reconstructing, transforming, filtering, and resampling the image. To a large extent, this decision should be based on considerations of computational efficiency. When possible, we exploit the geometric features of the warp transformation in search of the optimal implementation.

In this section we will cover the following topics related to the computation of the warping map:

- approximating the pixel geometry;
- direct mapping;
- inverse mapping;
- decomposable mappings.

15.5.1 Approximating the Pixel Geometry

When we reconstruct an image, each pixel corresponds to a region of the image support plane, which represents the pixel shape. Therefore, the warping map performs a region transformation when moving pixels. In general, the pixel shape is defined by some polygonal boundary, but the polygon edges get distorted during the warping. This fact is ignored in most of the computations: polygonal shapes are transformed into polygonal shapes by transforming their vertices. Since the pixels are small, we will not be making a great error by approximating curved edges of the transformed polygon by straight-line segments.

The most common case occurs when using uniform lattices associated to the matrix representation of an image. In this case the pixel shape is a rectangle that is transformed into an arbitrary quadrilateral shape by the warping map. Figure 15.12 shows the pixel shape (shaded rectangle on the left), the transformed pixel (shaded region on the right), and the approximation used for the transformed pixel (the dashed quadrilateral).

Therefore, in the continuous universe the discretization of the image domain U consists of a rectangular mesh. When transforming the image domain this mesh is mapped into an irregular mesh, as illustrated in Figure 15.13. Each quadrilateral of this irregular mesh is called a *cell*.

Since we are working in the continuous domain U, each point $p \in U$ has a color attribute. In particular, each point in the pixel shape has a color, including the vertices. Whether we consider the color of the pixel shape constant or variable depends on the reconstruction process used.

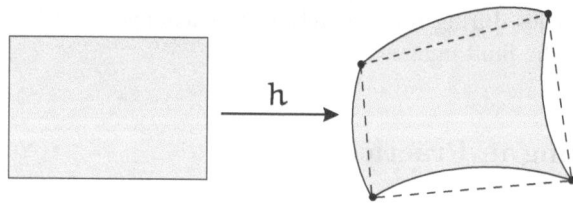

Fig. 15.12. Approximating the warped pixel by a quadrilateral.

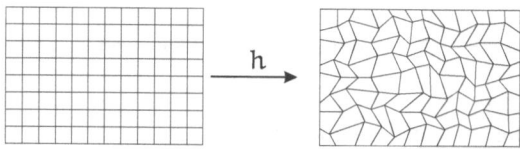

Fig. 15.13. Image of the original pixel mesh under the warp map.

We should stress that the methods we are about to study apply equally well to pixel geometries other than rectangles.

15.5.2 Warping Using the Direct Map

The technique we will introduce to compute the warp is called direct mapping because it does not consider the warp as a coordinate change. The warping map is directly applied to the source image elements in order to compute the elements of the target image.

Thus, the original rectangular mesh of the reconstructed image is mapped into an irregular mesh by the warping transformation, as seen in Figure 15.13. This irregular mesh represents the warped image in the continuous domain. Indeed, from the colors of each mesh vertex we can reconstruct the color at any point of the domain.

Several reconstruction techniques are possible here. A common, easy-to-compute technique is to take for each cell the average of the colors at the cells' vertices and applying the resulting color to every point of the cell. This is clearly a poor reconstruction method because it corresponds to a reconstruction with an adaptive, and irregular, box filter, which has discontinuities at the boundaries of the cells. A better reconstruction uses bilinear interpolation in defining the color at each point of the cell. This is illustrated in Figure 15.14: the colors c_1 and c_2 are interpolated linearly to obtain c'; the process is repeated with the colors c_3 and c_4 to obtain the color c''; finally, the colors c' and c'' are interpolated to obtain the color at the point p. As we know, this corresponds to the use of an adaptive, irregular, Bartlett filter for reconstruction. We will use both Bartlett and box reconstruction techniques in what follows.

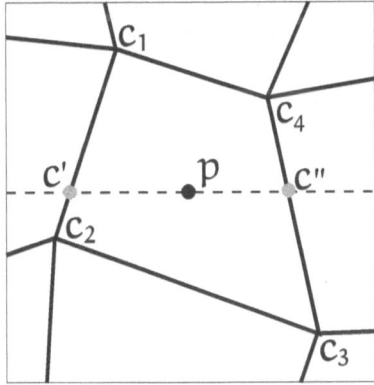

Fig. 15.14. Reconstruction using bilinear interpolation.

In order to discretize the warped image, we superimpose the rectangular mesh from the regular lattice of the target image to the irregular mesh and sample the image at the vertices of the uniform lattice. Of course, we should use a sampling technique that minimizes aliasing problems in the target image. This will be explained below.

Let p be a rectangular pixel of the target image to which we need to assign a color. The pixel p is superimposed to the irregular mesh. If we are lucky, p is entirely contained in a cell of the irregular mesh, as shown in Figure 15.15(a). The color at pixel p in this case is obtained by interpolating the colors of each of the vertices of the cell. An easy method is to use bilinear interpolation as illustrated in Figure 15.15(b).

Generically, the pixel p intersects several polygons of the irregular mesh, as shown in Figure 15.16, left. Let $p_1, p_2, \ldots, p_n$ be the pixels of the source image such that $h(p_i) \cap p \neq \varnothing$. We have a partition of the pixel p,

$$p = \bigcup_{i=1}^{n} \big(h(p_i) \cap p\big).$$

The area of each partition set is small compared to the area of the pixel p. Therefore, we can consider each cell to have a constant color (as we know, this amounts to reconstructing the warped image using an adaptive, irregular, box filter). This is illustrated by Figure 15.16, right. Some filtering must be done before sampling the pixel p, in order to minimize aliasing artifacts. A natural choice here is area sampling. To do this, we compute the fraction w_i of the area of p that overlaps with each $h(p_i)$ and use these fractions to compute a weighted average. Thus, we set

$$g(p) = \sum_{i=1}^{N} f(p_i) w_i,$$

where $g(p)$ is the color of the pixel p, and f is the reconstructed warped image.

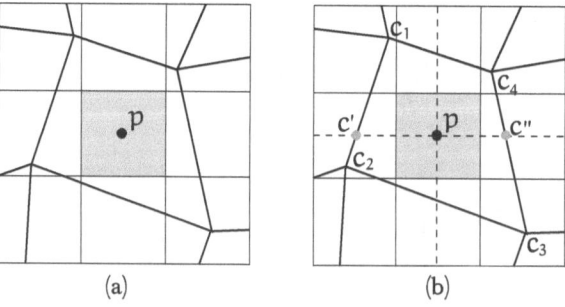

(a)   (b)

Fig. 15.15. Target pixel entirely contained in the warped image of a source pixel.

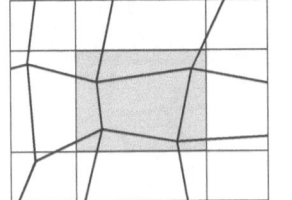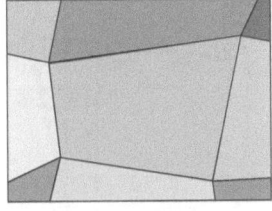

Fig. 15.16. Target pixel that overlaps several warped source pixels.

A more accurate approach to compute the color of pixel p is to reconstruct the warped image using bilinear interpolation and apply supersampling to compute the color of the pixel p.

We divided the computation of the color of pixel p into two cases mainly for pedagogical reasons. In fact, the reader should note that the first case is a particular case of the second situation, when we have only one set in the partition.

15.5.3 Warping Using the Inverse Map

This approach to compute the warping map considers the mapping as being a change of coordinates: for each point p in the domain of the warped image, we find its original coordinates $h^{-1}(p)$ in the domain of the original image, in order to compute its color.

Applying the inverse map h^{-1} to the regular lattice of the warped image, we obtain an irregular mesh superimposed to the regular lattice of the source image. This is illustrated in Figure 15.17.

If we reconstruct the original image on the irregular mesh, we obtain the color values for every pixel of the target image.

Generically, each cell of the irregular mesh intersects several rectangles of the regular lattice of the source image, as illustrated by Figure 15.18, left. The exceptional case where the cell is contained in a rectangular pixel is a particular case of a partition with one set, the cell itself; see Figure 15.18, right.

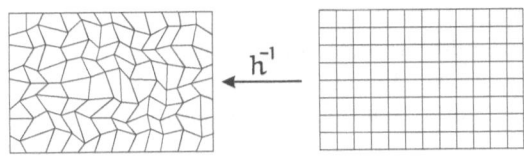

Fig. 15.17. Image of the target pixel mesh under the inverse warp map.

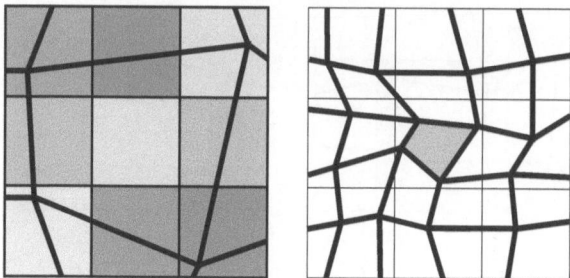

Fig. 15.18. Inverse warp image of a target pixel.

The color of each cell is computed using supersampling or area sampling, as we did for the direct method approach.

It is interesting to observe that in the direct map approach the target image is easily computed on the vertices of an irregular grid, and we must reconstruct it in order to sample on a regular lattice. On the other hand, in the inverse map approach the source image must be reconstructed in order to be sampled on an irregular grid. From these irregular samples, the pixel values of the target image are computed without additional effort.

The reader can verify that the two approaches lead to the same results to first-order approximation, assuming that the same sampling method is used. The choice between them should therefore be guided by implementation concerns, such as whether it is easier to compute the image of a vertex under h or h^{-1}.

15.5.4 Decomposable Transformations

Sometimes one can exploit specific properties of a warp map, or class of warp maps, in order to gain a more efficient implementation than can be achieved using the general methods of the preceding two sections. We have already seen how zooms by integer factors can be easily implemented. We now turn to another important special case, that of decomposable maps. We say that a map $h : U \subset \mathbb{R}^2 \to \mathbb{R}^2$ is *decomposable* if it can be written as a composition $h = h_n \circ \cdots \circ h_2 \circ h_1$ of simpler transformations. The term "simpler" is intentionally vague and depends on the application, but subsequent examples will clarify the situation.

(Warning: in the literature, decomposable maps are often called "separable." We avoid this term because it conflicts with its meaning, introduced in Section 7.3, of a filter that can be written as a product of filters in each variable.)

When the desired warp transformation is decomposable, the image can be transformed in stages: one first applies h_1, then h_2, and so on. This may be more efficient than applying h directly, and it may have other advantages such as reduced approximation errors, when the component maps are particularly simple.

An important case of "simple" component maps consists of shears. A *vertical shear* is a map $f : U \subset \mathbb{R}^2 \to \mathbb{R}^2$ that preserves the x-coordinate of each point: $f(x, y) = (x, f_2(x, y))$. In other words, f maps each vertical line into itself. Similarly, a *horizontal shear* preserves y-coordinates, mapping each horizontal line into itself: $g(x, y) = (g_1(x, y), y)$. When the map is a shear, the warping process is effectively reduced to a number of independent one-dimensional operations, since each row or column, as the case may be, can be considered separately. This means that the resampling concerns discussed earlier are much simplified.

We now consider the problem of writing a general warp map $h = (h_1, h_2)$ as a composition $g \circ f$, where f is a vertical shear and g is a horizontal shear. Since h will be acting in two steps, we have three coordinate systems, as shown in Figure 15.19: the original system (x, y), the system (u, v) after the application of f, and the system (r, s) after application of g to (u, v). Thus we have

$$(r, s) = h(x, y) = (h_1(x, y), h_2(x, y)), \tag{15.1}$$

and we wish to determine $f(x, y)$ and $g(u, v)$. By the assumptions on f and g, we have

$$f(x, y) = (x, f_2(x, y)) \quad \text{and} \quad g(u, v) = (g_1(u, v), v). \tag{15.2}$$

Therefore,

$$(r, s) = g(u, v) = g(f(x, y)) = g(x, f_2(x, y)) = (g_1(x, f_2(x, y)), f_2(x, y));$$

comparing with (15.1) we get

$$f_2(x, y) = h_2(x, y),$$

so that $f(x, y) = (x, h_2(x, y))$. To compute $g(u, v)$, we use the fact that

$$g(u, v) = h(x, y) = (h_1(x, y), h_2(x, y)).$$

Comparing with the expression of $g(x, y)$ in (15.2), and taking into account that $x = u$, we obtain

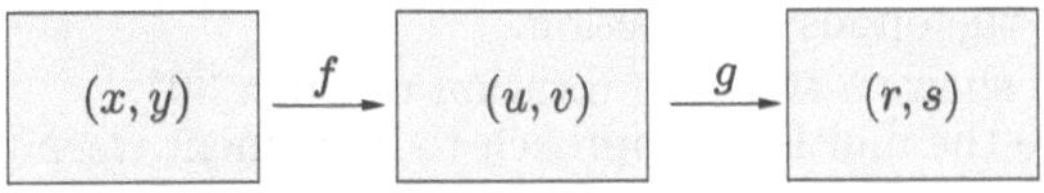

Fig. 15.19. Decomposing a general warp map into shears.

$$g_1(u, v) = h_1(x, y) = h_1(u, y). \tag{15.3}$$

Thus, we need to determine a function φ such that $y = \varphi(u, v)$. Using this function we can write

$$g_1(u, v) = h_1(u, \varphi(u, v)), \tag{15.4}$$

so that $g(u, v) = (h_1(u, \varphi(u, v)), v)$.

The function φ is guaranteed to exist because g is bijective, but finding it is not always easy. We consider a simple example.

Example 15.2 (Two-step rotation). Suppose the warp map h is a plane rotation through an angle θ, so that

$$h(x, y) = \begin{pmatrix} \cos\theta & -\sin\theta \\ \sin\theta & \cos\theta \end{pmatrix} \begin{pmatrix} x \\ y \end{pmatrix}.$$

This implies

$$h_1(x, y) = x\cos\theta - y\sin\theta,$$
$$h_2(x, y) = x\sin\theta + y\cos\theta.$$

Substituting $h_2(x, y)$ into (15.5.4), we get

$$f_2(x, y) = x\sin\theta + y\cos\theta,$$

which takes care of the vertical shear f. Turning to the horizontal shear g, we substitute $h_1(x, y)$ into (15.3), obtaining

$$g_1(u, v) = u\cos\theta - y\sin\theta. \tag{15.5}$$

We must compute y as a function of u and v, that is, we must determine the function φ from (15.4). We have

$$v = f_2(x, y) = x\sin\theta + y\cos\theta,$$

so that

$$y = \frac{v - u\sin\theta}{\cos\theta} \qquad \text{if } \cos\theta \neq 0.$$

Substituting into (15.5) gives

$$g_1(u, v) = u\cos\theta - \frac{v - u\sin\theta}{\cos\theta}\sin\theta,$$

which determines g.

Notice that g is not defined when θ is an odd multiple of $\pi/2$. Moreover, when θ is near such a value, the horizontal shear distorts the image very severely. Fortunately we can avoid these values altogether: if we wish to turn an image through $87°$, for example, we can first perform a $90°$ rotation, then a $3°$ rotation in the opposite direction.

Figure 15.20 shows a two-step rotation through $30°$.

When we use the multistep approach to implement warp transformations, there may occur problems during the intermediate step. A common case, known as a *bottleneck*, happens when one of the component maps is not injective, so certain points collapse together.

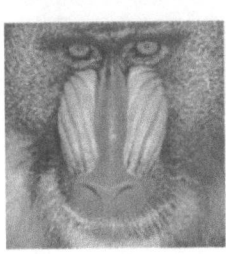 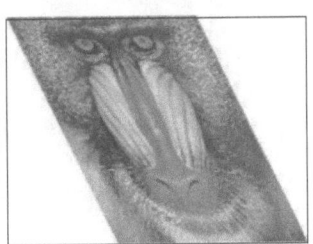 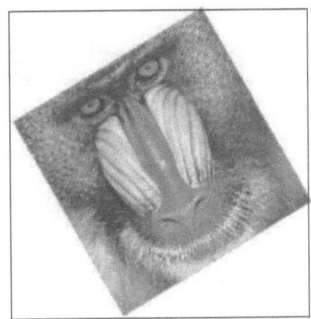

Fig. 15.20. A rotation can be achieved through a horizontal shear followed by a vertical one.

15.6 Morphing

As we saw in the preceding chapter, one important class of filters consists of *amplitude filters*, those that act only on the color space of the image and perform a color change. In mathematical terms, this corresponds to post-composition of the image with a transformation $T : C \rightarrow C$ of the color space: $g(x, y) = T(f(x, y))$, where g is the new image and f is the original one.

Amplitude filters are often used to create color transitions between two images. Thus, the *cross-dissolve* operation of Section 14.2.1 performs a linear interpolation in color space between two images:

$$h_t(x, y) = (1 - t)f(x, y) + tg(x, y). \qquad (15.6)$$

For $t = 0$, the result is the initial image $h_0 = f$ of the transition, and for $t = 1$ it is the target image $h_1 = g$. In order to regard a cross-dissolve as a filter (which is a unary operation), we make the target image g part of the definition of the filter and make only f its input.

The cross-dissolve operator performs a transition between two images without taking into account the transition between the different objects contained in the image. The result is generally very poor. This is illustrated by Figure 15.21(c), which represents a cross-dissolve between the images in (a) and (b).

The shapes of the objects contained in an image are given by the color of the pixels that define them. Certainly, changes of these colors imply changes of the associated shapes. But transforming an object shape into another by simply transforming the colors of one object into the colors of the other is not effective, because we distinguish shapes by their boundaries. Therefore, to obtain good shape transition between objects in two distinct images, we should transform the boundaries from one shape into the boundaries of the other shape, while simultaneously changing the colors from one shape into those of the other.

(a)   (b)

(c)   (d)

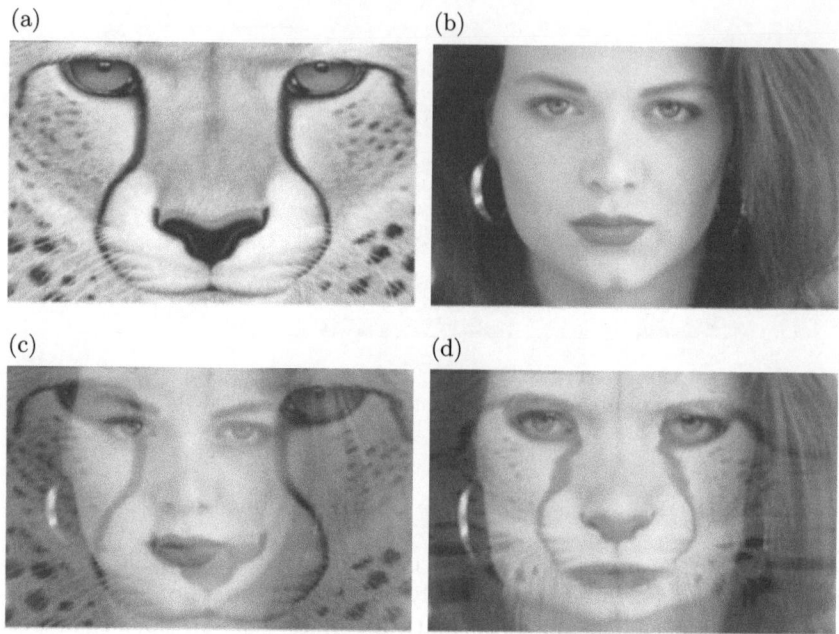

Fig. 15.21. Morphing and cross-dissolve. Image (c) is an intermediate frame of the cross-dissolve between (a) and (b), where (b) has weight $t = 0.6$. Image (d) is a frame from a morphing sequence between (a) and (b).

Mathematically, a transition that changes the boundaries and the colors of the objects in an image can be achieved by transformations that act both on the image domain (warping) and on the image range (color space). These transformations are called *morphing transformations*. As stated before, morphing transformations allow us to obtain much better transitions between images than can be obtained using only color transformations. This is illustrated in Figure 15.21(d), which shows a morphing transformation applied to the images in (a) and (b). The reader should notice the perfect alignment of the eyes, nose, and mouth, in contrast with the image in (c), obtained by using only cross-dissolve. A sequence of frames of the morphing between the woman's face and the cheetah is shown in Figure 15.24.

The morphed image in Figure 15.21(d) has two important characteristics:

- We have applied a warping filter to the original images, so as to make the geometry of the woman's face align with that of the panther (note how the mouth, the nose, and the outline of the face are made to coincide). The warping is carried out at the same time as the dissolve.

- The dissolve is carried out adaptively, that is, the weight functions are not constant from pixel to pixel. For example, the woman's hair on the right-hand side has been attenuated more than other features such as the earring.

The following diagram synthesizes the different maps involved in a morphing transformation:

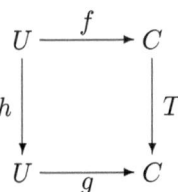

T is the amplitude transformation, h is the warp transformation, f is the initial image, and g is the final image. The morphing transformation M, when applied to f, results in

$$g = M(f) = T \circ f \circ h^{-1}. \tag{15.7}$$

If h is the identity, there is no change of coordinates in the domain, and we're back to the case of an amplitude filter, $g = T \circ f$, which changes only the image color:

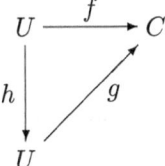

If, on the other hand, T is the identity, we're back to the case of a pure warp map, $g = f \circ h^{-1}$, which deforms the shape boundaries without changing colors:

$$\begin{CD} U @>f>> C \\ @VhVV @AAgA \\ U \end{CD}$$

15.7 Continuous Families of Transformations

Warping, morphing, and color transformations are used to correct distortions in the geometry and in the color space of an image, and also to create smooth transitions between images. In this latter case these operations usually occur in families, which are defined as follows.

Given a subset $V \subset \mathbb{R}^n$, an *n-parameter family of transformations* is a map $h : V \subset \mathbb{R}^n \to \mathcal{F}$, where $\mathcal{F}$ is a space of transformations. We call V

the *parameter space*. To each $(k_1, \ldots, k_n) \in V \subset \mathbb{R}^n$, there corresponds a transformation in the family, which we denote $h_{(k_1, \ldots, k_n)}$. An example of a one-parameter family is the cross-dissolve (15.6) between two images.

Consider an n-parameter family $h_{(k_1, \ldots, k_n)}$ of warp maps and an n-parameter family $T_{(k_1, \ldots, k_n)}$ of color transformations, both defined on the same parameter space V. From (15.7), we obtain an n-parameter family of morphing transforms $g_{(k_1, \ldots, k_n)}$ of an image f, by setting

$$g_{(k_1, \ldots, k_n)} = T_{(k_1, \ldots, k_n)} \circ f \circ h^{-1}_{(k_1, \ldots, k_n)}.$$

An n-parameter family of images can be regarded geometrically as an n-dimensional parametrized surface in the space of images. For $n = 1$ it is natural to regard the parameter as time, and we get a curve $h(t)$ in image space, as shown in Figure 15.22. As the parameter t varies from 0 to 1, we obtain a continuous transition between the images $h(0)$ and $h(1)$; such a continuous sequence of images constitutes an *animation*. One classical example is the cross-dissolve operation discussed earlier.

The case of two parameters is also important. Here we have a surface $h(u, v)$ in image space, as indicated in Figure 15.23. Setting $v = 0$ and varying u, we get a one-parameter family $h(u, 0)$, which transforms the image $h(0, 0)$ into the image $h(1, 0)$. Setting $v = 1$ and varying u, we get another animation sequence $h(u, 1)$, going from $h(0, 1)$ to $h(1, 1)$. Thus, the full two-parameter family provides a transition between the two animation sequences $h(u, 0)$ and $h(u, 1)$, for $0 \leq u \leq 1$. Picking a path $c : [0, 1] \to \mathbb{R}^2$ in parameter space, where

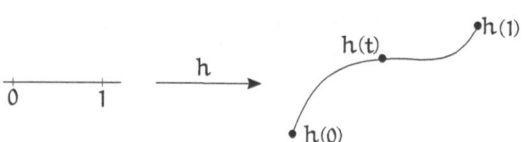

Fig. 15.22. One-parameter family in image space.

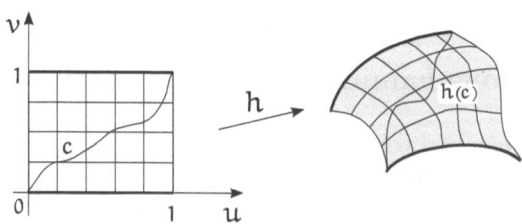

Fig. 15.23. Two-parameter family in image space.

$c(0) = (0,0)$ and $c(1) = (1,1)$, we obtain a one-parameter family (animation sequence) $h_{c(t)}$, which combines the two animations in a certain way. This process is known as *animation morphing*.

Figure 15.24 shows eight frames from a morphing transformation between a woman's and a panther's face. Note that the position of the main features (nose, mouth, outline of face) changes gradually by means of a warp transformation, to mediate between the initial and final positions; at the same time, the colors are averaged.

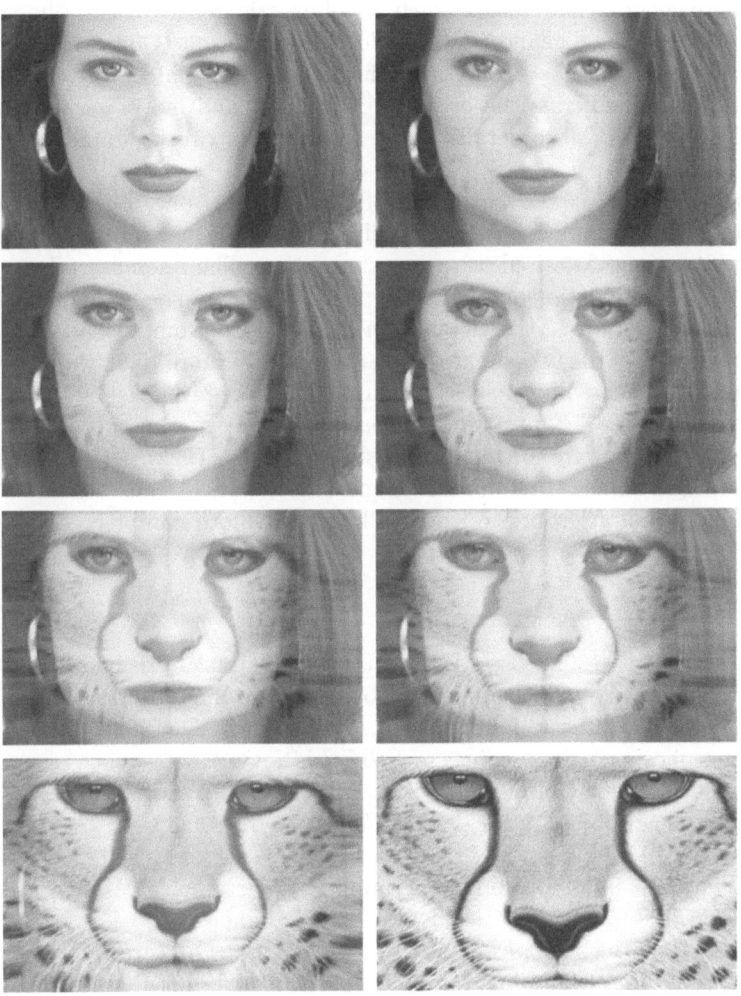

Fig. 15.24. Morphing animation sequence. See Plate 10 in color insert.

15.8 Comments and References

The concept of resampling first appeared explicitly in the literature in (Heckbert 1989). This is a pioneering work in image warping that has never been published.

For a fairly complete book on warping filters and morphing, see (Wolberg 1990). This work covers sampling and reconstruction techniques and is strongly directed toward implementation, containing several examples with source code.

Digital image warping is a particular case of deformations of graphical objects. The n-dimensional warping problem deals with deforming graphics objects more general than images (such as volumetric images or two-dimensional geometric models). The interested reader should consult (Gomes et al. 1996a). This reference contains a rigorous definition of a graphical object and considers, in a unified way, the problem of warping and morphing of arbitrary graphical objects. More details about these topics can be found in (Gomes et al. 1996b).

As mentioned at the beginning of this chapter, one major problem in implementing warping and morphing filters lies in the user interface, including the specification of desired warp transformation. The reference (Costa et al. 1992) has a classification of the various methods that can be used for specifying such transformations. More details can be found in (Gomes et al. 1996b).

The images in the two-step rotation example of Figure 15.20 were produced by George Wolberg. The images in Figure 15.21 were produced by Lucia Darsa and Bruno Costa. The morphing sequence of Figure 15.24 was produced using the morphing software *Visionaire* (Costa and Darsa 1992).

The original image in Figures 10.3 and 10.9 is a detail from "Market Place", by Alfons Rudolph, from the Kodak PhotoCD, Photo Sampler.

References

[Costa and Darsa 1992]Costa, B. and Darsa, L. (1992). Visionaire users and reference manual. Technical report, Impulse Inc., Minneapolis, MN.

[Costa et al. 1992]Costa, B., Darsa, L., and Gomes, J. (1992). Image metamorphosis. In *Proceedings of SIBGRAPI V*, 19–27.

[Gomes et al. 1996a]Gomes, J., Costa, B., Darsa, L., and Velho, L. (1996a). Graphical objects. *The Visual Computer*, 12(6):269.

[Gomes et al. 1996b]Gomes, J., Costa, B., Darsa, L., and Velho, L. (1996b). Warping and morphing of graphical objects. Notes from *Colóquio Brasileiro de Matemática*. IMPA, Rio de Janeiro.

[Heckbert 1989]Heckbert, P. S. (1989). *Fundamentals of Texture Mapping and Image Warping*. Master's thesis, Dept. of Electrical Engineering and Computer Science, University of California, Berkeley.

[Wolberg 1990]Wolberg, G. (1990). *Digital Image Warping*. IEEE Computer Society Press, Los Alamitos, CA.

16

Image Systems

The end product of a computer graphics process is usually an image or sequence of images. This is so, by definition, in the area of image synthesis, where the goal is to generate an image according to certain criteria: photo-realism, simulation, animation, and so on. The image may be displayed on a variety of devices, depending on the application. This chapter discusses certain problems arising in computer systems for image manipulation.

16.1 Image Characteristics

An *image system* is a graphics device, or set of devices, used for the storage, processing, display, and transmission of images. Three types of images may be present in an image system:

1. An *optical image* is one where the signal consists of visible electromagnetic radiation. Ultimately, this is the only type of image that the eye can perceive, and therefore the system must use this type of image to communicate with the user.
2. An *analog-electronic image* is generally characterized by a physical magnitude varying continuously in time. The modulation of the physical magnitude encodes the image. The image produced on the screen of a CRT monitor is of this type.
3. *Digital images* and their models were the subject of Chapter 6. They arise from analog-electronic images by discretization. From the viewpoint of image systems, we can distinguish four types of discretization:
   - *Spatial discretization* is the discretization of the domain into pixels, studied in Chapter 6.
   - *Spectral discretization* is the representation of the spectral color space by means of a finite-dimensional space, obtained by sampling the visible spectrum, as discussed in Chapter 4.

L. Velho et al., *Image Processing for Computer Graphics and Vision*,
Texts in Computer Science, DOI 10.1007/978-1-84800-193-0_16,
© Springer-Verlag London Limited 2009

- *Amplitude discretization* is the discretization of the color vector components, also known as quantization; see Chapter 6.
- *Time discretization* is necessary when one considers images that vary in time, as in the case of an animation. For example, time discretization underlies the various digital formats used by the video and TV industries.

16.1.1 Matrix Representation of a Digital Image

In this chapter we consider the spatial model of an image, where the domain is a rectangle

$$R = [a, b] \times [c, d] = \{(x, y) \in \mathbb{R}^2 : a \leq x \leq b \text{ and } c \leq y \leq d\}.$$

In practice, we work with digital images, where the domain is a discrete regular lattice of the rectangle R:

$$R_\Delta = \{(j\Delta x, k\Delta y) \in R : j, k \in \mathbb{Z}\}, \qquad \text{with } \Delta x, \Delta y \in \mathbb{R} \text{ fixed.}$$

Each pixel of the image can therefore be assigned integer coordinates (j, k), and the image is given by the matrix (c_{jk}), where each entry is a vector in color space, representing the color at the pixel (j, k). For a monochrome image, c_{jk} is a real number representing the luminance of the pixel. See Figure 16.1.

The quotient $(d - c)/(b - a)$ is called the image's *aspect ratio*. This ratio varies greatly in practice, depending on the application and the target display device. Images for television have aspect ratio $\frac{3}{4}$; images for 35-mm film should have aspect ratio $\frac{2}{3}$.

The quotient $\Delta y/\Delta x$ is called the *pixel aspect ratio*. It determines the shape of the pixel, which can be rectangular or square, as shown in Figure 16.2. The pixel aspect ratio depends on the device that stores the image (frame buffer). Most commonly, it equals unity ($\Delta x = \Delta y$), but for many matrix devices it

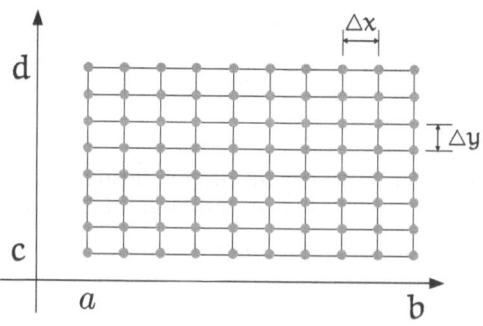

Fig. 16.1. Matrix model of a digital image.

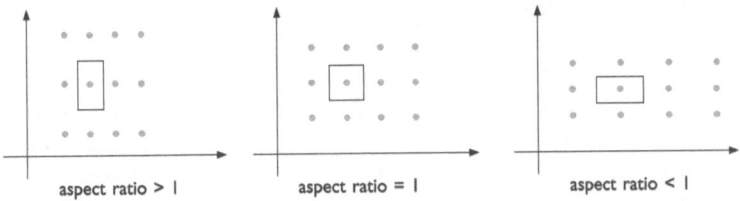

aspect ratio > 1   aspect ratio = 1   aspect ratio < 1

Fig. 16.2. Pixel aspect ratio and geometry.

has a different value, and this must be taken into account in generating and displaying an image.

We recall some terminology introduced in Chapter 6. The order $m \times n$ of the matrix (c_{ij}) representing a digital image is called the image's *spatial resolution* (or *geometric resolution*); the number m of rows is the *vertical resolution*, and the number n of columns is the *horizontal resolution*. The number of bits used to represent the color of a pixel is the image's *color resolution*. The set $\{c_{ij}\}$ of all colors occurring in the image is the *gamut*.

16.1.2 Pixel Geometry

In the matrix model, as we have seen, each pixel determines a rectangle in the image plane, as shown in Figure 16.2. In fact, one can also use other pixel configurations, where the plane region assigned to each pixel (pixel shape) is not a rectangle but some other compact, convex subset of the plane. Hexagonal cells are a relatively common choice, as shown in Figure 16.3.

There are many ways of encoding a digital image for storage, manipulation, and transmission. These different encodings give rise to the many existing image file formats (PCX, GIF, TIFF, PhotoCD, and so on). We will not attempt to cover these formats in this book. There are also a great many graphics programs and libraries to convert among the various formats. See the references listed in Section 16.7.

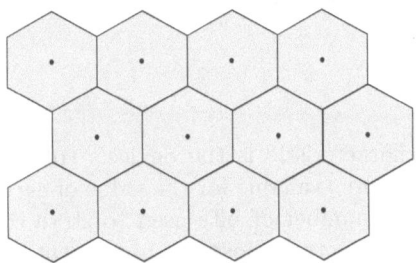

Fig. 16.3. Hexagonal pixel lattice.

16.2 Image Display

When an image is displayed on a matrix device, such as a CRT-based device, its matrix representation is mapped to the device's dot matrix. The device's physical pixel shape is given by the pixel *spread function*, which is the pixel's response curve to the impulse that generates the color of one pixel. The impulse signal here consists of an image with maximum luminance value in one pixel and zero intensities on the other pixels. The quality of the displayed image depends on several characteristics of the pixel spread function. Four of these characteristics are related to the geometry and the color space of the device: pixel size, pixel density, geometric resolution, and color resolution.

Pixel Size

The physical dimensions of the pixel are determined by the distance between dots in the device matrix. In video monitors this value is known as the *dot pitch* and ranges from 0.26 to 0.31 mm (0.01 to 0.012 inches) for commercial CRT monitors.

Pixel Density

Pixel density is the number of pixels per unit length in the matrix device. It is usually expressed in dots per inch (dpi), and it may be different in the horizontal and vertical directions. When only one density is mentioned, it applies to both directions.

Geometric Resolution

The device's geometric resolution is the order of the display matrix defining the number of pixels of the image display. It must be compatible with the resolution of the image to be displayed.

Color Resolution

Another important characteristic is the device's color gamut, which determines the number of colors available for the value of each pixel. This number usually depends on the number of bits used to store color (number of "bit planes") and on the architecture chosen to implement color space. The color of each pixel must be quantized to the number of bits available in the device's color space.

16.2.1 Support Media

To display an image on a *support medium* means to reconstruct the color values stored in the pixels. There are many types of support media: a monitor screen, paper, film, video, and so on. The technologies used for displaying the image vary widely: some support media use digital technology; others, analog technology; yet others a hybrid analog-digital technology. The most appropriate support medium to use depends, of course, on the type of application as well as the cost.

The most commonly used display device is the video monitor. It uses a cathode ray tube (CRT) to realize the image, which is of the analog-electronic type.

To display a color image on paper, one can use a color printer or one of the traditional printing processes (offset printing, rotogravure, and so on). There are many printing technologies, such as thermal wax transfer and dye sublimation. Choosing the best process depends largely on the application. Generally speaking, thermal wax transfer is appropriate for vector-based images ("line art"), while dye sublimation gives much better results when the gamut is wide, as in the case of color photographs.

Offset printing, traditionally used in the graphics industry for large print runs, is completely analog. With the development of the market and the dissemination of electronic publishing systems, the production process is now almost entirely in the digital realm. Offset printing and its interface with digital image processing will be studied in more detail later in this chapter.

16.2.2 Tone Maps

When an image is displayed, the values of the color components at each pixel are used to generate colors in the color space of the display device. Ideally, there should be a linear relation between the intensity of the pixel value and the color produced, so that, for example, doubling the red component also doubles the amount of energy emitted by the red phosphor on the CRT screen. This is important in order to obtain a faithful color balance. The transformation that maps color intensities of the image into color intensities of the display device is called the *tone map*. Computation of the tone map is a complex task, as will be clear from the discussions of this section.

Unfortunately, the response of most display devices is nonlinear. It is possible to get around this problem by compensating for this nonlinearity before sending the color values to the device. Typically, the user's application has to do this, based on the type of device or other information supplied by the manufacturer.

A general rule to carry out this correction is to create an image with a linear gradient of color and to plot the graph of the function associating the intensity of each pixel of this image to the intensity of color actually displayed. This requires the use of an instrument that can measure color intensities accurately.

If this graph is linear, no correction is necessary; if not, the function is used as the basis for the correction. When the display device has a *lookup table*, that is, a user-controlled map between input values and values sent to the actual display, this table can be used to compensate for the nonlinearities. Otherwise, it is necessary to actually change the values of the pixels in the image.

Laser Printers

Laser printers are usually markedly nonlinear, due to the type of paper used, to the imperfect shape, dimensions, and placement of pixel dots, and to interference among pixels due to dot overlap.

CRT Devices

In a CRT device, the components of a pixel determine the voltage that drives the electrons toward the phosphor on the screen, creating an electron flow (current). The relation *input intensity* × *voltage* is roughly linear, but the relation *voltage* × *pixel luminance* is not, and a correction is needed. We discuss this in detail now.

Gamma Correction

In a CRT monitor, the voltage V driving the electron flow is related to the intensity I of light emitted by the phosphor hit by the electrons according to the approximate formula

$$I \sim V^{\gamma}, \tag{16.1}$$

where γ is a constant that depends on the phosphor. This nonlinearity causes serious distortions in the colors of the image displayed: some areas are too dark, while in other, the colors are too saturated. To avoid this problem, we must use an amplitude filter that corrects the intensity of each pixel before that information is passed to the CRT. Because of the conventional use of the letter γ as the exponent in (16.1), this step is generally called *gamma correction*. We can rewrite (16.1) as

$$V \sim I^{1/\gamma},$$

so we know what voltage to use in order to create a desired intensity. In Figure 16.4, the black line is the graph of Equation (16.1). The gray line shows the correction function that must be used to obtain a linear relationship between intensity and voltage, which is indicated by the dotted line.

The correction given by the equation $V \sim I^{1/\gamma}$ must be applied to the pixel intensities before the signal is converted into voltage at the CRT. As already mentioned, if the device has a lookup table, this table can be used to store

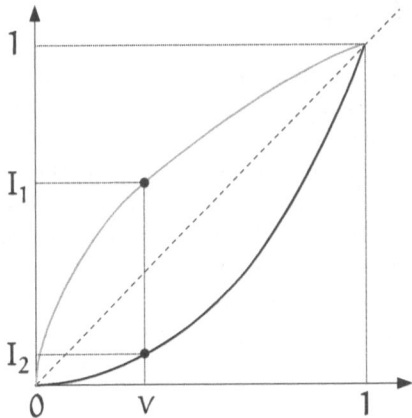

Fig. 16.4. Monitor gamma correction curve.

the correspondence, avoiding the need to recompute the gamma correction for every pixel. However, there is a disadvantage to this procedure: usually, the precision of the values that can be stored in the lookup table is low; typically only eight bits are available. In any case, when there is no lookup table, one must carry out the correction separately for each image pixel.

In the case of synthetic images, another alternative is to perform the correction immediately after the calculation of the value of each pixel. However, this means the image is now optimized for a particular display device, and one cannot use it on a device having a different response, or process it in other ways, without first undoing the gamma correction already introduced. In general, it is better to store the image with its "natural" color values and apply a gamma correction only when it is being sent to a particular device.

To carry out the gamma correction we must of course know the value of γ. If this value is not specified by the manufacturer, it must be experimentally determined. The literature contains procedures for this determination. For best results one should determine separately the γ values for each of the monitor's three primary colors.

16.2.3 Calibration

Gamma correction is only one component in the process of calibration of the various devices in an image system, a process that is essential for the correct display of the image. Calibration is often complicated and may involve even mechanical adjustments, in the case of systems with analog components. One important step in calibration is adjusting the system's color space. In general, this is done by comparison with some standard color space.

16.3 Cross Rendering

An image system often involves more than one display device or display subsystem, and the physical processes involved in the realization of the image on each device may be completely different. One of the main problems in such systems is to make an image look the same when displayed on the various devices. This problem is known as *cross rendering*. For example, one may have a workstation connected to a printer and expect images to come out on the printer looking as closely as possible to how they look on the screen. Clearly, "as closely as possible" involves some notion of distance, or metric, for comparison between images; this metric should take into account that, ultimately, it is our perception of the images that matters.

To give another example, the reader may have seen in a video and electronics store a number of TV sets tuned to the same show simultaneously. Almost certainly, the image was noticeably different on sets from different manufacturers. This is a simple but significant example of the problem of cross rendering: monitors from different manufacturers use different phosphors and sometimes even different technologies to turn on the pixels on the screen. This makes the color spaces of the devices different.

In general, the solution to the problem of cross rendering is quite complex, since it involves the particular characteristics of each device. A real-life example, namely electronic publishing systems, will be studied later in this chapter.

16.3.1 Gamut Transformations

Making an image look the same in different display devices is a difficult task, because in general the gamut of these equipments is quite different, and we must take into consideration the gamut of the image and of each device. Thus, it is necessary to carry out a color transformation, called a *gamut transformation*, with the purpose of making the gamuts of the various color systems compatible. In order to carry out the gamut transformation, we should consider a "universal space" that contains the gamut of the different devices and of the image. In this space we determine the gamut transformation in a way that makes the various gamuts compatible by means of appropriate corrections. This is essentially the task to be accomplished by *color management systems*.

The object of a gamut transformation is to make the image look the same when displayed on any device of the system. The right choice of gamut transformation involves psychophysical considerations and is, in general, extremely difficult when the devices involved use very different technologies.

16.4 Color Correction

In image synthesis, it is often impossible to display a given color on an RGB monitor for two reasons:

- The luminance of a pixel's color may be outside the range that can be displayed on the device; this is known as *luminance overflow*.
- The chromaticity of a pixel's color may be outside the color gamut of the device; we say that the color is *unrealizable*.

16.4.1 Luminance Overflow

Assuming that the color components are normalized to be in the interval $[0, 1]$, luminance overflow occurs when one of the color components computed for a pixel falls outside this interval. An attempt to solve this problem should try to maintain the color information, changing only the luminance. For example, reducing to 1 each color component that is greater than 1 is easily seen to change the chromaticity, and is therefore an inappropriate method. An appropriate method is to divide each component by the maximum value of the components; thus, if $C = (r, g, b)$ is the original color vector, with $\max\{r, g, b\} > 1$, we replace it by the new color vector

$$C' = \left(\frac{r}{\max\{r, g, b\}}, \frac{g}{\max\{r, g, b\}}, \frac{b}{\max\{r, g, b\}} \right).$$

16.4.2 Unrealizable Colors

A color is unrealizable when its chromaticity point, given by the radial projection of the color vector onto the chromaticity plane, does not belong to the device's gamut. Figure 16.5(a) illustrates this problem when the gamut is a triangle.

The existence of unrealizable colors is very common in image synthesis, especially when one uses a spectral model of color space in computing lighting and shading. The same problem occurs when we apply certain filters. One such example is the laplacian filter (see Chapter 7 on image operations).

There are several possible solutions, and the choice among them depends on the application. If the correction is to be made to the final image, one should use a method that preserves hue and luminance, changing only the saturation. In other words, one should add white to the color, so as to bring it into the device's color gamut: see Figure 16.5(b). We describe two methods to do this: by a *color space contraction*, and by *color clipping*.

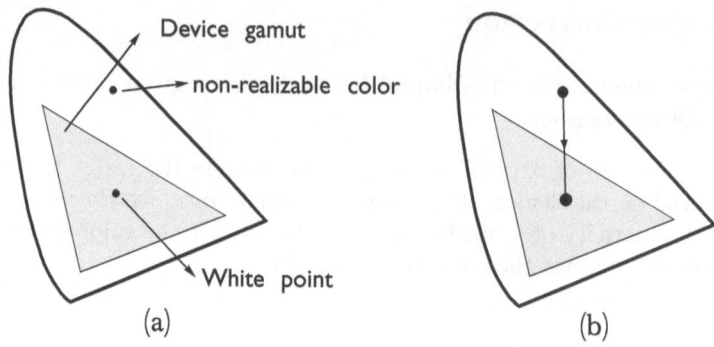

Fig. 16.5. Correcting an unrealizable color.

Color Space Contraction

One can make a color contraction in the chromaticity diagram of the CIE-XYZ system in order to guarantee that the colors to be displayed will belong to the device's color gamut. An easy way to do this change of coordinates is to take three (unrealizable) primary colors in such a way that the triangle they determine contains the gamut of colors to be reproduced. We then compute the chromaticity coordinates in this new color system and use these values as the chromaticity coordinates in the CIE-RGB system. Geometrically, this corresponds to applying a contracting affine transformation of the chromaticity diagram, leaving fixed the color white. One disadvantage of this method is that a significant part of the device's color gamut will then never be accessed.

Color Clipping

Another disadvantage of the contraction mapping method just described is that all colors have their saturation decreased, and this can change the whole image perceptually. The method of color clipping changes the saturation only of those colors that cannot be displayed; it is therefore a less intrusive and more local correction method. Here is a simple clipping method. Start by changing coordinates to HSV. An unrealizable color has saturation $S > 1$; replace this value by $S = 1$ and convert back to RGB coordinates. The color thus obtained is obviously realizable and differs from the original color only in saturation.

16.5 Display Models

After studying some technical aspects of the image display process, our goal in this section is to introduce a conceptual model for the process. Such a model

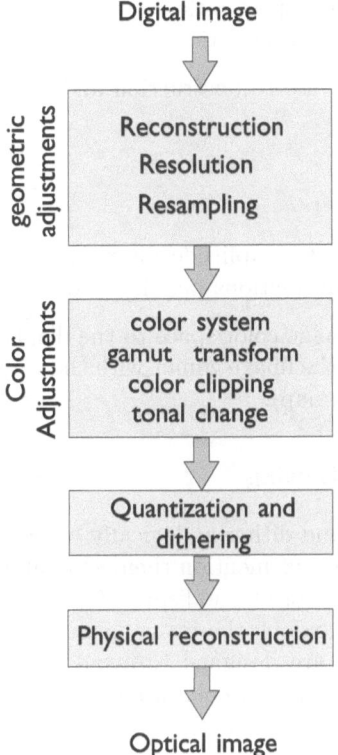

Fig. 16.6. Display model for digital images.

allows the comparison of different systems with greater objectivity and helps in the development of algorithms for solving image display problems.

Figure 16.6 shows a conceptual model for a digital image display system. The input consists of a digital image, and the output is an image displayed on some graphics device, which is seen by the observer as an optical image. We turn to each box in this diagram.

Geometric Adjustments

The first box represents geometric filters that transform the image domain with the goal of making it geometrically compatible with the characteristics of the display devices. The necessary transformations may include, for example

- a change in spatial resolution,
- a change in the image representation (from a hexagonal lattice to a rectangular one, say),

- a change in the image aspect ratio,
- a change in the pixel aspect ratio.

Sometimes one must use reconstruction and resampling to carry out the desired adjustment.

Color Adjustments

The second box represents amplitude filters used to transform the image's color gamut. Such transformations may include

- a change from the image color space to the display device color space,
- compatibilization of the image gamut with the gamut of the display device,
- hue correction (tone map).

Quantization and Dithering

Although quantization and dithering logically belong to the color adjustment operations just discussed, we mention them separately because of their great importance in the process of image display. Quantization is one of the transformations that aim at making the image gamut compatible with the device gamut; dithering filters are designed to avoid perceptual quality loss after quantization. Dithering is of particular importance in the case of bitmap (two-level) display devices.

Physical Reconstruction

Finally, after the preprocessing operations, the image is physically reconstructed in the support of the graphics display device. After being displayed, the image can be seen by the observer as an optical image. The filters that reconstruct the image at this step can be quite complex. We describe one model for this filtering process.

16.5.1 Physical Reconstruction Function

The physical reconstruction function of a graphics device uses the values of color intensity of the image to reconstruct the image physically. It is important to have specific knowledge of the device's physical reconstruction function in order to optimize the quality of the image displayed; this knowledge should influence one's choice of device-dependent filters to be applied to the image.

Let p_i be a physical pixel of the device, such as a dot on the screen of a video monitor, and let δ be a unit impulse at p_i. The system's impulse response function (the response to δ) is called the *pixel spread function* and will be denoted by h_i. If the physical reconstruction process were linear in terms of the various image parameters (image color, image dimensions, pixel

size, etc), the reconstructed image would be given by the convolution of the
digital image I with the pixel spread function:

$$(I * h)(n) = \sum_{j=-\infty}^{+\infty} I(j)h(n - j).$$

In practice, the process introduces nonlinearities, both spatial and of amplitude, and their modeling is more complicated than we have written above.
One must take into account several factors, such as the following:

- randomness in the spatial placement of the device's pixels (this is called *positional noise*, and it occurs quite markedly in the case of low-resolution laser printers, for example);
- randomness in the pixel's impulse response function;
- influence of the support medium on the appearance of the reconstructed image;
- the nonlinearity of the values of the reconstructed function with respect to the pixel values of the digital image.

Figure 16.7 shows a general conceptual model of the physical reconstruction function, taking into account the factors just mentioned. The first box in the figure represents the reconstruction of the image function in the continuous domain. Positional noise is modeled by a unit impulse $\delta(x - \varepsilon(x))$, where

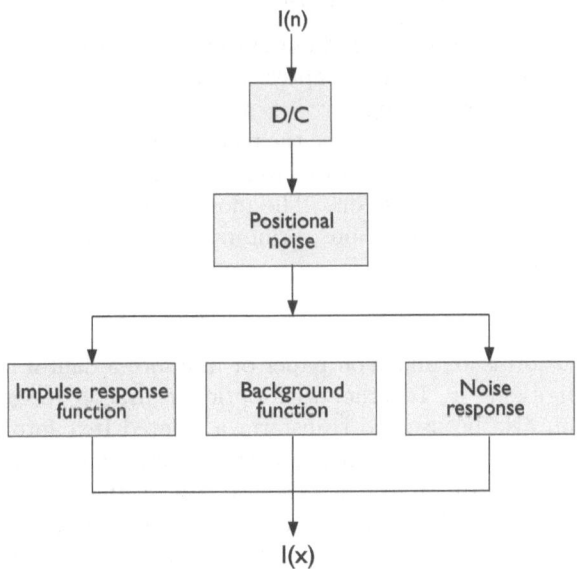

Fig. 16.7. Conceptual model of the device's physical reconstruction function.

$\varepsilon(x)$ is a random perturbation. The variation of the spread function at each pixel is modeled by a noise spread function $\omega(x)$, to be added to the average impulse response function $d(x)$. The influence of the support medium on the perception of the reconstructed image is modeled by the *background function* $b(x)$. The final reconstructed image is therefore given by

$$I(x) = \left(\sum_n I(n)\delta(x - \varepsilon(x)) \right) * (d(x) + \omega(x)) + b(x).$$

This equation does not include a correction for amplitude nonlinearity, because we included such a correction among the preprocessing operations (Figure 16.6).

For each specific device, we can determine δ, ε, d, ω, and b. With this knowledge, we can apply specific filters to the image at display preprocessing time, in order to improve the perceptual quality of the final result.

16.6 Electronic Publishing Systems

We could use different image systems to illustrate the above problems in a real-world environment. Several of these systems exist for different areas of application. We could mention

- image systems for video production,
- image systems for film production,
- image systems for multimedia, and
- image systems for electronic publishing.

A complete study of each of these systems is outside the scope of this chapter. In the remainder of this chapter we will give an overview of an image system for electronic publishing.

Electronic publishing systems use the techniques of computer graphics and image processing toward the goal of producing printed publications (books, magazines, newspapers, and so on). The most common devices used in such a system are scanners, workstations, printers, and phototypesetters.

Scanners

A scanner transforms an image on paper or film into a digital image by sampling it and digitizing it. Together with optical character recognition (OCR) software, it can also be used to transform a printed text into its electronic equivalent. Other types of software can extract vector data (line art) from the digital image. Figure 16.8 illustrates the use of a scanner.

Digital cameras avoid the need to take a photograph of a scene and then scan it; their output is already a digital image. Traditional analog video cameras can also be used to capture a scene digitally when they are coupled with a *frame grabber*, which takes one frame of the video signal and transforms it into a digital image.

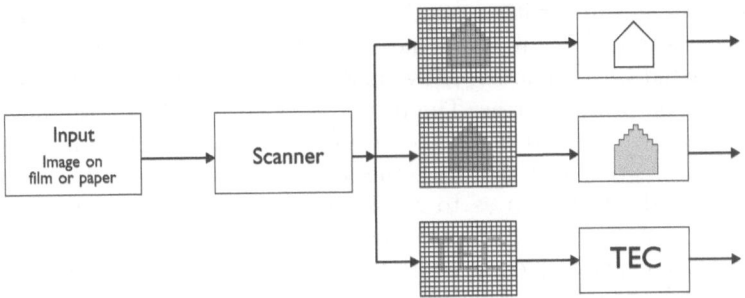

Fig. 16.8. Use of a scanner in an electronic publishing system.

Workstations

Typically, an electronic publishing system includes a workstation (desktop computer) with a matrix-based monitor, mouse, keyboard, and frame buffer. The spatial and color resolutions should be compatible with the needs of the publications to be produced. The software used generally has interaction capabilities that allow the user to manipulate directly the various elements of the publication.

Printers

A printer is part of any electronic publishing system, so that proofs (samples) of the material being composed can be made at low cost on paper. The resolution densities typically range from 300 to 1200 dpi (dots per inch) for black-and-white printers, and from 120 to 600 dpi for color printers. Printers can be used to produce final copy for jobs where high resolution is not essential.

Phototypesetters

Phototypesetters are monochrome devices that use laser or infrared technology to produce very high-resolution output (up to 3000 dpi). They can typically output either onto photographic paper or directly onto film. The per-page cost is relatively high, so phototypesetters are not used for mass printings; instead, their output is used as input for offset printing, which we turn to now.

16.6.1 Offset Printing

For large print runs, the most economical alternatives are still the two traditional methods of offset printing and rotogravure. Here we discuss offset printing briefly, concentrating on its characteristics that matter to electronic publishing.

The process starts with a positive image on paper or a negative on film. The image is transferred to a thin metal plate by a photochemical process: the plate is coated with a photosensitive material, and exposure to light hardens the coating on printing areas. The rest of the coating (on nonprinting areas) is washed away. The plate is rolled onto a cylinder and a greasy ink is applied; it sticks only to the areas that have the coating, the nonprinting areas having been previously wetted so as to repel the ink. The rotating plate cylinder comes into contact with a rubber cylinder, to which it transfers the ink; finally, the traveling paper is pressed against the rubber cylinder, and receives the printed image. The basic steps are shown in Figure 16.9.

This process has been used for decades, employing analog techniques at each step. With the development of electronic publishing, it became possible to go digital in several ways:

- As discussed earlier, for small print runs and low-to-middle quality, the final output may be obtainable from a low-end digital device (laser printer). For a single printout, a high-end digital device (phototypesetter) provides better quality than offset printing. However, the use of digital printers instead of offset printing is far from satisfying the combined requirements of time, quality, and cost for large print runs.
- There now exist graphical output devices that produce a metal plate directly, thus allowing the elimination of the film step in offset printing. However, the cost–quality equation is not yet favorable to the use of this shortcut. Moreover, the material from which the plate is made is not very resistant, so that large print runs tend to wear it out.
- Image processing techniques and laser printer technology allow the creation of film directly from a digital image, using high-resolution phototypesetters, as explained earlier. This is the path most commonly followed. Thus, the production of good-quality film can be carried out efficiently by digital means.

We now consider separately the cases of monochrome (grayscale) and color printing, and how they affect the production of film.

Monochrome Images

The offset printing process is essentially a one-bit affair: a point on the metal plate either accepts ink or does not. Thus, in order to reproduce a grayscale image, one must use digital halftoning, or dithering, as discussed in Chapter 12. The halftone images in this book can serve as examples; the film

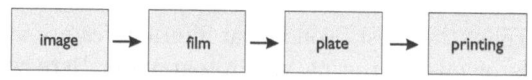

Fig. 16.9. Flowchart of the offset printing process.

for them was produced from digital images using a cluster ordered dithering algorithm, on an 1800-dpi phototypesetter.

The raster image processor (RIP) associated with a phototypesetter in general has built-in dithering algorithms. The user has the choice of using his or her own algorithm or relying on the one built into the RIP software.

Color Images

Because, perceptually, three primary colors are enough to generate color space, color images can be obtained by offset printing if the same sheet of paper is exposed to inks of the three primary colors. Thus, color offset printing theoretically reduces to three monochrome printings: we create a piece of film for each color channel; we then transfer each channel to a metal plate; and we print the image from each plate in succession. To make this actually work, one must answer two basic questions:

- What color system should be used?
- How do the colors from each channel combine to produce the final color seen on the paper?

Color Systems

The perception of color printed on paper is reflective, that is, light falling on the paper is reflected (diffused) and so reaches the eye. Starting from a white sheet of paper, we have maximal luminance before adding any color. As we add color, there is a loss of luminance, until when we reach black there is (ideally) no reflection of the incident light. Thus, color formation takes place subtractively. From our study of color systems in Chapter 5, we know that the most appropriate system to use in this context is that of the complementary primary colors cyan, magenta, and yellow, or CMY.

As shown in Figure 5.7 of Chapter 5, in the CMY system zero luminance (black) is obtained by superimposing the three primary colors. From the additive point of view, cyan is a mixture of the primaries green and blue, so it has the effect of eliminating the red component of light reflected from the paper. An analogous process of elimination happens for the other two complementary colors.

The Black Component

Although, from the point of view of color theory, all shades of gray can be produced from cyan, magenta, and yellow, in practice it is more convenient to treat gray (including black) separately, for the following reasons.

- *Registration problems*, that is, slight misalignments in the mechanical setup, cause the cyan, magenta, and yellow components of the image to be printed at slightly different positions. Thus, a black dot made from dots of the primary colors invariably has colored edges. The better (and more expensive) the printing press, the less perceptible these registration errors are, but they are always present. Therefore, for printing text, the use of black is essential.

- It is extremely difficult to combine the three primary colors consistently in exactly the proportions that yield black; most often the combination has a brownish or purplish hue.
- Making black from colors requires much more ink, and color ink is more expensive to begin with. The excess of ink also delays the drying.
- The use of black ink allows a better balance in the printing density in areas of low luminance.

For these reasons, instead of using three channels CMY, we use also a fourth channel, black, abbreviated K (since B already means blue). This leads to the CMYK system, which is universally used in offset printing. Figure 16.10 shows a color image produced by offset printing (in fact, the whole book was printed this way, the films having been created digitally). Figure 16.11 shows, from left to right, the results of printing separately the cyan, magenta, yellow, and black channels of the same image.

The introduction of the K component solves the practical and technical problems we have mentioned, but it creates another: how to convert from the image color space to the CMYK system, and specifically how to compute the K component, which, we observe, is linearly dependent on the components C, M, and Y. This is known as the *color separation* problem, and we will return to it later.

Fig. 16.10. Reproduction of a color image. See Plate 11 in color insert.

Fig. 16.11. CMYK components of the image of Figure 16.10. See Plate 12 in color insert.

Note that in some special applications it may be desirable to go beyond four colors and directly apply inks of other colors that are difficult or impossible to obtain as a combination of cyan, magenta, yellow, and black. This is known as a *spot color* application, while the use of the four primaries only is known as *process color*. Spot colors can be used, for instance, to obtain metallic and fluorescent effects and a wide range of textures that complement color. A photograph with very dark regions will nonetheless stand out against a black background if the background is matte and the photograph's colors are glossy.

Color Reconstruction

As we have seen, the physical reconstruction of color in offset printing happens when the C, M, Y, and K color channels are successively printed on the same piece of paper. The details of this reconstruction process are somewhat complex.

In producing the film for each channel, the images must be dithered, as explained above; cluster dithering algorithms are preferable, since they are more robust with respect to errors in the placement of individual pixels on the phototypesetter. Next, when combining the channels, we want to ensure that small variations in registration affect as little as possible the final perceived color. Ideally, this would be achieved by having the distribution of dots be totally independent for each channel, so that, when combining a 50% magenta with a 50% cyan, exactly 25% of the area should be covered by both inks, 25% by each ink separately, and 25% by neither. In practice we can only achieve partial independence; we will return to this point later.

Note that, in terms of color theory, the effect of nonoverlapping dots of different colors is *additive*, while the effect of overlapping dots is *subtractive*. The overall dependence of the final color on the intensity of each channel, is, therefore, neither purely additive nor purely subtractive; the formulas expressing it are complicated, and it is usually more practical to use an empirical approach to calibrate the press. Hence the use of *match prints*, which are high-quality proofs produced from the negatives by means of some other technology and are used for comparison to ensure reasonably faithful tones.

Figure 16.12 shows an enlarged detail of the floor from the image in Figure 16.11. Notice the clusters of each one of the colors CMYK, and look at how they vary in size to control the intensity of each channel.

The choice of the lattice (screen) that underlines the dithering of each channel is very important. The most important consideration is that the screens should be placed *at different angles*. If they were aligned, a small registration error might cause all yellow dots (for example) to overlap with magenta dots, while an equally small error in the opposite direction might cause all dots to be disjoint; the final perceived color would be significantly different from the desired one in either case. In practice, one chooses screen angles widely spaced in the interval $(0, 90°)$: for example, $0°$, $15°$, $45°$, and $75°$. (Notice that a screen rotated $90°$ is aligned with one rotated $0°$.)

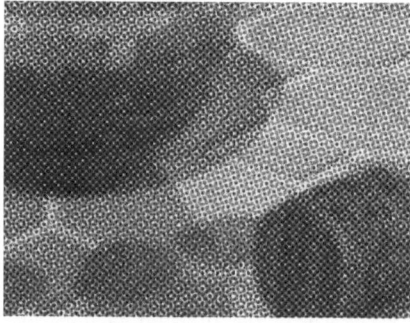

Fig. 16.12. Enlargement of an offset-printed image. See Plate 13 in color insert.

Fig. 16.13. Moiré pattern arising from interference between two screens.

Even with this precaution, there is a certain amount of interference among the screens, due to periodicity. Figure 16.13 illustrates this: on the left we have an image dithered on a (very coarse) screen tilted 5°, and on the middle the same image is dithered on a screen tilted 25°. The image on the right is the superposition of the first two. Notice the interference patterns (moiré patterns) in the form of regularly distributed, darker than average, clusters. Compare to the images in Section 7.5.1.

The most common choices for screen angles lead to the type of "rosette patterns" seen in Figure 16.12 rather than the type of pattern seen in Figure 16.13. Rosette patterns don't stand out, and screen interaction tends not to be a problem at screen densities of 150 lpi (lines per inch) and higher.

More recently, there has been a move toward the use of nonperiodic dispersed algorithms for color printing. This has been possible because of the high precision in the dot placement attained by modern phototypesetters. The Peano curve dithering algorithm studied in Chapter 12 is well suited for this purpose. For more information, consult the references given in Section 16.7.

Color Separation

We now turn to the question of *color separation*, that is, conversion to CMYK color space of a digital image whose image space is, most likely, quite different. Figure 16.14 shows the main steps in this conversion, assuming that the image color space is RGB and that we use the standard color space XYZ to perform the conversion. Another commonly used standard is CIE-Lab. Both of these systems allow the specification of color independently of the device and are used in order to mediate between image color space and device color space because their gamut contains the gamut of all graphical output devices.

In principle, color separation with the computer should be very simple, since cyan, magenta, and yellow are complementary to red, green, and blue. Assuming that colors are normalized to lie in the interval $[0, 1]$, we can write (see Section 5.4.4)

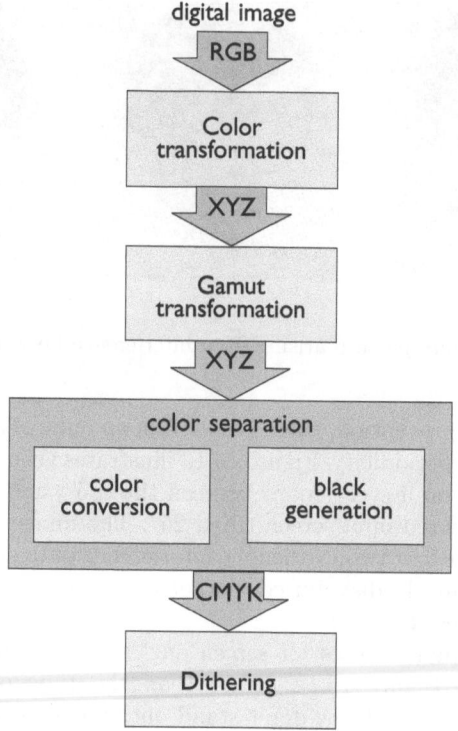

Fig. 16.14. Steps in the digital generation of film.

$$C = 1 - R, \quad M = 1 - G, \quad Y = 1 - B.$$

The next step is to remove the presence of gray that is produced via a combination of the primaries CMY and replacing it directly by black, K. The simplest method to do this is known as *gray component replacement* (GCR). To compute the intensity of the K component at a point, we find the least of the three values CMY at that point; call it d. We then take a certain fraction of d, say $p\%$, where p is a number determined empirically, and set $K = pd/100$. Finally, we subtract K from the other components, setting $C' = C - K$, $M' = M - K$, and $Y' = Y - K$. See Figure 16.15.

For simple applications, this procedure works well enough. However, for more complex images, having a wide gamut, it may lead to unexpected results, and the printed image may be perceptually quite different from the one seen on the monitor screen. This is because the map from the monitor color gamut to the offset printing gamut is nonlinear.

Another method for replacing "composite" gray by a black component is known as *undercolor removal* (UCR). It acts only on areas where the CMY components are approximately equal, that is, areas that are really grayish

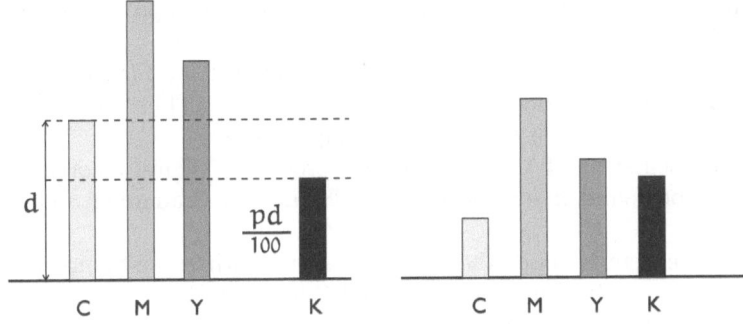

Fig. 16.15. Calculation of the black component.

and not merely dark in color. This method can also be combined with GCR, leading to more sophisticated algorithms involving complicated heuristics and designed to counteract the nonlinearities of the process.

The replacement of process gray by a black channel allows greater control over the ink density deposited on the paper during printing. This density may vary from 0% (absence of ink) to 400% (all four components at maximum). Control over ink density is important because this density should be suited to the type of paper being used.

We stress that our discussion of offset printing and color separation was meant as an illustration of the problems inherent in image systems and is far from a complete treatment of the subject.

16.7 Comments and References

For a fairly complete reference containing technical information about a great variety of image formats, see (Murray and Ryper 1994). Another good source is (Kay and Levine 1992).

In comparing renderings of an image on different devices, one must take into account perceptual factors, so it is important to be aware of the various mechanisms of visual perception. An elementary but reasonably comprehensive discussion can be found in (Rosenfeld and Kak 1976). For a more detailed discussion, see (Wyszecki and Stiles 1982).

A concise but excellent reference on the use of color in image systems is (DeMarsh and Giorgianni 1989).

A detailed discussion of color clipping and compression algorithms can be found in (Hall 1989). This book also contains a good discussion about image systems for video production, with emphasis on the NTSC composite video standard.

A more complete reference on image systems for video production is (Winkler 1992), a comprehensive, if somewhat terse, reference.

The conceptual model for the physical reconstruction function presented in this chapter is the one given in (Ulichney 1987). This book gives more details on this model, for the case of bitmap devices.

Good discussions of the problem of reproducing digital color images on paper can be found in (Stone et al. 1988) and in (Lamming and Rhodes 1990).

Complete details about the use of dithering with space-filling curves for color printing can be found in (Velho and Gomes 1996).

The subject of gamma correction for monitors is well covered in (Catmull 1979). The procedures discussed there can be adapted to the case of tone maps for other display devices. A brief discussion about tone maps for printers can be found in (Ulichney 1987). Detailed procedures for the calibration of individual devices are set forth in their technical documentation, and more general treatments are sometimes found in manuals for image manipulation software. A good example is the Photoshop manual, from Adobe Systems.

The original image used in Figures 16.10, 16.11, and 16.12 is taken from the Strata Vision tutorial demo.

References

[Catmull 1979]Catmull, E. (1979). A tutorial on compensation tables. *Computer Graphics (SIGGRAPH '79 Proceedings)*, 13(3):1–7.

[DeMarsh and Giorgianni 1989]DeMarsh, L. and Giorgianni, E. (1989). Color science for imaging systems. *Physics Today*, September, pp. 44–52.

[Hall 1989]Hall, R. A. (1989). *Illumination and Color in Computer Generated Imagery*. Springer-Verlag, New York.

[Kay and Levine 1992]Kay, D. C. and Levine, J. R. (1992). *Graphics File Formats*. Windcrest/McGraw-Hill, Blue Ridge Summit, PA.

[Lamming and Rhodes 1990]Lamming, M. G. and Rhodes, W. L. (1990). A simple method for improved color printing of monitor images. *ACM Transactions on Graphics*, 9(4).

[Murray and Ryper 1994]Murray, J. D. and Ryper, W. V. (1994). *Encyclopedia of Graphics File Formats*. O'Reilly and Associates, Sebastopol, CA.

[Rosenfeld and Kak 1976]Rosenfeld, A. and Kak, A. C. (1976). *Digital Picture Processing*. Academic Press, New York.

[Stone et al. 1988]Stone, M. C., Cowan, W. B., and Beatty, J. C. (1988). Color gamut mapping and the printing of digital color images. *ACM Transactions on Graphics*, 7(3).

[Ulichney 1987]Ulichney, R. (1987). *Digital Halftoning*. MIT Press, Cambridge, MA.

[Velho and Gomes 1996]Velho, L. and Gomes, J. (1996). *Color printing, stochastic screening and space filling curves.* Preprint, IMPA, Rio de Janeiro.

[Winkler 1992]Winkler, D. (1992). Video technology for computer graphics. *SIGGRAPH '92 Course Notes.*

[Wyszecki and Stiles 1982]Wyszecki, G. and Stiles, W. S. (1982). *Color Science.* John Wiley & Sons, New York.

A

Appendix:
Radiometry and Photometry

This appendix deals with the photometric and radiometric variables that are useful in setting and understanding problems about color and energy exchange.

All eletromagnetic radiation transfers energy, called *radiant energy*. *Radiometry* is the science of the measurement of the physical variables associated with the propagation and exchange of radiant energy. *Photometry* is the branch of radiometry that deals with these variables from the viewpoint of visual responses; that is, it studies how radiant energy is perceived by an observer. Thus, radiometry in general deals with physical processes, while photometry deals with psychophysical ones.

A.1 Radiometry

Although an understanding of the physics of electromagnetic waves is important in the study of light–matter interaction, we need not be concerned with the nature of electromagnetism in the study of radiometry and photometry. It is enough to know that radiant energy flows through space; the fundamental variable involved is the *flux* through a surface, that is, the rate at which radiant energy is transferred through a surface. This notion is therefore analogous to the notion of an electric current or to the flow of matter in fluid dynamics.

In contrast with the situation in fluid dynamics, however, there exist point sources of radiant energy, and indeed they are of great importance in radiometry. To define the radiometric variables associated with point sources, we introduce the notion of solid angles.

Solid Angles

In the plane, angles can be regarded as a measure of apparent length from an observer's viewpoint. To compute the angle subtended by an object when

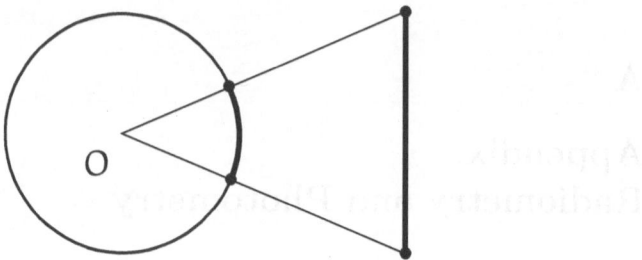

Fig. A.1. Angles and radial projection.

observed from a point O, we project the object radially onto the unit circle centered at O and measure the arc determined by this projection; this is how big the object looks to someone stationed at O (Figure A.1).

We can also use a circle of radius other than 1, but in this case we must divide the length of the projected arc by the radius of the circle. Thus, angles are dimensionless quantities. *Radians* and other units of angle measurement are a notational device to avoid confusion when working with such measurements: a radian is simply the number 1; a degree is the number $\pi/180$; and so on.

This way of looking at angles can be extended to higher dimensions, leading to the definition of a *solid angle*, that is, a measurement of the apparent area as seen from a point. Consider a subset A of space and a viewpoint O. The *solid angle* ω determined by A (with respect to O) is the area of the radial projection of A onto the unit sphere centered at O, which we call the *visual sphere* (Figure A.2). As in the plane case, we can take as the visual sphere

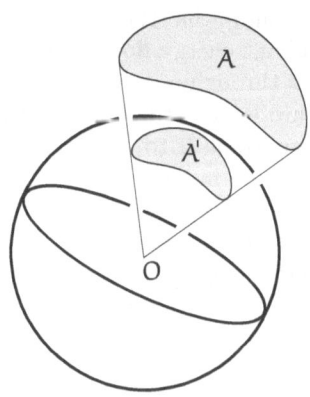

Fig. A.2. Measuring a solid angle.

a sphere of arbitrary radius $r \neq 1$, but then we must divide the area of the projection by the square of the radius. Thus, the solid angle is given by

$$\omega = \frac{\text{Area}(A')}{r^2}. \tag{A.1}$$

The radial projection of A onto the visual sphere determines a *cone* with vertex O and base A, that is, the solid formed by all the rays (half-lines) in space starting at O and going through points of A (Figure A.3). The solid angle defined by (A.1) can also be regarded as a measurement of this cone.

Two subsets of the plane that subtend the same angle (from a fixed point of view O) are called *perceptually congruent* (with respect to O); this is illustrated in Figure A.4, left. Similarly, subsets of space that determine the same cone (or, which is the same, the same projection on the visual sphere) are called perceptually congruent; see Figure A.4, right.

Since (A.1) is a ratio of two areas, solid angles are dimensionless. However, just as in the case of plane angles, it is comforting to be able to use a unit when

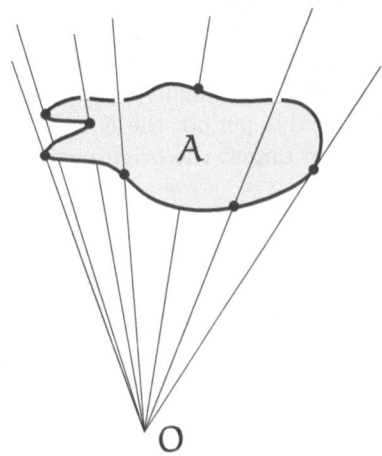

Fig. A.3. Cone with vertex O and base A.

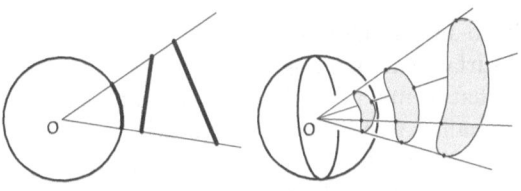

Fig. A.4. Perceptual congruence.

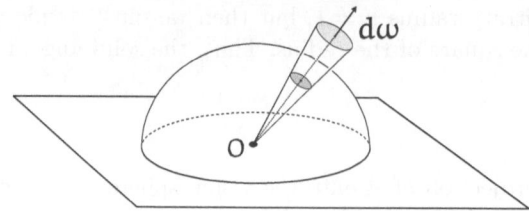

Fig. A.5. Element of solid angle.

discussing measurements of solid angles. The standard unit, representing the number 1, is the *steradian*, abbreviated sr. The whole visual sphere has solid angle 4π sr, since the area of a sphere of radius r is $4\pi r^2$.

When integrating over the visual sphere (or part thereof), we will consider an infinitesimal element of solid angle $d\omega$. Pictorially, we represent $d\omega$ by a vector pointing radially away from O; see Figure A.5.

A.1.1 Radiometric Magnitudes

Suppose a light bulb is turned on, left on for a while, then turned off. There are several measurements that one might be interested in: the total energy emitted by the bulb during this period; the energy emitted per second; the energy that reaches a certain target; the brightness as seen from that target; and so on.

Radiant Flux

The total energy emitted is denoted by Q_e, and it is measured in units of energy: *joules* in the MKS system. The *radiant flux* Φ_e is the rate at which energy is being emitted:

$$\Phi_e = \frac{dQ_e}{dt}.$$

In the MKS system it is measured in joules per second, also known as *watts*.

Irradiance

The energy emitted by the light source can also be considered to be going through a closed surface surrounding the source, so it makes sense to consider the *flux density*—that is, flux per unit area—at points of such a closed surface. Flux density, also called *irradiance* and denoted E_e, is vector-valued: given a point P and an element of surface containing P, having area dA, the flux through that surface element is $E_e \cos\theta dA$, where E_e is the magnitude of the irradiance vector at P and θ is the angle between the irradiance vector and the normal to the area element (see Figure A.6, where the flux through the

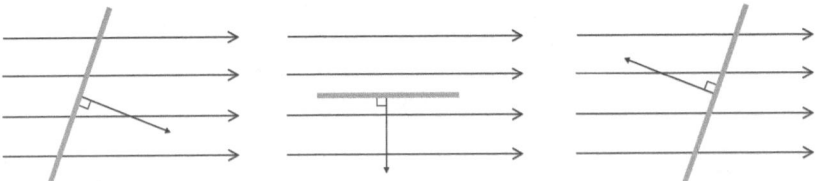

Fig. A.6. The amount of radiant energy crossing a surface per unit time depends on which way the surface faces.

surface is negative on the left, zero in the middle, and positive on the right). Loosely, we can write

$$E_e = \frac{d\Phi_e}{dA}.$$

Irradiance is measured (in the MKS system) in watts per square meter.

Radiant Intensity

The irradiance depends, of course, on how far the point of measurement is from the source (and usually also on the direction as seen from the source). It is often useful to work instead with a magnitude that is associated with the source itself. For a point light source, this is easy: we just consider flux per solid angle instead of flux per area. The *radiant intensity* I_e of a point source is defined as

$$I_e = \frac{d\Phi_e}{d\omega},$$

where $d\omega$ is the element of solid angle as seen from the source and is measured in watts per steradian. (Note that this does not make sense unless the light source has negligible extension, since the notion of solid angle depends essentially on the choice of an origin.) The radiant intensity is a function of the direction as seen from the source. When we integrate it over all directions, we recover the total flux, $\Phi_e = \int I_e d\omega$. For a source that sheds light uniformly in all directions, I_e is constant and $\Phi_e = 4\pi I_e$.

Clearly, the flux density an observer perceives at a distance d from the source is $E_e = I_e/d^2$.

Radiance

This concept can be adapted to the case of nonpoint light sources, as follows. Consider a surface S that delimits the source in question—the surface of a light bulb, say, or a sphere around it. (The light need not be generated on S; it is the flux through S that concerns us.) If we take an element of the surface, of area dA, we can regard it as a point source and look at its radiant intensity

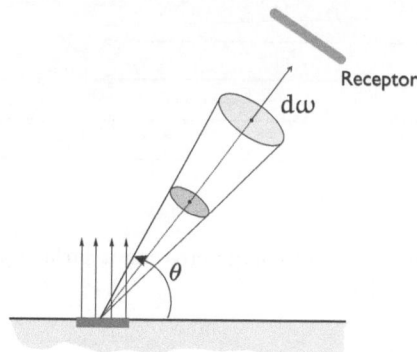

Fig. A.7. Radiance of a light source.

dI_e in a certain direction. The ratio dI_e/dA is the density of radiant intensity at the given point of the source surface, in the given direction. Actually, we must take into account that the area of the surface element as seen from the chosen direction is not dA, but $dA \cos \theta$, where θ is as in Figure A.7. This leads us to the following definition: the *radiance* (not to be confused with the irradiance defined earlier) is

$$L_e = \frac{dI_e}{dA \cos \theta} = \frac{d^2 \Phi_e}{d\omega \, dA \cos \theta}.$$

We stress that this is a function of the chosen point P on the source surface S and of the chosen direction as seen from P. (In mathematical terms, it is a function on the unit tangent bundle to S.) Radiance is measured in watts per square meter per steradian.

Integrating the radiance over the points of the source gives the flux density. More precisely,

$$E_e = \int_S u_P \frac{L_e(P, u) \cos \theta \, dA}{r^2},$$

where P ranges over the surface S, u is the unit vector in the direction from P to the observer, θ is the angle between u and the normal to S at P, and r is the distance from P to the observer. This equation is a generalization of the earlier formula $E_e = I_e/d^2$ for point sources.

A.1.2 Spectral Distribution

So far we have ignored the fact that light is composed of many wavelengths. When it is necessary to study the dependency on wavelength, we can define a spectral version of each of the variables studied above, called a *spectral distribution function*. For example, recall that Φ_e denotes the radiant flux (through some fixed surface). By writing $\Phi'_e(\lambda) \, d\lambda$ for the contribution to this

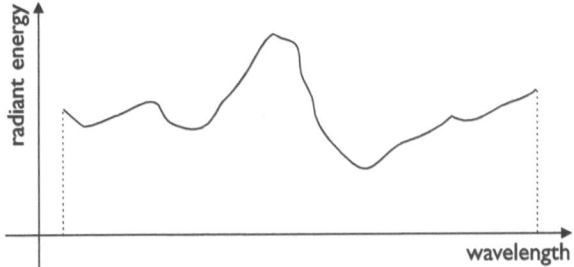

Fig. A.8. Spectral distribution of a radiometric variable.

flux that has wavelength between λ and $\lambda + d\lambda$, where $d\lambda$ is an infinitesimal wavelength, we obtain the *spectral distribution of flux* $\Phi'_e(\lambda)$. Clearly, we have

$$\Phi_e = \int_{-\infty}^{+\infty} \Phi'_e(\lambda)\, d\lambda.$$

Figure A.8 shows a possible spectral distribution.

The radiometric variables we have introduced are also functions of time, and some are also functions of position and/or direction, as we have seen. These dependences are essential in image synthesis and animation. In colorimetry, however, we usually concentrate on wavelength dependence.

Black-Body Radiation

Every material emits radiant energy, at a rate that increases rapidly with the temperature. The spectral distribution of this radiation depends on the temperature and on the nature of the emitter, but at each frequency and temperature the radiant flux of a physical body is bounded by a certain limit predicted by quantum mechanics. A *black body* is an ideal object that emits exactly the predicted maximum amount of energy at each temperature. The radiance of a black body is given by *Planck's equation*,

$$L_e(\lambda) = \frac{2c^2 h}{\lambda^5 \left(e^{hc/(kT\lambda)} - 1\right)},$$

where T is the temperature (in degrees Kelvin), $h = 6.6260755 \times 10^{-34}$ joule-second is *Planck's constant*, $c = 2.99792458 \times 10^8$ meters per second is the speed of light, and $k = 1.380658 \times 10^{-23}$ joule per degree is *Boltzmann's constant*. Figure A.9 shows the graph of this function for several temperature values.

It is possible to construct, for experimental purposes, devices that approximate very well the emission of a black body, at least within a certain range of frequencies and temperatures.

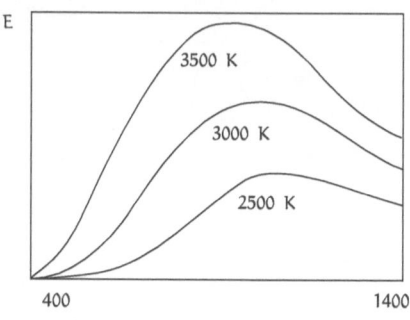

Fig. A.9. Spectral distribution of black-body radiance.

Standard Illuminants

Paint manufacturers give fancy names to dozens of shades that mere mortals would call white. Because the designation "white" is applied so loosely and subjectively, it is essential (for example, in calibration procedures or in the specification of color systems) to specify exactly the spectral distribution of certain colors, taken as standard whites, or *standard illuminants*. The CIE, or International Commission on Illumination, defines illuminant A as the spectral distribution of a black body at 2856°K; this spectrum can be approximated by the light of an incandescent tungsten filament. Illuminant B, which corresponds approximately to direct solar light, has by definition the spectral distribution of a black body at 4874°K, whereas illuminant C has the spectral distribution of a black body at 6774°K. A number of illuminants attempt to approximate daylight under different conditions: D_{55}, D_{65}, and D_{75} correspond to temperatures of 5500°K, 6500°K, and 7500°K, respectively. Finally, Illuminant E is defined by an ideal source whose spectral distribution is flat in terms of energy; for this reason it is also called the *equal-energy white*. The color of such a source is perceptually the same as that of a black body at around 6000°K, but its spectral distribution is of course different.

    More details on these illuminants and other standards can be found in the literature cited in Section A.3.

A.2 Photometric Variables

In photometry our interest shifts from purely physical characteristics of light to the question of how a human observer perceives light. To a first approximation, this means that radiometric variables, such as the energy flux reaching the observer, should be weighted according to the human eye's sensitivity to light of that wavelength, which is encoded in the *light-efficiency function* $V(\lambda)$, discussed in Section 4.6. Recall that this function measures the eye's

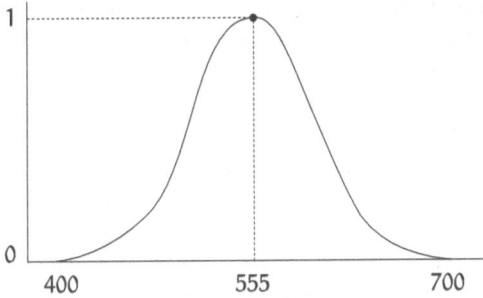

Fig. A.10. Graph of the light-efficiency function $V(\lambda)$.

relative sensitivity to each wavelength of the visible spectrum (and is zero outside the visible range); by convention, it has value 1 at $\lambda = 555\,\text{nm}$, the wavelength at which sensitivity is maximal. See Figure A.10.

Thus, each radiometric variable has a photometric counterpart. Each radiometric variable defined in the preceding section was denoted by a letter with the subscript e (for "energy"); the corresponding photometric variable will be denoted by the same letter, with the e replaced by v (for "visual"). Each photometric variable is commensurable with its radiometric counterpart but is traditionally expressed in a different unit, to avoid confusion and perhaps also because photometry predates radiometry.

Thus, the *luminous flux* Φ_v is the photometric counterpart of the radiant flux Φ_e. For monochromatic light of wavelength λ, we can write $\Phi_v = V(\lambda)\Phi_e$. For light that is not monochromatic, we need to consider the spectral distribution $\Phi'_e(\lambda)$ of Φ_e; then the spectral distribution $\Phi'_v(\lambda)$ of Φ_v is given by $V(\lambda)\Phi'_e(\lambda)$, and we can obtain the total luminous flux Φ_e by integrating over the visible spectrum. In symbols,

$$\Phi_v(\lambda) = \int_{\lambda_a}^{\lambda_b} \Phi'_v(\lambda)\,d\lambda = \int_{\lambda_a}^{\lambda_b} \Phi'_e(\lambda)V(\lambda)\,d\lambda,$$

where λ_a and λ_b are the bounds of the visible spectrum. The luminous flux is measured in *lumens*, abbreviated lm; there are approximately 680 lumens per watt at the wavelength 555 nm, while for an arbitrary wavelength we can write

$$1 \text{ watt} = 680\,V(\lambda) \text{ lumens}.$$

Integrating radiant flux over time, we obtain radiant energy, which is expressed in joules. In the same way, integrating luminous flux over time we obtain *luminous energy*, which is measured in lumens-second. A lumen-second is also also called a *talbot*.

The counterpart of the radiant intensity I_e is the *luminous intensity* I_v, measured in *candelas* (cd). Thus, 1 candela equals one lumen per steradian.

It is in fact the candela that is taken as the fundamental unit of photometric magnitudes: by definition, one candela is the luminous intensity in the perpendicular direction of a surface of 1/600,000 of a square meter of a blackbody at the temperature of fusion of platinum (approximately 1773°C, 2046°K, or 3223°F).

The photometric counterpart of irradiance is *illuminance*, measured in lumens per square meters. A lumen per square meter is also called a *lux* (lx).

The photometric counterpart of radiance is *luminance*, L_v. Luminance is a photometric variable that corresponds most closely to the notion of brightness perceived by the eye. It is measured in candelas per square meter.

Other photometric and radiometric variables are used in the literature, including some that are specifically geared toward computer graphics needs. Moreover, for the variables discussed here, there are other units in use, the most important of which is the *foot-candle*, a unit of illuminance equal to one lumen per square foot, or 10.7639 lux.

Table 12.1 summarizes the preceding discussion.

Table 12.1. Photometric and radiometric variables.

Radiometric variable	Symbol	Unit
radiant energy	Q_e	J (joule)
radiant flux	Φ_e	W (watt)
irradiance	E_e	W/m^2
radiant intensity	I_e	W/sr
radiance	L_e	W/sr·m^2
Photometric variable	**Symbol**	**Unit**
luminous energy	Q_v	lm·s (talbot)
luminous flux	Φ_v	lm (lumen)
illuminance	E_v	lm/m^2 (lux = lx)
luminous intensity	I_v	lm/sr (candela = cd)
luminance	L_v	cd/m^2 = lx/sr

Example: Spectral Luminance

Consider a light source whose radiance spectral distribution function is known; let it be $C(\lambda)$, in units of W/sr·m^2. The corresponding luminance distribution function is therefore

$$680 C(\lambda) V(\lambda),$$

in units of cd/m^2. The total luminance of the source can then be computed by integrating over the visible spectrum:

$$L_v = 680 \int_{\lambda_a}^{\lambda_b} C(\lambda) V(\lambda) \, d\lambda,$$

in units of cd/m^2.

In performing integrals such as this one in practice, we must be aware the $V(\lambda)$ is known from tabulated values at a discrete set of points $\lambda_a = \lambda_0 < \lambda_1 < \cdots < \lambda_n = \lambda_b$. The integral therefore must be approximated numerically; the simplest method is to use the trapezoid rule, so that

$$L_v = 680 \sum_{i=1}^{n} \tfrac{1}{2}\big(C(\lambda_i)V(\lambda_i) + C(\lambda_{i-1})V(\lambda_{i-1})\big)(\lambda_i - \lambda_{i-1}).$$

Better results can be obtained by using, for example, Gaussian quadrature.

A.3 Comments and References

The purpose of this appendix is simply to give self-contained definitions of the main variables of interest in radiometry and photometry, such as the luminance of a color, without interfering with the exposition in Chapter 4. It does not attempt to be a complete exposition of the subject. In particular, we have omitted any mention of the *illumination equation*, which is of fundamental importance in image synthesis. A concise but good exposition of radiometry and photometry geared toward computer graphics can be found in (Kajiya 1990).

A comprehensive treatment of radiometry and photometry, describing physical experiments and including quantitative information on standard illuminants, can be found in (Wyszecki and Stiles 1982).

There are whole books devoted to colorimetry and photometry in general; a good one is (Walsh 1958). The subject is also covered in many optics books, such as (Klein and Furtak 1986).

References

[Kajiya 1990]Kajiya, J. (1990). Radiometry and photometry for computer graphics. *SIGGRAPH '90 Course Notes*.

[Klein and Furtak 1986]Klein, M. and Furtak, T. (1986). *Optics*, 2nd ed. John Wiley and Sons, New York.

[Walsh 1958]Walsh, J. T. (1958). *Photometry*. Dover, New York.

[Wyszecki and Stiles 1982]Wyszecki, G. and Stiles, W. S. (1982). *Color Science*. John Wiley & Sons, New York.

Index

support medium, 417
dot dispersion, 327
three-dimensional image, 137
two-dimensional image, 137

abstraction paradigms, 3, 9
achromatic color point, 98
achromatic line, 98
ACM, 133
acuity visual
    angle de, 314
adaptive filter, 32
addition of signals, 32
Adobe Systems, 436
algorithm
    Floyd–Steinberg, 330
    median cut, 305
    populosity, 303
algorithms
    cluster ordered dithering, 429
    digital halftone, 314
aliasing, 191
    and reconstruction, 211
    error, 192
alpha channel, 369, 370
alpha-channel compositing, 374
amplitude discretization, 413
analog signal, 18
analog-electronic image, 413
analytic sampling, 195
angle of visual acuity, 314
animation, 410
animation morphing, 411

Antunes, André, VIII
area sampling, 31, 145, 195
atlas
    of color, 127
Author
    Adelson, E., 185
    Anderson, C., 185
    Barnsley, Michael, 357
    Barsky, Brian, 384
    Bayer, B., 342
    Beatty, J. C., 436
    Bergen, J., 185
    Bloomenthal, James, 384
    Burt, P., 185
    Buzo, A., 311
    Carpenter, Loren, 384
    Catmull, Edwin, 184, 436
    Clark, R., 357
    Cole, A., 343
    Cook, Rob, 184
    Costa, Bruno, 215, 384, 412
    Cowan, W. B., 436
    Crow, Frank, 215
    Darsa, Lucia, 215, 384, 412
    Daubechies, Ingrid, 185
    Davidson, J., 184
    DeMarsh, LeRoy, 435
    Dubois, Eric, 216
    Duff, Tom, 384
    Fishkin, K. P., 384
    Fiume, Eugene, 215, 311, 384
    Floyd, R., 342
    Foley, James, 133